Numerical analysis in geomorphology

an introduction

John C. Doornkamp
Lecturer in geography, University of Nottingham

Cuchlaine A. M. King
Professor of physical geography,
University of Nottingham

EDWARD ARNOLD

First published 1971 by
Edward Arnold (Publishers) Ltd
41 Maddox Street
London WIR OAN

ISBN: 0 7131 5589 2

Printed in Great Britain by
Butler & Tanner Ltd, Frome and London

Preface

Land forms and drainage patterns are being studied by more and more elaborate methods all the time. The increasingly accurate methods of measuring land form and geomorphological processes are providing a vast amount of quantitative data. This has to be analysed by numerical methods so that an orderly behaviour may be discerned from amongst the mass of accumulated data. This book sets out to provide an introduction to those numerical methods which we have found to be most useful in the analysis of geomorphological data. In it we have tried to show how even the simplest statistical tests can bring out significant points concerning the landscape. But land forms are sometimes complex, having many measurable properties, and this demands of the geomorphologists an appreciation of the value of multivariate statistical analysis. At its most complex such methods demand the use of a computer. Nevertheless, many of the methods in this book can be applied with no more than a set of mathematical tables or a slide-rule. These are well within the grasp of students in the upper forms of schools, at colleges of higher education and polytechnics as well as those at a university.

Where a computer is required we have tried to bring out the potential of these methods by concentrating on the results that can be obtained. In this way we hope that some may be stimulated to pursue their familiarity with the landscape up to and beyond this level, while those who are already well past the position reached by this book may still find in it some stimulus and guidance.

The book concentrates on four major fields of geomorphological study, namely, drainage basins, slopes, and coastal and glacial land forms; its theme is that of the analysis of form. Some differences in approach have been adopted under each of the main headings. This is partly because the subject matter dictates which techniques and methods are most appropriate, but these differences of approach are also designed to illustrate some of the different ways in which numerical analysis can be used.

We are indebted to many of our colleagues at the University of Nottingham, but in particular we owe our thanks to P. M. Mather and M. J. McCullagh who kindly assisted us by making available computer programs which have been used in some of the analyses presented. We are also grateful to J. A. Dawson and A. Hughes for discussions over earlier drafts of the manuscript. For cartographic assistance we are indebted to M. Cutler who has drawn the maps and diagrams.

Department of Geography
University of Nottingham
July 1970

JOHN C. DOORNKAMP
CUCHLAINE A. M. KING

Acknowledgements

The authors and publisher gratefully acknowledge permission given by the following to reproduce or modify material from copyright works:

The editor and the Institute of British Geographers for eight figures (2.1, 4.1, 5.1, 5.2, 5.3, 5.4, 5.5, 5.6)[1] from 'Multivariate analysis in geography, with particular reference to drainage basin morphometry' by P. M. Mather and J. C. Doornkamp, *Transactions* 51; the editor, the author and the University of Liège for a figure (6.9) from 'The analysis and classification of slope profile forms' by R. A. G. Savigear in *L'Evolution des Versants, Les Congrès et Colloques de l'Université de Liège* 40, 173–93; the editor and the Geologists' Association for a figure (9.2) from 'The development of hillside slopes' by A. Wood, *Proceedings* 53, 128–40; Gebrüder Borntraeger and the editor for two figures (9.4, 9.5) from 'A hypothetical nine unit land surface model' by J. B. Dalrymple, R. J. Blong and A. Conacher, *Zeitschrift für Geomorphologie* 12, 60–76; the editor, the author and the Göttingen Academy of Sciences for a figure (9.12) from 'Descriptive models of slope evolution' by Anthony Young in *Nachrichten der Akademie der Wissenschaften in Göttingen, II: Mathematisch-Physikalische Klasse* 5, 45–66; the McGraw-Hill International Book Company for three tables (appendix tables A, B, C) from *Nonparametric statistics for the behavioral sciences* by S. Siegel and also Oliver and Boyd Ltd, Dr. F. Yates and the Literary Executor of the late Sir Ronald Fisher for two of the three tables above (A and B) which were adapted by Siegel from *Statistical tables for biological, agricultural and medical research* by R. A. Fisher and F. Yates; the editor for two tables (appendix tables D, E) from 'Tables of percentage points of the inverted beta distribution' by M. Merrington and C. M. Thompson, *Biometrika* 33, 73 ff.

[1] Numbers in brackets refer to figures or tables in this book.

Contents

Introduction

Scales of measurement

The application of statistical methods to geomorphological data requires an understanding both of the methods of analysis and the nature of geomorphological data. Both are dependent upon the *scale of measurement* used when the data are collected. Measurement can be made on one of four scales which are equivalent to four levels of generalization. These are the nominal, ordinal, interval and ratio scales. Of these the *nominal scale* is the most generalized in that measurement values are only placed in classes. For example, in one area a hill could be placed in one of the three classes of low, intermediate and high. Each hill belongs to one class only, and every hill must belong to a class. On the *ordinal scale* the measurements can be ranked. For example, pebbles can be arranged in order according to their roundness, and an ordinal scale ranging from an index of 1 up to 10 can be used for this purpose. The *interval scale* is one where it is possible to state the actual difference between two measurements, but an arbitrary zero point has to be adopted for the scale. For example, sea-level is an arbitrary zero point from which to measure altitude. Some other starting point might have been taken instead. The most refined scale of measurement is known as the *ratio scale*, and it applies when a true zero exists. For example, the length of a river or the weight of a pebble can be measured on the ratio scale. In each case the scales used start from an actual (not an arbitrary) zero value.

Descriptive and inferential statistics

Statistical theory provides many of the techniques useful in the numerical analysis of geomorphological data. *Descriptive statistics* allow the range of values in the data to be summarized by means of simple diagrams or a few calculated values. In geomorphology it is seldom possible to study all examples of a particular feature or process. Instead of studying the total *population*, therefore, it is necessary to study a *sample*. Through descriptive statistics it is possible to estimate the size of an element in the total population from the sample.

During data collection hypotheses often arise in the mind of the investigator. These hypotheses may be tested through *inferential statistics*, by means of *non-parametric* and *parametric* tests. Of these two the non-parametric tests can be used with data measured on the nominal and ordinal scales. Parametric tests are more rigorous in their requirements and data have to be measured on the interval or ratio scales. The conditions which need to be fulfilled before a parametric test can be used are:

1   observations must be independent and random

2   the data must be drawn from a population for which the descriptive statistics are known.

If the conditions are not met then the data can be reduced to the ordinal or nominal scale so that non-parametric tests can be applied. This means, however, that less rigid tests are being applied to more generalized data, with a corresponding decrease in the precision of the inferences made.

Geomorphological data

There are varying degrees of complexity associated with geomorphological data. In its simplest form interest may be centred upon a single characteristic of certain land forms. This *variable* can be measured and its proportion analysed through *univariate statistics*. A special kind of univariate statistics involves the analysis of changes which take place in one variable over time. In many instances, however, the geomorphologist needs to know the relation between two, or more, variables. This requires the use of *bivariate* or *multivariate statistics* respectively. For example, bivariate statistics may be used to discover the numerical relationship between the size of a drainage basin and the length of the river which it contains. The analysis can be made more complex, when it will require the use of multivariate statistics, by adding other variables, such as the width or height characteristics of the basin. Skill, however, lies not so much in applying the multivariate technique as in the geomorphological interpretation of the results.

Many of the techniques used in this book are described in textbooks on statistics. Some of these are listed below. In the following chapters statistical techniques are applied, and described, as required by the problem or area under investigation. The use of a particular technique in the text can be found by reference to the index. However, not all of the numerical methods applicable to geomorphology involve statistical tests. Slope form analysis, for example, includes methods of numerical classification. *Model building* allows standards of reference to be set up with which future case studies can be compared, or through which a problem may be simplified for further analysis. Some models are statistical, but others such as simulation models are not. For example, geomorphological events can be simulated by a model through which processes may be isolated and their respective influences on land-form development studied.

Useful statistical texts

ALDER, H. L. and ROESSLER, E. B. 1962: *Introduction to probability and statistics*. San Francisco and London: Freeman.

BAGGALEY, A. R. 1964: *Intermediate correlation methods*. New York: Wiley.

BLALOCK, H. M. 1960: *Social statistics*. New York: McGraw-Hill.

  1961: *Causal inferences in nonexperimental research*. University of North Carolina Press.

BROOKES, B. C. and DICK, W. F. L. 1969: *Introduction to statistical method.* London: Heinemann.

EZEKIEL, M. and FOX, K. A. 1959: *Methods of correlation and regression analysis.* New York: Wiley. (3rd edn.)

KRUMBEIN, W. C. and GRAYBILL, F. A. 1965: *An introduction to statistical models in geology.* New York: McGraw-Hill.

MORONEY, M. J. 1956: *Facts from figures.* London: Penguin. (3rd edn.)

SIEGEL, S. 1956: *Nonparametric statistics for the behavioral sciences.* New York: McGraw-Hill.

Part 1 Drainage basins

1 Drainage basins and stream networks

1.1 Defining a drainage basin[1]

Ordering systems; Morphometric properties; Problems in stream ordering; Basin morphometry and geomorphological processes; Basin sampling; Delimitation of drainage basins

1.2 The laws of drainage composition

The analysis of drainage basins, either as single units or as a group of basins which, taken together, comprise a distinct morphological region, has particular relevance to geomorphology. Fluvially eroded landscapes are composed of drainage basins, and these provide convenient units into which an area can be subdivided. The development of a landscape is equal to the sum total of the development of each individual drainage basin of which it is composed. The fact that morphological regions can be recognized suggests not only that within each region the drainage basins have forms similar to each other but also that these basins are evolving in a similar way to each other. Thus, by analysing the development of each drainage basin, greater understanding of the landscape as a whole may be achieved. This is possible if there are definable relationships between the form of a drainage basin and the processes at work within it.

The drainage basin may be thought of as an open system in near steady state (Chorley, 1962; Strahler, 1964; Schumm and Lichty, 1965; Slaymaker, 1966; Morisawa, 1968). In an open system there is, ultimately, a balance between the rates of import and export of material and energy. This is a form of equilibrium. Although the basin and its streams are in this steady state it does not mean that there is no development taking place. On the contrary because sediment is leaving the basin, changes must be taking place within it. The condition of steady state is time-independent for it requires only that the rate of sediment supply from within the basin should be the same as the rate of sediment movement out of the basin. These conditions must be considered in the context of the processes and time periods involved (Schumm and Lichty, 1965). Once steady state has been reached the system becomes self-regulating, for any major changes in the environment (e.g. deforestation) will result in

---

[1] A summary of the section headings appears at the beginning of each chapter. Section titles are printed bold and sub-titles roman. In addition specific techniques referred to within each section are listed (printed italic and enclosed in square brackets).

compensating changes in the system. Under the condition of steady state it is likely that the morphometry of a drainage basin will display some recognizable regularity from one neighbouring basin to the next. The purpose of part I of this book is to illustrate how this type of orderly arrangement may be discerned through the application of numerical techniques. These techniques are here to be used in order:

1   to study the relationships between the morphometric properties of drainage basins
2   to examine the relationships between basin morphometry and other characteristics of the environment (e.g. rock type, stream discharge)
3   to find an objective grouping of drainage basins having similar morphometric properties, and to see if this grouping coincides with morphological regions originally defined by ground inspection
4   to consider drainage basins as convenient units within which to study slope form (part II).

1.1 Defining a drainage basin

Ordering systems

The problem of defining a drainage basin does not lie in a definition of the term itself but in locating the boundary of the basin on the ground, or on a map. Fundamental to any numerical analysis of drainage basin characteristics is the concept of stream (or valley) ordering (Fig. 1.1). A system of channel ordering was suggested by Gravelius (1914), but the work of Horton (1932, 1945) marked the beginning of the widespread use of channel ordering systems in geomorphology. Since that time numerical analysis of drainage basins has relied on an ordering system. The usual method of ordering used today is that suggested by Strahler in 1952 (Fig. 1.1B). In this system all streams which have no tributaries are known as first-order streams. When two first-order streams join together they form a second-order stream; when two second-order streams join they form a third-order stream, and so on. If a first-order stream enters a second-order stream then there is no change in the order of the second-order stream. An increase in stream order only occurs when two streams of like order join each other. In this system the head of each second-order stream occurs at the junction of two first-order streams. In the system used prior to 1952 (Fig. 1.1A) it was usual to extend the second-order designation back to the head of the longest of its first-order tributaries, and so on for higher orders. In this study the Strahler system of ordering is employed (Fig. 1.1B). This can be used equally well for the ordering of valley networks as it can for streams.

Alternative systems of ordering streams have been suggested by Scheidegger (1965), Woldenberg (1966) and Shreve (1967). These alternatives were suggested because of weaknesses apparent in the systems

of Horton and Strahler. Scheidegger (1965) points out that in the Strahler system where two streams of order u join they form a stream of order $u + 1$. No change in order takes place, however, when a stream of order u is met by one of a lower order. Scheidegger suggests that account should be taken of all tributaries (Fig. 1.1C). This can be achieved by defining

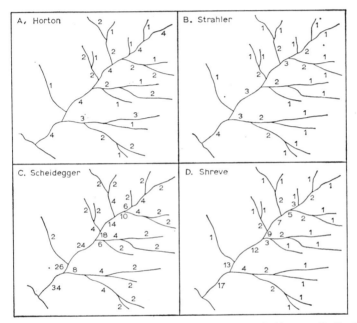

Fig. 1.1   Channel ordering systems as proposed by **A** Horton; **B** Strahler; **C** Scheidegger; **D** Shreve.

the order (x) after the junction of two streams (of order v and u respectively) by:

$$x = \log_2 (2^u + 2^v)$$

This means that in the Scheidegger system all of the extreme outward branches must be of order 2 and not 1 as in the Strahler system.

Shreve (1967) refers not to stream order but to the 'magnitude of a link'. This is defined (Fig. 1.1D):

1   each exterior link has magnitude 1
2   if links of magnitude μ_1 and μ_2 join, then the resultant downstream link has magnitude $\mu_1 + \mu_2$.

By this means account is still taken of all tributary junctions (as demanded by Scheidegger), and the magnitude of a link is also a direct statement of the number of sources ultimately tributary to it. As Shreve points out, magnitude is a purely topological concept. Networks with equal magnitudes have an equal number of links, forks and sources.

The mathematical relationship between stream order (x), as defined by Scheidegger, and its magnitude, as defined by Shreve, is:

$$x = \log_2 2\mu$$

Woldenberg (1966) finds that successive orders in the Strahler system are new logarithmic cycles to the base of the bifurcation ratio (this is a measure of the amount of branching in the network, see Table 1.1). He suggests, therefore, that a new absolute order should be derived by raising the bifurcation ratio to successive integer powers equivalent to stream order minus one.

Neither the Scheidegger, Woldenberg, or Shreve systems have been widely used, and they are not used in this book for, by so doing, little or no direct comparison could have been made with results previously obtained in studies using the Strahler system. Nevertheless, there seem to be many advantages in the Shreve system and further research should be directed towards examining its application more fully.

Given that an ordering system, such as the Strahler system, is to form the starting point for the collection of numerical information concerning a drainage basin, there are certain obvious characteristics which can be defined and measured. For example, as a development of the ordering system it is possible to speak not only of first-order streams but also of first-order drainage basins. These are the areas drained by first-order streams. Likewise it is possible to have a second-order drainage basin, but in this case the parallel with the stream ordering system no longer holds good for a second-order basin also includes the area occupied by all of the first-order basins which drain into it, and not just the area around the stream channel which has been designated as a second-order stream. A third-order drainage basin includes not only the slopes which supply water directly to the third-order stream but also the area occupied by the first- and second-order streams which drain into it.

Morphometric properties
Once a drainage basin watershed and its stream (or valley) pattern has been defined, measurements can be made of some of its morphometric properties. These variables are defined in Table 1.1, and the symbols used in this account are also listed. The variables fall into several distinct groups. Some of them are measured directly, usually from a map; these include basin area (A), length of first-order streams (L_1), total length of all streams (ΣL), number of first-order streams (N_1), and so on. Other variables are derived from such direct measurements and include measures of basin shape, such as basin circularity (R_c), as well as drainage density (D), bifurcation ratio (R_b), and basin relief ratio (R_h). The variables also differ according to their dimensional characteristics. Thus measures of stream length are one-dimensional ($\mathbf{L}$, not to be confused with stream length), while basin area is a two-dimensional measure ($\mathbf{L^2}$). On the other hand drainage density is the reciprocal of a one-dimensional

Table 1.1 Morphometric and related variables, their dimensions and symbols used

Variable	Symbol	Units	Dimensions
Drainage network			
Stream order (used as subscript)	u	enumerative	0
Number of streams of order u	N_u	"	0
Total number of streams within basin order u	$(\Sigma N)_u$	"	0
Bifurcation ratio	$R_b = N_u/N_{u+1}$	"	0
Total length of streams of order u	L_u	miles	L
Mean length[1] of streams of order u	$\bar{L}_u = L_u/N_u$	"	L
Total stream length within basin of order u	$(\Sigma L)_u = L_1 + L_2 \ldots + L_u$	"	L
Stream length ratio	$R_l = \bar{L}_u/\bar{L}_{u-1}$		0
Basin geometry			
Area of basin	A_u	square miles	L^2
Length of basin	L_b	miles	L
Width of basin	B_r	"	L
Basin perimeter	P	"	L
Basin circularity	$R_c = A_u$/area of circle having same P		0
Basin elongation	$R_e =$ diameter of circle having same P/L_b		0
Measures of intensity of dissection			
Drainage density	$D_u = (\Sigma L)_u/A_u$	miles per miles^2	L^{-1}
Constant of channel maintenance	$C = 1/D_u$	miles^2 per mile	L
Stream frequency	$F_u = N_u/A_u$	number/mile^2	L^{-2}
Texture ratio	$T_u = N_u/P_u$	number/mile	L^{-1}
Measures involving heights			
Stream channel slope	θ_c	feet/mile or degrees	0
Valley-side slope	θ_g	degrees	0
Maximum valley-side slope	θ_{max}	"	0
Height of basin mouth	z	feet	L
Height of highest point on watershed	Z	"	L
Total basin relief	$H = Z - z$	"	L
Local relative relief of valley-side	h	"	L
Relief ratio	$R_h = H/L_b$		0
Ruggedness number	$R_n = D \times H/5280$		0
Stream flow			
Discharge	Q	cubic feet/second	L^3

[1] The superscript bar indicates (here and throughout) a mean value.

measure, being equal to total stream length divided by basin area ($\mathbf{L} \div \mathbf{L}^2 = \mathbf{L}^{-1}$). Stream numbers and the ratios such as bifurcation ratio are dimensionless. In the following chapters these variables will, very largely, be referred to by their symbols; it may be necessary, therefore, constantly to return to Table 1.1 for reference.

Problems in stream ordering

Problems which are a result of local peculiarities become apparent when stream ordering is undertaken according to the Strahler system. These should be noted, for they can indicate a great deal about the particular character of the geomorphology of the area under examination. For example, it is frequently the case that stream networks are denser in areas of exposed well-jointed bedrock on steep slopes than on neighbouring vegetated slopes. These bedrock slopes therefore introduce many additional first- and even second-order streams into the network. This has a direct consequence on the ordering of the rest of the system.

Figure 1.2A shows another example. Here the ordering of a system is seen to change entirely by the development of one small first-order tributary, and this can lead to an increase of one in each of the higher stream orders. This will only occur, however, in those cases where the first-order stream develops as a tributary to another first-order stream, no change in stream ordering takes place if it develops as a tributary to a stream of higher order. In similar manner, the convergence of two third-order streams just before their junction with a fourth-order stream turns the latter into a fifth-order stream. If, on the other hand, the two third-order streams had managed to take independent paths to the fourth-order stream then no fifth-order would have been involved. This is particularly critical when the two third-order streams meet the main stream (fourth-order in Fig. 1.2D) on a wide alluvial plain. Under such circumstances they might as readily have joined just before reaching the main stream, as shown by the broken line in Figure 1.2D.

Discontinuous drainage lines introduce problems of another kind. Within an area of karst landscape they present special circumstances which require careful assessment (Williams, 1966). Discontinuous flow can also occur on non-limestone areas. Such may be the case where hillside drainage passes below the surface on meeting a scree or footslope predominantly composed of waste material, and the measurement of stream length has to be approximated to the straight-line distance down the footslope or scree. In areas where river capture has recently taken place there may be difficulties over stream ordering. Taken to its extreme ordering becomes impossible where capture is actually taking place today. This is the case in the Kandekye–Koga drainage basins of the Buhweju Mountains in southern Uganda (Fig. 1.3). The Kayania Swamp drains to the south, along the early course of the Kandekye, but it has also been tapped by a tributary of the Koga so that this swamp also drains away

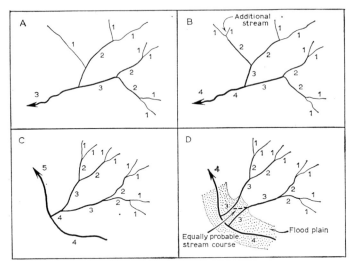

Fig. 1.2 Some problems in stream ordering:
A and **B**: by adding one tributary (**B**) the main stream increases from third to fourth order;
C and **D**: by delaying a stream junction (**D** as opposed to **C**) the main stream fails to increase in order. Had the equally probable course been taken in **D** the main stream would have attained fifth-order status (as in **C**).

to the north-east, over a high waterfall. The resulting network defies ordering by any of the systems outlined above.

Other environmental problems may also create difficulties. For example, heavy forest cover may prevent the observation and mapping of all

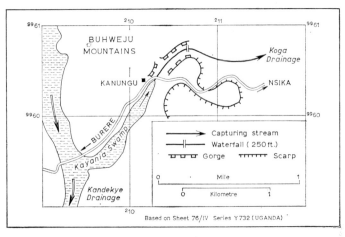

Fig. 1.3 The present capture of the Kayania Swamp.

members of a drainage network. Elsewhere the natural system may have been reorganized by man, for instance by the addition of drains or ditches, to such an extent that a natural system of streams can no longer be recognized. Despite these problems, however, many areas remain where ordering can be carried out and sensible analysis undertaken.

Basin morphometry and geomorphological processes
It has to be appreciated that the morphometric characteristics measured, such as those listed in Table 1.1, are not by themselves a definition of processes at work within a drainage basin. Through a consideration of stream frequency, however, a comparative study can be made of the intensity of drainage dissection from one basin to another, and drainage density gives a measure of stream lengths per unit area. Nevertheless many of the variables listed relate to basin form and are the result of denudational processes. However, they are just as likely, if not more likely, to be the result of processes which have operated in the past than those which can be seen in operation today. In a sense the processes of the past have given rise to the forms which we measure today, and the processes at work today will give rise to the forms of tomorrow. It has been suggested, though, that changes in basin morphometry in response to changes in processes may take place rapidly to provide a new 'steady state' (Strahler, 1964). In reality the situation may be even more complex than this. Many landscapes have been affected by climatic change. In extreme cases this may range from a period of glaciation to one of temperate conditions, or from a period of ample rainfall in the past to one of arid conditions today. This means that the present form of a drainage basin may be the result of a sequence of different processes, or of a different domination among the processes. Thus it is possible that forms produced under very wet conditions may be modified by the onset of aridity; if there is then a change to cooler and more moist conditions, the landscape, and thus the drainage basins of which it is composed, may carry features resulting from each of these three climatic phases. It is not known how long it takes one set of processes to eliminate all signs of a previous period of landscape development. The change in processes need not be a complete change in kind; instead it may be a change in intensity. Thus an area experiencing movement of soil and regolith on its slopes may be subject both to debris avalanches and to soil creep. At one period in time it might be the movement of regolith as debris avalanches which is the dominant process, but at another soil creep might become dominant. This change could be a response to a change in the climate, but it can also result from internal changes within the area under consideration. For example, debris avalanches may reduce the gradient of the hillside so much that the slope is no longer sufficiently steep for new avalanches to occur and soil creep may take over, under the same climate, as the main process affecting land-form development. Thus variations in process domination within a drainage basin may be the result of external in-

fluences or of internal adjustments. Clearly this discussion is also relevant to the study of slopes (part II).

It is unlikely that an analysis could be completed of the nature, sequence, duration or effect of all of the processes that have operated within a specific drainage basin. Certainly it would be an impossible task to assign numerical values to these processes. The only reflection of past processes which can be enumerated with any degree of certainty are the forms and basin characteristics which are present in the landscape today. Some of these may reflect past processes rather more clearly than others. For example a high density of valleys, within which there now exists a low density of stream channels, may reflect a period when a much greater amount of surface water was available for drainage basin development. On the other hand, there are characteristics of a drainage basin which, once established, tend to be more permanent and may persist from one period to the next regardless of the dominant process at work. Such features include, for example, the alignment of the main valleys.

The numerical analysis of the form of a drainage basin and of its stream network must be undertaken, therefore, with the knowledge that the history of basin development may be a complex one. At each stage of its development the forces within the drainage basin will have been active on the hillsides and within the valley net to establish a state of dynamic equilibrium. This may or may not have been achieved. Nevertheless it is only through an approach towards equilibrium between form and forces that regularity will be introduced into the measurable characteristics of a drainage basin.

Basin sampling
Before order can be sought through a numerical analysis of drainage basin characteristics, it is necessary to consider both some of the difficulties which arise in obtaining data for numerical analysis and also some of their limitations. To be effective many of the statistical techniques employed in numerical analysis require the measurement of data from many individual drainage basins. The results obtained when numerical analysis is applied to only ten basins may differ a great deal from those that result from measuring 50 basins. Once 100 basins have been measured, however, the addition of more basins to the data seems to make little difference to the results. Thus in the numerical analysis of drainage basins it would appear a desirable thing that at least 50 but, better still, 100 basins should be measured. Problems of sampling arise when it has to be decided which 100 basins are to be measured within a particular area.

Not only in the case of drainage basins but also in most other geomorphological investigations, it is not possible to measure the total population of features being studied. Instead any analysis has to be based on a sample drawn from the total population. One of the aims of statistics is to consider the likely nature of the population by analysing the characteristics of the sample. For this purpose it is necessary for the sample to

be drawn *at random* in such a way that each member of the population has an equal chance of being included in the sample. Random sampling is not always possible in geomorphology. For instance some drainage basins may be beyond analysis for one of the following reasons:

1   they are inaccessible in the field
2   there is no air photograph cover
3   there is no base map available.

Where these limitations apply a sample cannot be properly random. Where they do not apply, a random selection of drainage basins can be made by first defining all possible basins and locating these on a base map. Each basin is numbered and the required set of basins can be drawn at random by reference to a table of random numbers. If, on the other hand, regular selection across an area is required then the investigator can dispense with the table of random numbers. Thus if 150 basins are available and 50 basins spread evenly across the area are required for measurement then every third basin is included.

One frequently encountered problem, is that, although it may be desirable to exceed a certain number of drainage basins for a particular analysis, that number does not exist in nature. For example if a particular study is to be made of the drainage basins on an outcrop of granite, and there are only seven such basins in existence, what is to be done if the statistical tests require that a minimum of ten basins should be used to provide the basic data? The answer depends on the statistical test in question, and frequently it will have to be adapted to allow for only seven basins instead of the required minimum of ten. On the other hand this provides the sort of instance where it is possible to analyse the total population of basins (on this granite outcrop) rather than a sample, as is more usually the case.

Delimitation of drainage basins
In order to be effective a particular statistical test may require that a large number of drainage basins are measured, and where a large number of basins is available, it is desirable that the most speedy methods of measurement are adopted. If, however, accuracy is sacrificed for the sake of speed, then the whole point of using numerical analysis, and in particular some of the more refined statistical techniques, has gone. The easiest and most rapid way of making the required measurements of drainage basin characteristics is from good quality and detailed relief maps. The published maps, such as the Ordnance Survey 1 : 25,000 series, may be sufficiently accurate for some purposes, but where precise relationships and characteristics are being defined they may not prove to be adequate. Where this is the case these maps will have to be adapted to incorporate the data required. This can either, and frequently best, be done in the field, or from high-quality and large-scale air photographs. The funda-

mental failing of most published relief maps is that they do not allow a precise definition of the number or lengths of first-order valleys and streams. Since the whole of the ordering system is dependent upon this one feature it is important that it should be correctly defined, both in number and extent. Once this has been done the definition of many other characteristics follows automatically. For example, the number of higher order streams is dependent on the number and location of first-order streams, as is the assessment of total stream numbers and lengths. The bifurcation ratio between first- and second-order streams depends on the number of first-order streams, and the drainage density cannot be calculated unless the correct lengths of all valleys, or streams, are known. Stream frequency is derived from the number of streams in the basin, and thus depends on a correct assessment of the number of first-order streams. Once first-order streams have been correctly defined a great deal of information can be derived from the base map. When the drainage network has been established basins of the required order can then be defined. If, on examination of the relief map, doubt exists about the precise position of a watershed then this too will have to be ascertained either in the field or from air photographs.

When taken from a base map the measures of basin area and all measurements of length are the horizontal equivalents of the ground measurements. The actual area of the basin will differ from the horizontal equivalent by a significant amount only if the terrain contains some very steep slopes. If this is likely to be a serious difference then a precise value for the surface area of the drainage basin can be obtained by the construction of a morphological map of the type described in section 6.2. Such a map defines both the area and steepness of each slope within the drainage basin. As a result, the map area of each slope (which is in two dimensions) can be converted, by dividing by the cosine of the angle of slope, to give the actual area of the slope on the ground. The compilation of such a morphological map involves an immense amount of time if it is carried out in the field, especially if many basins are involved. It may prove to be an impracticable task in many circumstances, and so the horizontal equivalents will have to be employed as the nearest approximation available to the actual values. This may not be serious when all of the basins studied come from the same type of terrain and have the same kind of relief characteristics. It will, however, provide a source of error when basins are being compared from an area of steep slopes in one case and of gentle slopes in another.

Many other complications can arise when drainage basins are being defined and delimited on a base map. The limitations of the published topographic maps have already been touched upon. The blue lines which appear on many topographic maps, including the Ordnance Survey 1 : 25,000 maps, cannot be taken as an absolute definition of the stream network, let alone the valley network, within an area. The limitations of using such blue lines are discussed by Morisawa (1957). A slightly better

picture of the true state of affairs is obtained if the network suggested by the contour crenulations is analysed. These do not, however, include every valley. This in itself might not be too serious if on every map and in every area those valleys omitted were consistently of the same order. Unfortunately this is seldom the case, for while two adjacent valleys may both in fact be of second-order one may be shown on a map (e.g. because it is deeper) whereas the other may not, thus leading to inconsistencies. Another difficulty lies in the fact that the detail presented by the contours on these maps may vary from one area to the next. For example a map of a lowland with coarse-grained relief may contain few valleys and each of these may be clearly represented by the relief map. A mountain area, with many mountain-slope streams, may have these less adequately represented simply because there is not sufficient space on the map to show all of the tributaries within the drainage system. Thus in the former case all of the valley network, right down to the first-order streams, will be shown, whereas in the latter, no stream of third- (or lower) order may appear on the map. Thus if contour crenulations were taken as definitive in each case and a comparison made between the characteristics of what appeared to be the first-order basins in each case, the analysis would be made on two unlike things. In the lowland example they would in reality be first-order basins, but in the mountains they would be fourth-order basins. When two areas are being compared it is vital that the basis for comparison is the same in each case. This frequently requires the checking of map data in the field or on good air photographs.

When air photographs of the same quality and scale are used for the delimitation of valley networks, uniformity is more likely to result than when only topographic maps are employed. The value of air photographs is increased if sample areas can be compared with valley networks defined in the field.

Photographs at a scale of 1 : 30,000 were used to define the valley network of a part of Uganda (chapter 2). A comparison of the network mapped in this way with the network on the ground revealed several important points. Valleys more than 20 feet deep were easily detected on the photographs, and were consistently defined for all types of terrain devoid of forest cover. Interesting differences occur, however, between stream networks and valley networks in the same areas. In the Rwampara Mountains (Fig. 2.1) a single valley may contain several streams within its floor. Such a valley (Fig. 1.4) may have a distinct upper limit, but beyond this head the crestslope of the mountain may carry a network of gullies. These make little impression on the surface, but can be mapped from photographic enlargements of the air photographs. The valleys within the Rwampara Mountains open out towards the footslope where distributaries are formed by the streams draining the valley floor. In addition new streams may start on the footslope. Under these conditions the valley would be designated first-order, but higher orders are involved if the stream network within that valley is being ordered. Under these

circumstances it is extremely difficult to define the entire stream network, especially if a large number of such valleys were to be analysed. As a result, for general morphometric studies it is probably better to accept these differences and to concentrate on an analysis of valley rather than stream networks. On the other hand if the purpose of analysis is strictly

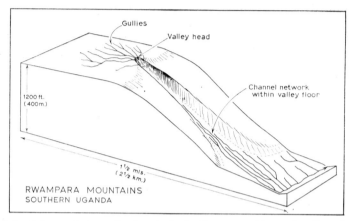

Fig. 1.4 Valley and channel networks in the Rwampara Mountains, Uganda.

Table 1.2 Drainage density and stream frequency values derived from field and map measurements in part of south-east Devon (after Gregory, 1966)

	Stream channel net (field)	Valley net (field)	Stream net based on maps
Stream frequency (Otter basin)	3·39	10·03	1·86
Drainage density (Otter basin)	2·01	3·44	1·68
Average drainage density for five basins		4·88	2·63

hydrological, then the stream network is much more relevant and these complications must be considered.

A comparison has been made by Gregory (1966) of the results obtained from an analysis of the morphometric properties of drainage basins when measured in the field and from topographic maps. He defined separately the valley net on the ground and the stream nets both in the field and on the 1 : 25,000 topographic maps, within a part of south-east Devon. When Gregory calculated the value of stream frequency and drainage density,

for example, he found that there was a real difference in the values obtained according to the way in which the data had been collected. These values are reproduced in Table 1.2.

It is important that whenever the results of a morphometric analysis are presented a clear statement should be made of the method used to obtain the data. In addition, since results are to some degree dependent on the method of data collecting, comparisons made on data collected from two different areas will only be meaningful if the data are collected in the same way, and from the same type of source material, in both areas.

1.2 The laws of drainage composition

In addition to his ideas on stream ordering Horton (1945) showed that there is a close relationship between the number of streams and the length of streams of each order when compared with the order value (Fig. 1.5). The first law of drainage composition (the *law of stream numbers*) was stated by Horton (1945, p. 291) to be:

The number of streams of different orders in a given drainage basin tend closely to approximate an inverse geometric series in which the first term is unity and the ratio is the bifurcation ratio.

The second law (the *law of stream lengths*) given by Horton (*ibid.*) was:

The average lengths of streams of each of the different orders in a drainage basin tend closely to approximate a direct geometric series in which the first term is the average length of streams of the first order.

Put in simpler terms, Horton showed that in a drainage basin there is a progressive decrease in the number of streams as the numerical value of the stream order increases. The number of first-order streams is greater than the number of second-order streams, and so on. This relationship follows a geometrical series which can be demonstrated by plotting the number of streams of each order on a logarithmic scale against stream order on an arithmetic scale (Fig. 1.5). The second law suggests that as stream order increases so does the mean length of the streams. This too is apparent from a graph of mean stream length, on the log scale, against stream order, plotted on the ordinal scale. More recently suggestions have been made that 'cumulative mean length' should replace the 'mean length' term originally used by Horton (Broscoe, 1959; Bowden and Wallis, 1964). These two laws of drainage composition have been substantiated by several people for different areas, whether the Horton or the Strahler method of stream ordering has been used (Chorley, 1957; Morisawa, 1962; Chorley and Morgan, 1962; Strahler, 1964; Gregory, 1966; Slaymaker, 1966; Selby, 1967 & 1968; Ghose *et al.*, 1967).

Horton (1945) also found that there is a close relationship between the

gradient of a stream and its order (the *law of stream gradients*). This is demonstrated by the straight-line graph which results from plotting the logarithm of mean stream gradient against basin order. Chorley (1957) describes these laws with reference to examples drawn from Exmoor, Pennsylvania and Alabama. He also includes in this account other laws,

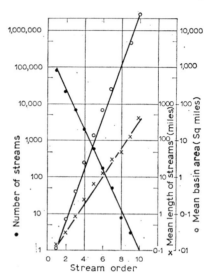

Fig. 1.5 Relation of stream numbers, mean stream length and mean basin area to stream order for the Susquehanna River basin. *After Brush (1961) using Horton's ordering system.*

such as the *law of basin areas*, developed by Schumm (1956), which states that:

the mean area of drainage basins at each of the different orders tends to approximate closely to a geometrical series in which the first term is the mean area of the first-order basin.

Thus basin area, total stream length and the number of streams are all related to the magnitude of the order of the basin in which they occur. Since each of these change in value with basin order it is likely that they will also change in response to changes in each other. Such relationships are examined more closely in chapter 3.

Perhaps no great surprise should be expressed at the consistent manner in which these laws are duplicated. In any natural branching system, where all of the members are orientated towards a single stem, the same laws would have been obtained if the members of that system had been ordered in the same way as stream networks. For example, a similar analysis made on the branches of a weeping cherry tree (*Shidare zakura*), in one of the present writers' gardens, produced the same relationships. These results have also been obtained by Lawson and Downey (1966) from a simulation model derived from random numbers. This suggests that the laws of drainage composition may be the inevitable outcome of

the logic of the ordering system applied to a naturally branching network. This is not to suggest that the ordering system therefore has no
value. On the contrary it is a methodical and easily understood method
of stream and drainage basin classification. In more recent years this
system has been employed for defining drainage basins of similar order
within which the relationship between a number of variables can be
ascertained. For example it could be assumed that the larger drainage
basins of a group of third-order basins would also be those which contain
the greatest number of first-order streams. Or, it is likely that in those
third-order basins where the number of first-order streams is greatest
the number of second-order streams is also greatest. This type of approach
is rather different from that employed by Horton in producing his laws
of drainage composition. The precise study of the relation between two
morphometric variables within a group of drainage basins cannot be
undertaken without numerical analysis. Not only does numerical analysis
show whether or not a relationship exists between two variables, it also
defines the precision with which that relationship can be established.

The next chapter introduces a case study of the morphometric properties
of some drainage basins in Uganda. A summary of the geomorphology
of the area is provided, as it is within this context that the numerical
analysis is undertaken. The variables and their statistical properties are
also described. In chapter 3 the variables are examined two at a time, as
an illustration of the type of pair-wise relationships which exist within the
Uganda third-order basins. These results are compared with those
derived for fourth-order basins within the same area, and with those
obtained in other parts of the world. Chapter 4 is concerned with the same
variables but considers the relationships between them for combinations
greater than two at a time. The more complex multivariate analysis of
the whole mass of Uganda data by means of a form of factor analysis
known as principle components analysis, concludes that chapter. The
morphometric data are drawn from a number of different morphological
regions which are defined in chapter 2. In chapter 5 the data are re-
examined by means of cluster and multiple discriminant analysis to see if
basin morphometry can be used to distinguish between these different
morphological regions, or if other forms of grouping are indicated.

References

BOWDEN, K. L. and WALLIS, J. R. 1964: Effect of stream-ordering technique
    on Horton's laws of drainage composition. *Bull. Geol. Soc. Am.* **75**(8),
    767–74.
BROSCOE, A. J. 1959: Quantitative geomorphology of small drainage

basins of southern Indiana. *Columbia Univ., Dept. of Geol., Tech. Rept.* **18.** (73 pp.)

CARSON, M. 1966: Some problems with the use of correlation techniques in morphometric studies. In Slaymaker, H. O., editor, Morphometric analysis of maps, *British Geomorph. Research Group, Occ. Paper* **4,** 49–67.

CHORLEY, R. J. 1957: Illustrating the laws of morphometry. *Geol. Mag.* **94,** 140–50.

1962: Geomorphology and general systems theory. *U.S. Geol. Surv. Prof. Paper* 500-B. (10 pp.)

CHORLEY, R. J. and MORGAN, M. A. 1962: Comparison of morphometric features, Unaka Mountains, Tennessee and North Carolina, and Dartmoor, England. *Bull. Geol. Soc. Am.* **73**(2), 17–34.

CHOSE, B., PANDEY, S. and LAL, G. 1967: Quantitative geomorphology of the drainage basins in the central Lumi basin in western Rajasthan. *Zeit. für Geomorph.* NF **11**(2), 146–60.

GRAVELIUS, H. 1914: *Flusskunde: Goschensche Verlagshandlung.* Berlin. (176 pp.)

GREGORY, K. J. 1966: Dry valleys and the composition of the drainage net. *J. Hydrol.* **4,** 327–40.

HORTON, R. E. 1932: Drainage basin characteristics. *Am. Geophys. Union Trans.* **13,** 350–61.

1945: Erosional development of streams and their drainage basins. *Bull. Geol. Soc. Am.* **56,** 275–370.

LAWSON, M. and DOWNEY, G. 1966: A stochastic model of fluvial patterns. *The Monadnock* **40,** 36–40. (Clark Univ. Geogr. Soc.)

MORISAWA, M. E. 1957: Accuracy of determination of stream lengths from topographic maps, *Am. Geophys. Union Trans.* **38,** 86–8.

1962: Quantitative geomorphology of some watersheds in the Appalachian Plateau. *Bull. Geol. Soc. Am.* **73**(9), 1025–46.

1968: *Streams, their dynamics and morphology.* New York: McGraw-Hill.

SCHEIDEGGER, A. E. 1965: The algebra of stream-order numbers. *U.S. Geol. Surv. Prof. Paper* **525-B,** 187–9.

SCHUMM, S. A. 1956: Evolution of drainage systems and slopes in badlands at Perth Amboy, New Jersey. *Bull. Geol. Soc. Am.* **67,** 597–646.

SCHUMM, S. A. and LICHTY, R. W. 1965: Time, space and causality in geomorphology. *Am. J. Sci.* **263,** 110–19.

SELBY, M. J. 1967: Aspects of the geomorphology of the Greywacke ranges bordering the lower and middle Waikato basins. *Earth Sci. J.* **1,** 1–22.

1968: Morphometry of drainage in areas of pumice lithology. *Proc. Fifth N.Z. Georgr. Conf., N.Z. Geogr. Soc., Auckland,* 169–74.

SHREVE, R. L. 1967: Infinite topologically random channel networks. *J. Geol.* **75**(2), 178–86.

SLAYMAKER, H. O. 1966: Morphometric analysis—a general introduction. In Morphometric analysis of maps, *British Geomorph. Research Group, Occ. Paper* **4,** 1–4.

STRAHLER, A. N. 1952: Dynamic basis of geomorphology. *Bull. Geol. Soc. Am.* **63**, 923–38.

— 1964: Quantitative geomorphology of drainage basins and channel networks. In Chow, V. T., editor, *Handbook of Applied Hydrology*, New York: McGraw-Hill, section 4–11.

WILLIAMS, P. W. 1966: Suggested techniques of morphometric analysis of temperate karst land forms. In Morphometric analysis of maps, *British Geomorph. Research Group, Occ. Paper* **4**, 12–30.

WOLDENBERG, M. J. 1966: Horton's laws justified in terms of allometric growth and steady state in open systems. *Bull. Geol. Soc. Am.* **77**(4), 431–4.

2 The nature of morphometric data

2.1 Geomorphological background to the Uganda case study

Descriptive statistics can be applied to morphometric data in order to summarize the data collected, and from this summary it is possible to define the characteristics of the area being examined. In this chapter descriptive statistics are used in relation to some morphometric data from third-order drainage basins in southern Uganda. The data will be used in succeeding chapters to illustrate the application of numerical techniques to drainage basin studies. Before the descriptive statistics are discussed, however, it is necessary to examine the geomorphology of the area from which the Uganda examples are taken.

The third-order drainage basins are drawn from eight morphologically distinct regions (Fig. 2.1). The area as a whole has suffered from tectonic warping and, in part, a reversal of its drainage (Doornkamp and Temple, 1966). Two planation surfaces occur within the area (Doornkamp, 1968) but the higher of these, the upland landscape surface, is only preserved on some of the watersheds around the drainage basins to be analysed. The drainage either flows to Lake Victoria or into the western rift valley. Basins have been included for analysis from both

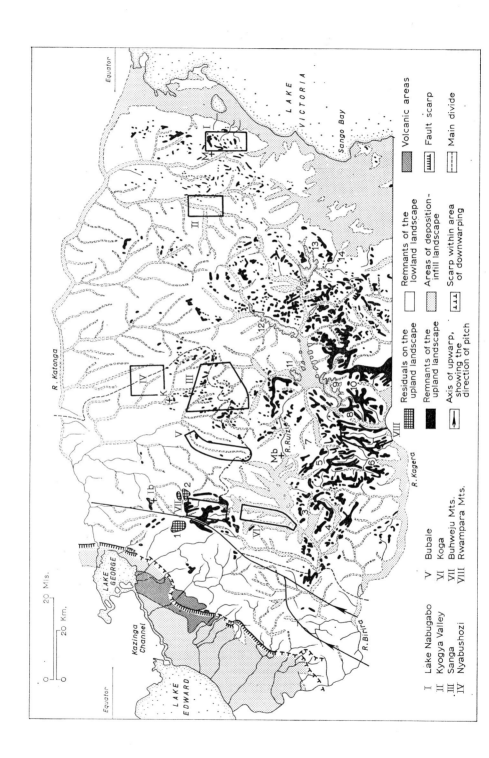

I Lake Nabugabo   V   Bubale
II Kyogya Valley   VI   Koga
III Sanga   VII   Buhweju Mts.
IV Nyabushozi   VIII   Rwampara Mts.

Residuals on the upland landscape

Remnants of the upland landscape

Axis of upwarp, showing the direction of pitch

Remnants of the lowland landscape

Areas of deposition-infill landscape

Scarp within area of downwarping

Volcanic areas

Fault scarp

Main divide

of these major drainage systems. In two instances, areas VII and VIII in Figure 2.1, the basins occur in mountain areas, and in three cases (areas IV, V and VI) they occur within lowlands. The other three areas (I, II and III) are intermediate in character between the mountains and the lowlands. These intermediate areas retain remnants of the upland landscape surface on some of their watersheds, as do most of the mountain basins. The lowland basins on the other hand do not: indeed, they are essentially related in form to the lowland landscape surface, which is the lower of the two main planation surfaces and is especially well developed across poorly resistant granite, gneiss and mica-schist. The mountain areas are underlain by resistant quartzites and conglomerates, as well as phyllites and shales run through with resistant bands of quartzite.

These third-order Uganda drainage basins therefore come from environments which have a wide range of differing characteristics. They include varying amounts of a summit planation surface; they are developed within different major drainage systems; and they have been cut into a variety of rock types. Although climatic data are scarce it is also probable that these basins occur within climatically quite different environments, with regions III and IV drier than the other areas, and the Buhweju Mountains (region VII) receiving more rain than all the rest (Atlas of Uganda, 1967). It has been possible to draw up a subjective classification of the area into morphological regions which has been based on an examination of the area in the field and an analysis of aerial photographs (Doornkamp, 1970). The regions from which the drainage basins were taken are delimited on Figure 2.1 and their main characteristics are summarized in Table 2.1.

Because the drainage basins have been selected from a number of contrasting areas any relationships between their morphometric characteristics will not be dependent on just a single environmental influence. It is more likely that the relationships which become apparent through numerical analysis result from forces governing the development of drainage basins in general. Nevertheless, in the analysis of results in chapter 3 it will become apparent that different morphological regions produce variations in the results obtained through numerical analysis. The importance of environmental control, and the fact that the basins analysed come from different morphological regions will be critically re-examined in chapter 5.

Fig. 2.1   Location map of the areas of south-west Uganda for which morphometric data has been obtained. (*Mather and Doornkamp, 1970*) Ib—Ibanda, K—Kiruhura, M—Masaka, Mb—Mbarara 1 Marangara, 2 Singiro, 3 Buzhaga swamp, 4 Munyere swamp, 5 Rwampara Mountains, 6 Chezho valley, 7 Masha arena, 8 Oruchinga valley, 9 Lake Nakivali, 10 Ngarama Hills, 11 Lake Mburo, 12 Lake Kachira, 13 Lake Kijanebalola, 14 Kibale River.

Table 2.1 Some geomorphological characteristics of the areas analysed

Region	Amount of upland planation surface within basins measured	Mean relative relief between upland and lowland planation surfaces	Dominant rock types	Position with respect to the main divide between the drainage to L. Victoria and that to the rift valley
I Lake Nabugabo (Masaka)	small areas on most watersheds	300 feet	gneiss	within L. Victoria system
II Kyogya Valley	very small areas on some watersheds	300 feet	gneiss	within L. Victoria system
III Sanga	small areas on most watersheds	300–350 feet	gneiss	astride main divide
IV Nyabushozi	none	—	gneiss	astride main divide
V Bubale Valley	none	—	schist and gneiss	within system draining to rift
VI Mbarara Lowland	none	—	quartz-mica and mica schist	within L. Victoria system
VII Buhweju Mts.	small areas on most watersheds	1,300 feet	shales, phyllite and quartzite	within L. Victoria system
VIII Rwampara Mts.	small areas on most watersheds	1,100 feet	shales, phyllite and quartzite	within L. Victoria system

2.2 Collecting the data

All the land occupied by the drainage basins which are analysed below, is shown on the Uganda 1 : 50,000 topographic maps (series Y 732). Heights are indicated on these base maps by form lines drawn from aerial photographs and with reference to spot heights established on the ground. Not all drainage lines and branches of the valley system are marked by blue lines on these maps. These were inserted for the basins measured, from the aerial photographs (scale about 1 : 30,000). Some of the maps produced in this way were checked against morphological maps (of the type described in chapter 6) drawn in the field. The actual delimitation of the valley net has been possible to a degree commensurate with the amount of detail shown by the aerial photographs. Field checking suggests that the result is the same as would have been obtained if the valley net had been mapped entirely in the field. For much of the year many of the lower order branches of the valley net are without running water. During the rainy seasons, however, every branch carries running water and is thus an integral part of the drainage system. In this area there is, therefore, no choice as to whether or not the dry valley sections of the basins should be included in such an analysis. During most rainy seasons the valley and drainage systems are one and the same thing.

Areas were measured from the 1 : 50,000 maps with a polar planimeter. These values are thus *not* ground values but their horizontal equivalents. Lengths and distances were all measured with dividers. These measurements too are horizontal equivalents of ground lengths. Numbers, of course, are actual values. Heights have been determined from the relief information provided by the 1 : 50,000 base maps. The values are only as accurate as the maps themselves. Although the absolute values may not be precise they are certainly of the right order of magnitude. For example, if one hill crest is mapped at 4,300 feet and another at 4,350 feet, the heights of these hill crests may not be absolutely correct but the second hill will certainly be higher than the first.

As a result of these measurements the available data for analysis consist of 25 variables, either measured or derived from those measured, for 130 third-order basins, i.e. 3,250 numbers. For ease of handling such information may be kept on filing cards, with one card carrying all of the values for one basin. Alternatively, if the data are to be processed by a computer they may be stored on punch cards or on paper tape.

2.3 Descriptive statistics

An important step in the numerical analysis of geomorphological data is its organization. When a very large number of observations have been made it is necessary to organize these in such a way that their main characteristics can be seen quickly. In addition, it is useful if a few values can

be quoted which summarize the main properties of the observations made on each variable. These values include *measures of central tendency* and *dispersion* (or spread of data values). For example, the area of the basins in Uganda, together with the drainage density and number of first-order

Table 2.2   Frequency (f), cumulative frequency (c.f.), relative cumulative frequency (r.c.f.) and relative frequency in proportion (r.f.p.)

A_3	f	c.f.	r.c.f. %	r.f.p. %	N_1	f	c.f.	r.c.f. %	r.f.p. %
0– 1	34	34	26·2	26·2	0– 5	5	5	3·9	3·9
1– 2	32	66	50·7	24·5	5–10	32	37	28·5	24·6
2– 3	24	90	69·2	18·5	10–15	50	87	67·0	38·5
3– 4	10	100	76·9	7·7	15–20	25	112	86·2	19·2
4– 5	13	113	87·0	10·1	20–25	10	122	93·9	7·7
5– 6	2	115	88·5	1·5	25–30	4	126	97·0	3·1
6– 7	4	119	91·6	3·1	30–35	2	128	98·5	1·5
7– 8	4	123	94·7	3·1	35–40	1	129	99·3	0·8
8– 9	2	125	96·2	1·5	40–45	0	129	99·3	0·0
9–10	1	126	97·0	0·8	45–50	1	130	100·0	0·8
10–11	2	128	98·5	1·5					
11–12	1	129	99·3	0·8					100·1 [1]
12–13	1	130	100·0	0·8					
				100·1 [1]					

D_3	f	c.f.	r.c.f. %	r.f.p. %
1– 2	16	16	12·3	12·3
2– 3	37	53	40·7	28·4
3– 4	34	87	67·0	26·3
4– 5	11	98	75·4	8·4
5– 6	7	105	80·8	5·4
6– 7	13	118	90·8	10·0
7– 8	9	127	97·8	7·0
8– 9	2	129	99·3	1·5
9–10	1	130	100·0	0·8
				100·1 [1]

A_3: Area of third-order basin (square miles)
N_1: Number of first-order streams in third-order basin
D_3: Drainage density (miles per square mile) within third-order basins (south-west Uganda).

[1] Rounding errors prevent the r.f.p. column from summing to precisely 100%.

streams in each basin, can be summarized by a *frequency distribution table* (Table 2.2). Here the range of values observed is subdivided into classes and the number of observations in each class is recorded. Thus, 34 basins have an area of between 0 and 1 square miles. The information on a frequency distribution table can be shown on a graph, or frequency

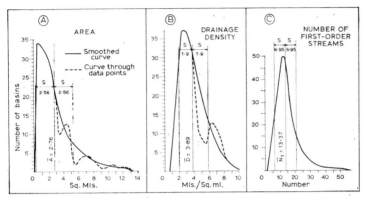

Fig. 2.2 Frequency distribution curves for **A** basin area (A_3), **B** drainage density (D_3) and **C** number of first-order streams (N_1).

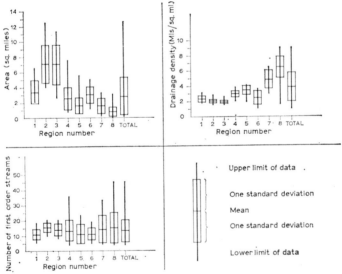

Fig. 2.3 Sample-range diagrams for area, drainage density and number of first-order streams in selected third-order basins in southern Uganda.

distribution curve (Fig. 2.2) where the number of observations in each class is plotted against the class value. An alternative form of diagrammatic representation of the data is by means of a *sample-range diagram* (Fig. 2.3), in which the range of data values is shown together with the mean and standard deviation (see Table 2.3). The former is a measure of central tendency in the data, the latter indicates its dispersion (see below).

There are three measures of central tendency, these are the mean, median and mode. The *mean* is well known as the average value. For the

130 third-order basins from Uganda the mean area ($\bar{A}_3$) is 2·76 square miles. The *median* is the value of the middle observation in the whole range of observations (i.e. the 65th value), and reference to Table 2.2 shows that this must be almost 2·0 square miles. (N.B. the 66th value is the last in the 1–2 square miles category.) The *mode*, on the other hand, is the value of the most frequently recurring observation or class. In the example of Table 2.2 the mode is best stated as the class which contains the most observations. For A_3 this is the 0–1 square miles class. Both the median and the mode indicate the importance of the two lowest classes

Table 2.3   Standard deviation calculated for the area of ten third-order basins, Kyogya Valley, Uganda

X	$X_i - \bar{X}$	$(X_i - \bar{X})^2$	
4·08	3·08	9·49	
4·53	2·63	6·92	$\bar{X} = \dfrac{\Sigma X_i}{n} = \dfrac{71·60}{10} = 7·16$
4·92	2·24	5·02	
6·18	0·98	0·96	
6·22	0·94	0·88	
6·54	0·62	0·38	$s = \sqrt{\dfrac{64·42}{10}} = \sqrt{6·442}$
7·55	0·39	0·15	
8·92	1·76	3·10	
10·16	3·00	9·00	$s = 2·54$ square miles
12·50	5·34	28·52	
Σ 71·60		Σ 64·42	

where X_i is the value of basin area in square miles.
Note: In calculating $X_i - \bar{X}$ negative signs are ignored.

in the total spread of values up to 13 square miles. The higher value of the mean takes into account all data values, whereas the median and the mode are obtained by reference to the number of observations rather than their magnitude.

The fact that most of the data observations are made in the first two size classes (for A_3) is also shown by the curve in Fig. 2.2. It produces an asymmetric curve with a high peak to the left-hand side of the histogram and a long tail to the right. This is described as a *skewed distribution*. A *normal distribution*, on the other hand, occurs when the curve is bell-shaped and symmetrical about its central axis. With a normal distribution the mean, median and mode all have the same value.

The spread of the data can be defined by measuring their dispersion (i.e. *standard deviation*). Knowledge of the standard deviation is most useful when the measured values are normally distributed. When this is the case two thirds of all the measurements lie within one standard deviation

of the mean, and 95% of the cases lie within two standard deviations of the mean.

Standard deviation (s) can be calculated from:

$$s = \sqrt{\frac{\Sigma\,(X_i - \bar{X})^2}{n}}$$

where Σ means "the sum of", X_i is each individual value, $\bar{X}$ is the mean of the observations, and n is the number of cases.

To show how the standard deviation may be calculated the area was measured for each of ten third-order basins tributary to the Kyogya Valley. These values are tabulated in Table 2.3. The mean area of the basins is 7·16 square miles and the standard deviation 2·54 square miles. The symbol σ is usually used to represent the standard deviation, but only when it relates to calculations based on the total population. In the example of Table 2.3 the standard deviation is calculated from only a sample of the population, thus s has been substituted for σ. It is possible however, to make an estimate ($\hat{\sigma}$) of σ from s by the equation:

$$\hat{\sigma} = \sqrt{\frac{n}{n-1}} \times s$$

This may be illustrated by reference to the example of Table 2.3:

$$\hat{\sigma} = \sqrt{\frac{10}{10-1}} \times 2\cdot54$$

$$= \sqrt{1\cdot111} \times 2\cdot54$$

$$= 2\cdot68 \text{ square miles}$$

Before discussing the value of these techniques in geomorphological studies it is important to note that the frequency distribution table (Table 2.2) can be expanded. Two more columns may be added relating to the *cumulative frequency* (c.f.) and *relative cumulative frequency* (r.c.f.) respectively. The former is derived by successively adding together the numbers in the frequency column (f), starting at the top of the table. The r.c.f. is the value of each cumulative frequency taken as a percentage of the total number of observations.

Not only does Table 2.2 provide a summary of the data, it also allows certain probability statements to be made. For example, as far as the values of A_3 are concerned the r.c.f. column shows that more than 50% of the basins have areas of less than 2 square miles.

Another, and final column, can be added to Table 2.2 and this is the *relative frequency in proportion* (r.f.p.) column. The values under this heading are stated as a percentage, and are the proportion that the frequency values in each class are of the total number of observations. For the 1-2 square miles class the calculation is:

$$\frac{32}{130} \times 100 = 24\cdot5\%$$

Table 2.4 Means and standard deviations (s) for basin area (A_3), drainage density (D_3) and number of first-order streams (N_1) in some third-order Uganda drainage basins

Area	n	Area		Drainage density		Number of first-order streams	
		$\bar{A}_3$	s	$\bar{D}_3$	s	$\bar{N}_1$	s
1 Lake Nabugabo	9	3·49	1·49	2·43	0·45	11·33	3·50
2 Kyogya Valley	10	7·16	2·54	2·14	0·33	15·90	3·21
3 Sanga	11	7·08	2·70	1·98	0·29	14·09	3·83
4 Nyabushozi	27	2·56	1·48	3·02	0·38	13·67	6·68
5 Bubale Valley	23	1·71	1·11	3·52	0·59	11·65	6·06
6 Mbarara Lowland	9	3·07	1·21	2·51	0·90	11·44	4·19
7 Buhweju Mts.	11	1·62	1·02	4·92	1·18	14·63	9·19
8 Rwampara Mts.	30	0·79	0·64	6·53	1·43	15·20	9·96
Whole sample	130	2·76	2·56	3·89	1·90	13·37	6·95

where n = number of basins in each morphological region

These relative frequency in proportion values can be divided by 100 in each case to allow a *probability* statement to be made. Thus there is a 0·015 probability that a basin chosen at random from the Uganda data will have an area of 8–9 square miles. Likewise the probability that a basin will have an area of 3–5 square miles is (0·077 + 0·101) = 0·178. This can be expressed another way. If A_i is the area of a randomly selected basin, then the probability (p) that it will have an area of between 3 and 5 square miles is:

$$p(3 < A_i \leqslant 5) = 0·178$$

(where $<$ means 'is less than', and $\leqslant$ means 'is less than or equal to'). The use of this type of probability statement is also described by Krumbein and Graybill (1965, 95).

Drainage density, within the Uganda third-order basins, ranges from 1 to 10, with 54·7% of the values lying between 2 and 4 miles per square mile. The drainage density values differ from one morphological region (Fig. 2.1) to another (Table 2.4). The Sanga region (region III), for example, has a mean value for D_3 of 1·98. The standard deviation of 0·29 indicates that 68% of the values of D_3, within the measured basins of the Sanga region, lie between 1·69 and 2·27 miles per square mile (i.e. 1·98 ± 0·29 miles per square mile). The Rwampara Mountains have the highest density, with a mean of 6·53 and a standard deviation of 1·43 miles per square mile. The whole set of 130 basins have a mean of $\bar{D}_3 = 3·89$ and $s = 1·9$ miles per square mile. These values may be compared with values obtained in other parts of the world (Table 2.5). They clearly lie near the lower end of these values, and conform to the tendency for the lower values of drainage density to occur on granite, gneiss and schist (regions I–VI) rather than on metasediments (regions VII–VIII).

Table 2.4 shows very little variation in the mean value of the number of first-order streams from one area to the next, though the standard deviations are high when compared with the mean values. Most (86·2 − 3·9 = 82·3%) of the individual basins (Table 2.2) have 5–20 first-order streams. The mean area of these basins varies from 0·79 square miles in the Rwampara Mountains (Table 2.4) to 7·16 square miles in the Kyogya Valley. Many of the standard deviations are again high compared with the mean values. This indicates a wide spread of the individual basin values within the data grouped by morphological regions.

The values shown in Table 2.4 indicate that $\bar{N}_1$ would not be of any value in discriminating between the morphological regions. However, if a variable is sought which distinguishes between one region and another, then on this limited evidence drainage density might be thought of as that variable. Likewise greater variations occur in the values of $\bar{A}_3$ than of $\bar{N}_1$, and $\bar{A}_3$ is thus the better discriminating variable. It is important, however, that a discriminating variable should show no overlap between its values for one area as compared with those of another. The amount of overlap is large for $\bar{N}_1$, as can be quickly detected from the sample

Table 2.5 Drainage density in selected areas (after Selby, 1967)

Location	Lithology	Climate (Köppen symbol)	Vegetation	Drainage density
Pennsylvania (USA)	horizontal resistant sandstone	humid continental (Cfa)	coniferous and deciduous forest	3 to 4
Dartmoor (UK)	granite	temperate maritime (Cfb)	heath, rough grass and deciduous woodland	3·5
Colorado (USA)	granite, gneiss and schist	humid montane (H)	montane forest	4 to 9
North Carolina (USA)	granite and gneiss	humid continental (Cfa)	coniferous and deciduous forest	5 to 8
Maryland (USA)	shale	humid continental (Cfa)	coniferous and deciduous forest	7
Volcanic Plateau (NZ)	pumice and ignimbrite	temperate maritime (Cfb)	scrub and grass	8·7
Ozark Plateau (USA)	cherts and cherty limestone	humid continental (Cfa)	deciduous forest and pasture grasses	14
New Mexico (USA)	dacite, basalt and pumice	humid montane (H)	montane forest	14 to 17
Arizona (USA)	basalt	semi-arid (BSh)	chaparral and grass	15 to 20
California (USA)	igneous and metasediments (deeply weathered)	Mediterranean (Csa)	chaparral	20 to 30
South Auckland (NZ)	greywacke (deeply weathered) overlain by volcanic ash	temperate maritime (Cfb)	formerly evergreen broadleaf forest now pasture grass	25·3
South Dakota (USA)	clays and shale	semi-arid (BSk)	sparse bunch grasses or none	80 to 260
Arizona (USA)	horizontally bedded shale	hot desert (BSh)	none	170 to 350
New Jersey (USA)	clay and sand fill	humid continental (Cfa)	none	550 to 1320

Note: These values are not all entirely comparable for they have been measured in a variety of ways.

range diagram (Fig. 2.3). None of these three variables, taken on its own, appears, for this reason, to be fully discriminatory for the morphological regions shown on Figure 2.1. Indeed, none of the variables examined for these basins proved to be discriminatory. In chapter 5 it will be shown that certain combinations of variables can be used to distinguish between basins drawn from differing regions.

2.4 Normality of data

In the introduction it was stated that many parametric statistical tests require that the data must be drawn from a population for which the descriptive statistics are known. Some tests specifically require, in fact, that the data should be *normally distributed*. Although the histogram for normally distributed data is bell-shaped, not all bell-shaped histograms indicate that the data, from which they are compiled, are normally distributed. One of the ways of testing for a normal distribution is to plot the data on arithmetical probability paper. This paper is specially ruled so that if a normal distribution is present then, when the class mid-values are plotted against their relative cumulative frequencies, the points fall along a straight line.

Methods exist, however, whereby data values can be modified, or *normalized*, to make them conform to the requirements of the statistical test (Krumbein and Graybill, 1965, chapters 5 and 6). The application of these tests is especially important with skewed data. When the majority of observations occur in the smaller size classes of the histogram then the curve is *positively skewed*. When most of the values are in the higher classes then the histogram is *negatively skewed*. The former can frequently be normalized by using the logarithm of the data values in the tests, and the latter by using their square roots. If no suitable method of normalization can be found then a non-parametric test may have to be substituted for the parametric one (see introduction).

Not all drainage basin properties have been found in the past to be normally distributed (Maxwell, 1960). As a result logarithmic-normalization has had to be performed on the variables:

channel length, area of basins of order $u + 1$,
basin diameter and perimeter, relief, drainage density.

With reference to the Uganda case study, Figure 2.4 has been compiled from the data of Table 2.2. These data have been plotted on arithmetical probability paper. For none of the three variables (A_3, N_1, D_3) do the data points lie on a straight line. This shows, as does the skewed form of the histograms (Fig. 2.2), that the data are not normally distributed. Normalization has been attempted by plotting the relative cumulative frequency against the logarithm of the class mid-values. This is represented

on Figure 2.4 by a broken line in the case of each variable. In no case has the log-normalization successfully produced a single straight line passing through the plotted points. In all cases, however, the points fall along more than one straight line, where the two (or more) straight lines have differing gradients. This is typical of those circumstances where the data being analysed have been drawn from two (or more) different populations. Thus, in the case of basin area (A_3) the break-point between the two straight lines occurs at 4·75 square miles, on both the absolute data and log-normalized data graphs. As far as N_1 is concerned the absolute data

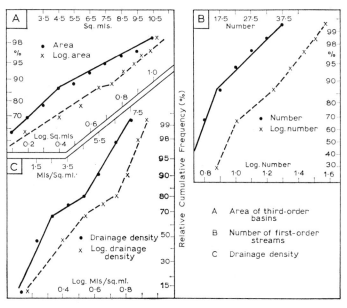

Fig. 2.4   Basin area, number of first-order streams and drainage density plotted on arithmetic probability paper.

provide a simpler relationship between the scatter of points than does the log-transformation, suggesting that the stream number data are normally distributed on either side of a break-point value of 17·5. Drainage density (D_3) provides almost identical graphs whether the data are log-transformed or not. In this context it is interesting to note that Krumbein and Graybill (1965) state that drainage density has a normal frequency distribution, while Maxwell (1960) found a log-normal distribution to apply. The Uganda data suggest that both may be possible, but in either case there is, in this case study, a distinct break-point at a value of 3·5 miles per square mile.

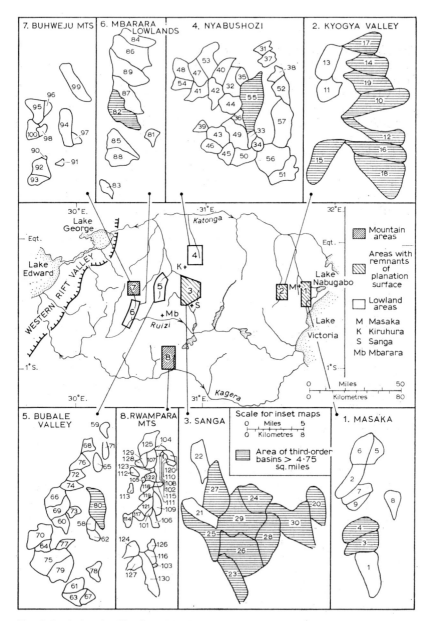

Fig. 2.5 A size classification of the Uganda drainage basins into those which exceed 4·75 square miles and those which do not.

2.5 Regional comparisons

Analysis of cumulative frequency graphs
The suggestion that the data shown on each of these graphs were drawn from two different populations is interesting because in fact regions VII and VIII are mountainous in character whilst the others are not. It is possible that the break in each of the graphs (Fig. 2.4) may separate the mountain basins from those of the lowlands. In the analysis of A_3 none of the mountain basins has an area which falls in the category that exceeds the break-point value of 4·75 square miles. In this sense the mountain basins are different from the other basins (Fig. 2.5). There are, however, differences between the other regions. As Table 2.1 indicated, some of the areas within regions I–VI carry remnants of the upland landscape planation surface, developed across gneiss bedrock which in part is deeply weathered. Where these planation surface remnants occur, the watershed areas are higher than where this surface no longer remains. At the same time it is found that the depth of the valley floors in these areas tends to be constant regardless of whether or not the watershed carries a planation surface remnant. This means that, where these remnants occur, the down-wearing of the ridge crests lags behind in the general denudation of the lowland areas. Changes in the position of divides is less likely as long as these remnants remain in the landscape. Once the remnant has been removed the watershed is rapidly lowered and a modification of the basin outline may well take place. The regional distribution of basin area (A_3) shown in Figure 2.5 suggests that where remnants of the upland landscape remain on the divides of the non-mountain basins, as in regions II and III, the areas occupied by the third-order basins tend to be significantly larger than elsewhere within this part of Uganda. The basins of region I, whose watersheds also carry remnants of the upland landscape, ought also to fall into this category, and indeed two of them do while two others come very close to it; but clearly other factors have to be taken into account. This illustrates the multivariate nature of geomorphological problems. A fuller consideration of multivariate analysis is deferred until chapter 4. In each of regions IV, V and VI there is just one basin which falls within the higher category, all of the others having a basin area of less than 4·75 square miles. Since these smaller basins no longer carry a remnant of the upland landscape, it would seem that third-order basins are developed more rapidly, that is to say within a smaller compass, once the divide has been lowered below the level of the upland landscape. Again, however, caution must be exercised, for a single answer to the problem of these regional differences may be a gross over-simplification of the real situation.

The regional distribution of the number of first-order streams in the two population groups, above and below 17·5 streams, shows little in the way of regional concentration (Fig. 2.6). All areas whether mountainous or not have members of both population groups. It is impossible to select

subjectively the determining factor behind this distribution, and the relationship between the number of first-order streams and other variables is reconsidered in chapters 3 and 4. Clearly N_1, on its own, is again shown to be useless as a discriminating variable.

The break-point for drainage density occurs at 3·5 miles per square mile. Most of the mountain basins (Fig. 2.7) are within the higher population category, and regions I, II and III (the areas intermediate between

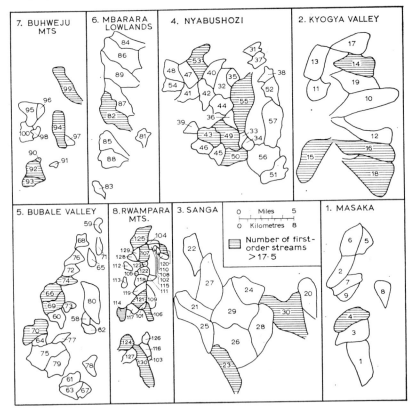

Fig. 2.6 The Uganda basins classified by the number of first-order streams they contain (above and below 17·5 streams).

mountain and lowland) are composed entirely of members of the lower population category. Regions IV and VI only contain one higher category member each. On this basis it would seem that the mountains can be successfully distinguished from the other areas by an analysis of drainage density values; drainage density is high in the mountain areas and low elsewhere. The one area which does not fall within this simple division is the Bubale Valley, where 14 of the 23 basins fall within the same category as the mountain basins. Drainage density is frequently dependent

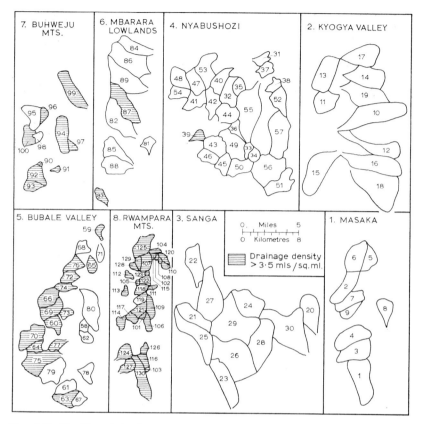

Fig. 2.7   The Uganda basins classified by their drainage densities (above and below 3.5 miles per square mile).

on the nature of the bedrock over which it has developed, and the combination of schists and gneiss within the Bubale Valley (Table 4.1) is unique amongst the lowland basins. The relatively high values of drainage density within some of the Bubale Valley basins may therefore indicate significant bedrock differences between these and the other lowland basins.

t-test for difference between group means

Table 2.4 records the mean value of basin area ($\bar{A}_3$) for the sample of basins drawn from each of eight morphological regions (Fig. 2.1). It is possible, through the *t*-test, to see if there is a statistically significant difference between the mean values of A_3 for any two of these regions, so long as the two samples have a normal distribution and their variances (s^2) are very nearly the same. For example, Table 2.4 shows that the 23 basins of the Bubale Valley (Fig. 2.1) have a mean area of 1·71 square

miles while the nine basins of the Mbarara Lowland have a mean area of 3·07 square miles. Their standard deviations, and therefore their variances, are nearly the same. Table 2.1 shows that the Bubale Valley and the Mbarara Lowland basins are developed on different rock types and within two different major river systems. It would be interesting, therefore, if a significant difference could also be detected in the mean area of the third-order basins drawn from these two areas. In both cases all but one of the basins belong to the group which is less than 4·75 square miles in area (Fig. 2.5).

In order to apply the *t*-test a *null hypothesis* (H_0) has to be set up, which states that the difference between the two mean values is due to chance and that there is no significant difference between them. The *t*-test is applied to see if H_0 can be rejected. First, however, a decision has to be made as to what can be accepted as a 'significant' difference. The symbol α is frequently used for *level of significance*, the commonly used values of which are 0·05 and 0·01. In the former case, for example, this is equivalent to saying there is a 95% probability that the result could not have occurred by chance, but that nevertheless there is a 5% (or 1 in 20) probability that the two means could have come from samples drawn from the same population.

There can be two types of error in making a decision in such a test. *Type I error* occurs if H_0 is rejected when in fact it is true. The probability of committing this error is given by α. Thus if α is small, i.e. 0·001, then there is little likelihood of committing a type I error. A *type II error* occurs if H_0 is accepted when in fact it is not true. Type II error is usually represented by β. The values of α and β can be specified before the tests are made, and when this has been done the number (n) in the sample that will give the required values can be determined. The smaller the errors that can be tolerated the larger the value of n must be. The probability of committing a type II error is called the power of the test, which may be defined as:

$$\text{Power} = 1 - \text{probability of type II error}$$
$$= 1 - \beta$$

Different statistical tests have different powers, so that it is possible to choose a test with the appropriate power. In general the parametric tests are more powerful than the non-parametric tests.

In order to apply the *t*-test the following values (Table 2.6) are required for both samples:

Table 2.6   Data required for a *t*-test

	No. of basins	$\bar{A}_3$	Variance[1]
Bubale Valley	27	1·71	1·11[2]
Mbarara Lowland	9	3·07	1·21[2]

[1] Variance is the square of the standard deviation.

To find the t-value the first step is to calculate from the two samples the best estimate $(\hat{\sigma}^2)$ of the variance of the population:

$$\hat{\sigma}^2 = \frac{27 \times 1.11^2 + 9 \times 1.21^2}{27 + 9 - 2}$$

$$= 1.37$$

In other words, this is calculated from the sum of the two basin numbers multiplied by the variance, and divided by the sum of the two basin numbers minus two. The latter number is known as the number of *degrees of freedom* (df). This is related to sample size and may be described as the number of observations that can be freely assigned before all the rest are completely determined. Thus, for each of the areas a mean value of basin area has been calculated, and this calculation uses up one degree of freedom. Looking at it in another way if all the data values except one are known, as well as the mean, then the last data value can be precisely determined. It can only have one value. In a sense, when the mean value is stated, although all other values cannot be calculated the last one can be. The number of degrees of freedom for each sample is therefore the number of data observations minus one. For the two samples taken together, as in the t-test, the number of degrees of freedom is equal to the sum of observations less two:

$$27 + 9 - 2 = 34$$

Next a value known as the *standard error of the difference between the means* $(\hat{\sigma}_w)$ has to be calculated:

$$\hat{\sigma}_w = \hat{\sigma}\sqrt{\frac{1}{27} + \frac{1}{9}}$$

$$= \sqrt{1.37} \times \sqrt{0.148}$$

$$= 0.445$$

The t-value is then:

$$t\text{-value} = \frac{\bar{A}_3 \,(\text{Bubale}) - \bar{A}_3 \,(\text{Mbarara})}{\hat{\sigma}_w}$$

$$= \frac{1.71 - 3.07}{0.445}$$

$$= (-)5.3$$

The t-value (the minus sign may be ignored) has now to be compared with tabled values of t to find out if the null hypothesis (that the two samples have been drawn from the same population) can be accepted or rejected. For this purpose it is again necessary to know the number of degrees of freedom, in this case 34. For $df = 34$ the tabled value of t at the 99.5% significance level is 2.73. The value of 5.3 calculated above is larger than this, and so the null hypothesis may be rejected. The probability of committing a type I error in this case is 0.005 (1−0.995), which is small. It is probable, therefore, that the difference in the mean

area of third-order basins in these two areas is not due to chance and reflects a real difference between them.

Mann–Whitney U-*test*

When data have been collected on at least the ordinal scale an alternative to the *t*-test is the Mann–Whitney *U*-test. This is a non-parametric test and is useful when the assumptions of the *t*-test cannot be fulfilled. Its purpose is to test whether or not two independent sets of data have been drawn from the same population.

The basis of this test is that the null hypothesis (H_0) is established which states that the two populations (**K** and **N** in the example given below) from which the samples are drawn have the same distribution. The alternative hypothesis (known as H_1) is that the samples are drawn from populations having different distributions. For example, the values of **K** may, in the main, be larger than those of **N**, in which case H_1 would be accepted. This situation arises when the probability is greater than one half that any value drawn from **K** is larger than a value drawn from **N**. In other words:

$$p(k > n) > \tfrac{1}{2}$$

where p is the probability, k and n are single values drawn from **K** and **N** respectively, and $>$ means 'is greater than'.

Table 2.4 shows that for the area of third-order basins the Kyogya Valley (**K**) and Lake Nabugabo (**N**) areas have differing variances (s^2). Since they are neighbouring regions (Fig. 2.1) with similar characteristics (Table 2.1) it could be argued on geomorphological grounds that the two sets of data are drawn from the same population. Doubt is thrown on this argument by the large difference in the mean area of third-order basins in these two regions (7·16 square miles for **K** and 3·49 square miles for **N**). The alternative hypothesis (H_1) would be, therefore, that the basins from **K** are significantly larger than those from **N**.

For the *U*-test the data are set out ranked from smallest to largest for each area (Table 2.7). The data for the two areas is then combined in

Table 2.7 The Mann–Whitney U-test for the area of drainage basins in the Kyogya Valley (K) and Lake Nabugabo (N) regions of south-west Uganda

A

K	4·08	4·53	4·92	6·18	6·22	6·54	7·55	8·92	10·16	12·50
N	2·11	2·18	2·26	2·47	2·60	3·78	4·56	4·85	6·62	

B

2·11	2·18	2·26	2·47	2·60	3·78	4·08	4·53	4·56	4·85
N	**N**	**N**	**N**	**N**	**N**	**K**	**K**	**N**	**N**

4·92	6·18	6·22	6·54	6·62	7·55	8·92	10·16	12·50
K	**K**	**K**	**K**	**N**	**K**	**K**	**K**	**K**

rank form, but the identity of the region from which the basin comes is maintained by the reference letters **K** and **N** (Table 2.7B). The value U is obtained by counting the number of times that the score in group **K** precedes a score in group **N** in the combined ranking table. Thus, for example, the **N** score of 2·11 is not preceded by a **K** score, nor indeed are any of the first six values of **N**. For the **N** score of 4·56 there are two preceding **K** scores and for the final score of **N** there are six preceding **K** scores. The value of U is given by summing these values, in this case $2 + 6 = 8$. In this example the number of observations is ten in the case of the **K** values and nine in the case of the **N** values. To test the significance of $U = 8$ it is necessary to compare this value with critical values of U as given in Appendix Table K of Siegel (1956). In the case where the number of observations in the larger group is ten and in the smaller group nine, the critical value of U is 8 at the 0·001 level of significance. In other words the null hypothesis may be rejected at the 0·001 level. The alternative hypothesis that the basins of the Lake Nabugabo area are significantly larger than those of the Kyogya Valley area may be accepted with considerable confidence.

The test has shown, therefore, that the assumptions of similarity between the two regions suggested by Table 2.1, are not verified when an analysis is made of the size of the third-order drainage basins (A_3) within each region. Indeed, A_3 may be thought of as a variable that discriminates between the two regions, and thus might be used to decide where the boundary between them lies. The cause for this difference is not indicated by the statistical test. The latter acts as a search procedure for isolating significant differences between regions. Causes have to be sought by further geomorphological investigation.

2.6 Summary

This discussion of the numerical properties of some drainage basins in Uganda indicates how a large amount of data can be reduced to manageable proportions by the use of descriptive statistics. These include the calculation of means and standard deviations, the drawing of histograms and the testing for normality in the statistical distribution of the data by the use of cumulative frequency curves plotted on probability paper. Three variables, namely, the area, drainage density and number of first-order streams in third-order basins, have been examined to illustrate the use of the t-test and the U-test as means of testing whether or not two samples are drawn from the same population. This account has touched upon the power of each of these variables to discriminate between different morphological regions, which will be discussed again in chapter 5.

References

DEPARTMENT OF LANDS AND SURVEYS, UGANDA 1967: *Atlas of Uganda.* (2nd Edn.)

DOORNKAMP, J. C. 1968: The nature, correlation and ages of planation surfaces in southern Uganda. *Geogr. Ann.* **50-A**, 151–61.

— 1970: The geomorphology of the Mbarara area. *Univ. of Nottingham, Dept. of Geogr. and Uganda Geol. Survey and Mines Dept., Geomorph. Rept.* **1**. (78 pp.)

DOORNKAMP, J. C. and TEMPLE, P. H. 1966: Surface, drainage and tectonic instability in part of southern Uganda. *Geogr. J.* **132**, 238–52.

KRUMBEIN, W. C. and GRAYBILL, F. A. 1965: *An introduction to statistical models in geology.* New York: McGraw-Hill. (475 pp.)

MAXWELL, J. C. 1960: Quantitative geomorphology of the San Dimas experimental forest, California. *University of California, Department of Geology, Technical Rept.* **19**.

SIEGEL, S. 1956: *Nonparametric statistics for the behavioral sciences.* New York: McGraw-Hill. (312 pp.)

SELBY, M. J. 1967: Morphometry of drainage basins in areas of pumice lithology. *Proc. Fifth N.Z. Geogr. Conf., N.Z. Geogr. Soc., Auckland* 169–74.

3 Pair-wise relationships among morphometric variables

A frequently recurring problem in geomorphology concerns an assessment of the relationship between two variables. The relationship between the area of a drainage basin and the total length of its stream network is a problem of this type, and there are several methods by which it may be numerically defined. These include product-moment and rank correlation methods as well as linear regression, and they will be dealt with here in that order. The data to which these techniques are applied in this chapter consist of the morphometric variables from Uganda described in chapter 2.

These variables are considered under the headings—basin geometry; basin area and stream lengths; relationships among the measures of stream length; stream numbers; drainage density, and valley gradient. The results obtained from an analysis of third-order basins are compared with those derived from the analysis of fourth-order basins, also measured in southern Uganda, and in some cases with the results of other studies.

3.1 Pearson product-moment correlation coefficient

Correlation coefficients are used to define the degree to which changes in the value of one variable are repeated in the behaviour of another variable. For example, let us suppose that a graph is drawn of basin area (A_3) against total stream length $(\Sigma L)_3$ for a set of third-order drainage basins. If all of the data points lie along one straight line then there is a perfect correlation between A_3 and $(\Sigma L)_3$. The two variables would be steadily changing together. In this case the correlation coefficient $r = +1$. If on the other hand A_3 increases in exact ratio to decreases in $(\Sigma L)_3$, then $r = -1$. Between these two extremes the value of r will indicate the level of correlation which exists between A_3 and $(\Sigma L)_3$. The intermediate situation of $r = 0$ occurs when there is absolutely no correlation between the two variables.

In general terms the *product-moment correlation coefficient*, r, may be calculated from:

$$r = \frac{\Sigma (X - \bar{X})(Y - \bar{Y})}{\sqrt{\Sigma (X - \bar{X})^2 (Y - \bar{Y})^2}}$$

$$= \frac{\Sigma xy}{\sqrt{\Sigma x^2 \Sigma y^2}}$$

(where $\bar{X}$ and $\bar{Y}$ are the mean values of the independent and the dependent variables respectively, and $x = X - \bar{X}$, and $y = Y - \bar{Y}$). Alternatively it may be applied in the re-written form:

$$r = \frac{\Sigma XY - n\bar{X}\bar{Y}}{\sqrt{(\Sigma X^2 - n\bar{X}^2)(\Sigma Y^2 - n\bar{Y}^2)}}$$

An example is presented in Table 3.1 where A_3 is the X term and $(\Sigma L)_3$ the Y term. The values are taken from the Kyogya Region (Fig. 2.1). The calculations (Table 3.1) show that

$$r = 0.929$$

which is very close to $r = +1$ (i.e. a perfect correlation), and means that basin area and total stream length tend to increase together. However, the significance of the r value is dependent on the number of observations upon which it is based. The level of significance can be assessed from t-tables, where:

$$t = \frac{r\sqrt{n - 2}}{\sqrt{1 - r^2}}$$

In the example of Table 3:1.

$$t = \frac{0{\cdot}929\sqrt{10-2}}{\sqrt{1-0{\cdot}929^2}} = \frac{0{\cdot}929 \times 2{\cdot}83}{1{\cdot}165}$$

$$t = 2{\cdot}26$$

This is significant at the 0·975 level. That is to say, there is a high probability that the correlation coefficient is significant. The level of correlation shown by a sample, in this case ten basins, cannot be applied directly to the population from which the sample is drawn. The larger the sample,

Table 3.1 The calculation of r, the product-moment correlation coefficient, for basin area (A_3) and total stream length ($\Sigma L)_3$ in the Kyogya drainage system

X A_3	Y $(\Sigma L)_3$	x $A_3 - \bar{A}_3$	y $(\Sigma L)_3 - \overline{(\Sigma L)}_3$	x^2	y^2	xy
4·08	9·40	−3·08	5·50	9·49	30·25	16·94
4·53	10·85	−2·63	4·05	6·92	16·40	10·65
4·92	10·69	−2·24	4·21	5·02	17·72	9·43
6·18	10·97	−0·98	3·93	0·96	15·44	3·85
6·22	11·94	−1·92	2·96	3·69	8·76	5·68
6·54	18·69	−0·62	−3·79	0·38	14·36	2·35
7·55	17·21	−0·30	−2·31	0·15	5·34	0·90
8·92	18·14	1·76	−3·24	3·10	10·50	5·70
10·16	20·14	3·00	−5·24	9·00	27·46	15·72
12·50	20·68	5·34	−5·78	28·52	33·41	30·87
$\bar{A}_3 =$ 7·16	$(\Sigma L)_3 =$ 14·90			Sum = 67·23	179·64	102·09

$$r = \frac{\Sigma\, xy}{\sqrt{\Sigma\, x^2\, \Sigma\, y^2}} = \frac{120{\cdot}09}{12077{\cdot}2} = \frac{102{\cdot}09}{109{\cdot}9} = 0{\cdot}929$$

however, the closer the sample correlation coefficient will come to the actual value of the coefficient which would result from a study of the total population. Ezekial and Fox (1959, Fig. 17.2) provide a diagram from which it is possible to convert the value of the sample coefficient to the value which would probably have been obtained if the population had been analysed. Thus, in the case of the relationship between A_3 and $(\Sigma L)_3$ for the Kyogya Region the correlation coefficient was 0·929 for ten third-order drainage basins. Comparison with the diagram in Ezekiel and Fox shows that this represents a probable correlation coefficient of at least 0·51 within the universe. The observed correlation comes closest to the true correlation when both the number of observations is large and the observed correlation coefficient is high. Some relationships may be illustrated by Table 3.2.

Difficulty occurs, however, in defining what the geomorphological 'total population' actually is. It could be all of the third-order drainage

basins in the world, or just the third-order drainage basins of the tropics, or indeed just those of southern Uganda. The precise definition of this type of 'population' must await much more research, for it requires a careful analysis of significant differences between pair-wise relationships in many areas. If the number of observations (n) is large then the calculation of the correlation coefficient (r) is a very long process best undertaken by a computer. This is especially true if correlation coefficients are

Table 3.2 Observed correlation in sample compared with the true correlation within the total population for differing sample sizes

No. observations in sample	Observed correlation in sample	True correlation of population
5	0·82	0·10
10	0·62	
20	0·46	
50	0·33	
5	0·88	0·40
10	0·76	
50	0·58	
5	0·95	0·70
10	0·89	
50	0·80	
5	0·98	0·90
10	0·965	
50	0·94	
100	0·927	

required between a large number of pairs of variables. Indeed it is possible for the computer to produce in one operation a matrix of correlation coefficients for a large number of variables. Although in reality the value of a correlation coefficient must lie between -1 and $+1$ it is more convenient when showing these values in tabular form to list them as $r \times 100$. A matrix of correlation coefficients is much tidier than a pair-wise list of variables with their correlation coefficients.

3.2 Uganda case study—product-moment correlation analysis

Results of an analysis of the correlation between morphometric variables, drawn from third-order basins in Uganda, are presented in matrix form in Table 3.3. The variables are listed along the margins of the table. When necessary these have been normalized by using log-values, for it is a

Table 3.3 Product-moment correlation matrix for 18 morphometric variables (Uganda, third-order basins). Data normalized by log-transformation in some cases

POSITIVE CORRELATIONS (upper‑right) / NEGATIVE CORRELATIONS (lower‑left)

	log A₃	log (Σ N)₃	log F₃	log L₁	log L₂	log L₃	log (Σ L)₃	log L̄₁	log L̄₂	log N₁	log N₂	log R₁ 1&2	log R₁ 2&3	log Rb 1&2	log D₃	Rh	log H
log A₃																	
log (Σ N)₃	44																
log F₃	83	70															
log L₁	85	55	74														
log L₂	87	51	79	66													
log L₃	92	68	91	85	86												
log (Σ L)₃	77	5	65	48	61	65											
log L̄₁	78	18	56	87	52	67	56										
log L̄₂	55	94	79	64	61	77	14	31									
log N₁	44	81	57	60	50	61	6	12	76								
log N₂	37	8	16	60	19	29	3	70	17	7							
log R₁ 1&2	39	41	48	12	70	44	24	40	43								
log R₁ 2&3	30	45	49	25	33	46	14	33	61			18					
log Rb 1&2																	
log D₃	78		92														
Rh	78		77												73		
log H	34		52												59	70	
log h	7														20	27	72

The numerical values listed = $r \times 100$; N = 130.
Coefficients in bold type are significant at better than the 99% level, while those in italics are significant at better than the 90% level.

condition of the product-moment technique that if the correlation coefficients are to be checked against significance tables, these coefficients should only be obtained from normalized data. The matrix of Table 3.3 is divided into two parts by a principal diagonal. The top right-hand corner includes all of the positive correlation coefficients, and the lower left-hand corner lists the negative coefficients. In each case their level of significance is also noted.

Basin area and stream lengths
The correlation coefficients between basin area and the variables of stream length are consistently significant at the 99% level, and many are significant at the 99·99% level. The correlation coefficient between basin area and the mean length of streams of a particular order is not usually as high as the coefficient between area and total stream length at that order. Nevertheless, the coefficients are still high enough for a significant correlation to be expressed, in most instances, between basin area and mean stream lengths. This implies that, within larger drainage basins, an increase in total stream length is due to the presence of longer streams and not necessarily because of an increase in the number of streams. This is also suggested by the fact that the correlation between A_3 and the total number of streams, or the number of streams at each order, fails either to be significant or to reach as high a level of significance as does the coefficient between basin area and stream length.

Stream lengths
There is nearly always a high level of correlation between the various measures of stream lengths (Table 3.3). The highest correlation coefficient between total stream length and the length of streams at any other order is always greatest with L_1. In fact, it is usually the case that, of all the streams within a drainage basin, it is the first-order streams which supply the greatest amount to the total length of the drainage network, despite the fact that the mean length of first-order streams is less than that of any other order. This implies that what the first-order streams lack in length of individual members they make up for in numbers. This is also indicated by the high correlation coefficient which usually occurs between $(\Sigma L)_3$ and N_1.

Stream number
The correlation table (Table 3.3) shows that there is a high correlation between the total number of streams and total stream length, between the number of streams and total stream length at each order, and between the number of streams at one order when compared with the number at the next highest order. All of these relationships bear logical examination and are consistent with the view that drainage basins tend towards a state of internal order and organization.

Drainage density

The correlation matrix (Table 3.3) shows that drainage density has no strong correlation with either of the two variables from which it is derived, namely basin area and total stream length. Other components of the drainage basin do, however, appear to have a relationship to drainage density. Positive correlations are to be found, for example, between $\log D_3$ and $\log F_3$, R_h, $\log H$ and $\log h$ (see Table 1.1 for meaning of abbreviations).

The relationship between drainage density and height values is of interest. In the analysis of the third- and fourth-order basins of Uganda, drainage density is greatest where there is the largest amount of basin relief. That is to say within the mountain regions. The implication that this is also associated with those basins having the steepest relief ratio is matched by the significant correlation, in the case of the third-order basins, between D_3 and R_h. Thus one can now suggest that mountain basins with steep slopes and considerable relative relief will have a greater drainage density than lowland basins of small relative relief. This may in itself be a reflection of rock type, for the mountains are composed of quartzites and phyllites while the lowlands consist of gneiss and schists. This conclusion supports the hypothesis raised by a different type of analysis in section 2.5.

Heights and basin area

The analysis of the fourth-order basins of southern Uganda indicates that the area of these drainage basins decreases with altitude. Thus the largest fourth-order basins tend to be found within the lowlands and the smallest occur within the mountains. This could be the result of a number of factors, including differences in lithology and amounts of precipitation. At the moment it remains a matter of observation and for comparison with other areas.

3.3 Spearman rank correlation test

If it is possible only to provide ranked data it is still possible to calculate a correlation coefficient. Only ranking may be possible, for example, where the scale or accuracy of the base map does not allow precise measurements to be made. Nevertheless it may still be possible to say, for example, which are the largest and which the smallest drainage basins, or which has the longest drainage network. It is possible of course to use rank correlation techniques where absolute data are precise, but in so doing some of the precision of the original data is ignored, for the rank correlation technique is designed for use with data measured on an interval scale. In the product-moment tests allowance is made, in the calculation of the correlation coefficient, for the interval between successive values. When the data are ranked, however, the interval between successive values in the ranking

table is unity, no matter what the difference between actual values happens to be. An advantage of rank correlation techniques is that they do not require the original data to be normally distributed. One ranking method is known as the *Spearman rank correlation test*, for which the data are set out as in Table 3.4. Here the drainage basin example has been

Table 3.4 Calculation of the Spearman rank correlation coefficient for the data of Table 3.1

Basin no.	Rank according to area	Rank according to stream length	Difference in rankings (d)	d^2
1	1 (smallest)	1 (shortest)	0	0
2	2	3	1	1
3	3	2	1	1
4	4	4	0	0
5	5	5	0	0
6	6	8	2	4
7	7	6	1	1
8	8	7	1	1
9	9	9	0	0
10	10 (largest)	10 (longest)	0	0

$$\Sigma\, d^2\ 8$$

ρ is calculated from:

$$\rho = 1 - \frac{6\,\Sigma\, d^2}{n_3 - n}$$

$$= 1 - \frac{6 \times 8}{1000 - 10} = 1 - \frac{48}{990}$$

$$\rho = 0.9515$$

used again so that the rank correlation coefficient (ρ (rho) $= 0.9515$) may be compared with the product moment correlation coefficient ($r = 0.929$).

This is a slightly higher coefficient than in the case of the product-moment correlation. Its significance has to be assessed by reference to a table such as that in Siegel (1956, 284). A ρ value of 0.56 for ten basins is significant at the 95% level, and $\rho = 0.74$ would be significant at the 99% level. The value obtained $\rho = 0.9515$ is in excess of both these values and is therefore highly significant.

3.4 Uganda case study—rank correlation analysis

In this section the discussion is limited to an analysis of the third-order basins of Uganda (Table 3.5) which were also the subject of the product-moment analysis of Table 3.3. A comparison of the results obtained by these two different techniques shows that high correlation coefficients tend

Table 3.5 Spearman's rank correlation coefficients for 14 morphometric variables (Uganda, third-order basins)

Upper-right region = POSITIVE CORRELATIONS; lower-left region = NEGATIVE CORRELATIONS.

	A_3	L_1	L_2	L_3	$(\Sigma L)_3$	$\bar{L}_1$	$\bar{L}_2$	N_1	$R_1\,1\&2$	$R_1\,2\&3$	$Rb\,1\&2$	D_3	Rh	H
A_3		**84**	**84**	**90**	**91**	**76**	**77**	**50**	**37**	**34**	**27**			
L_1			**75**	**82**	**97**	**64**	**56**	**80**	**19**	**39**	**49**			
L_2				**67**	**84**	**48**	**86**	**59**	**62**	07	20			
L_3					**89**	**63**	**53**	**55**	20	**61**	**40**			
$(\Sigma L)_3$						**64**	**65**	**75**	**30**	**40**	**41**			
$\bar{L}_1$							**58**	11	06	18	14			
$\bar{L}_2$								**28**	**71**	**71**	**31**			
N_1									**21**	**35**	**55**	03		**21**
$R_1\,1\&2$										18	19			
$R_1\,2\&3$											**31**	02		02
$Rb\,1\&2$														20
D_3	77	37	55	58	47	68	67		**33**	11	01		**78**	**48**
Rh	83	56	72	67	66	62	74	25	**42**	18	09			**63**
H	25	08	21	13	02	13	32		29	03				20

The numerical value listed = $r \times 100$; N = 130.
Coefficients in bold type are significant at better than the 99% level.

to occur between the same pairs of variables in each case, and significant correlations are the same in most of the pair-wise relationships. Since the calculations involved in the Spearman test are both less complex and less time consuming than those required for the calculation of the product-moment correlation coefficient, it must be concluded that in many cases, especially where a computer or even high-speed electronic calculator is not available, initial investigation of the data is probably better carried out by a rank correlation technique than by product-moment analysis.

3.5 Linear regression

When a significant correlation exists between two variables then an equation can be found which best defines the behaviour of one variable with respect to the other. This equation is known as a *simple regression equation*. The first step in regression analysis is to plot, on graph paper, one variable against the other. When the scatter of points tends to follow a distinct trend (i.e. tends to fall along a straight line) then it is worth calculating the regression equation.

The linear regression technique may be illustrated by reference to the relationship between the area of a drainage basin and the total length of

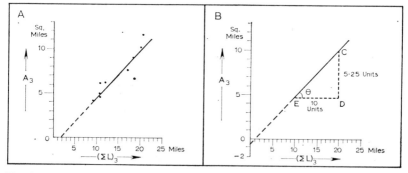

Fig. 3.1 The relationship between basin area (A$_3$) and total stream length (Σ L)$_3$ in 10 third-order basins, Kyogya Valley, Uganda.

the streams which it contains. A correlation analysis of ten third-order drainage basins in the Kyogya Region (Table 3.1) showed that the larger the basin the longer the total length of the streams which it contains. From the values of A$_3$ and (Σ L)$_3$ given in Table 3.1 a graph can be drawn (Fig. 3.1A) which supports the hypothesis that the greater the values of A$_3$, the higher the associated value of (Σ L)$_3$. The relationship between A$_3$ and (Σ L)$_3$ can be summarized by a regression line (which is itself described by a simple regression equation) drawn to pass as closely as possible to all points. Since this is a straight line all that is necessary to define its position is the value of A$_3$ (on the Y-axis) when (Σ L)$_3$

(on the X-axis) is zero (in this case -0.75), and its slope (0.525). The slope, or gradient, of any line is given by the tangent of the angle (θ) which the line makes with the horizontal. Thus in Figure 3.1B:

$$\tan \theta = \frac{5.25}{10} = 0.525$$

The fact that the gradient is positive means that the line slopes upwards from left to right, a negative gradient occurs when the line slopes upwards from right to left. If the point where the line crosses the Y-axis when $X = 0$ is called a, and the gradient of the line is represented by b, then the linear regression equation is:

$$Y = a + bX$$

In the example cited:

$$Y = 0.75 + 0.525X$$

The value b is also known as the *coefficient of regression* (regression coefficient). The value of b will be positive when both X and Y increase together, and negative when an increase in either X or Y is associated with a decrease in the value of the other variable. Since a is the constant value equal to Y when $X = 0$, the line can only pass through the origin [$X = 0$, $Y = 0$, or simply $(0, 0)$] when $a = 0$.

Figure 3.1B also shows that each point representing one of the ten basins does not fall exactly on the regression line. There may be a number of reasons for this. It is possible that they would have fallen on the line but for errors of measurement. On the other hand, it may also be that there are influences other than total stream length which determine the value of basin area, thus preventing a perfect relationship (correlation) from being demonstrated by means of a regression line between these two variables. It may also be that the precise relationship between A_3 and $(\Sigma L)_3$ does not follow a straight regression line but a curved one. In any investigation between pairs of geomorphological variables these possibilities have to be examined.

Dependent and independent variables

In the regression equation of the form $Y = a + bX$, the variable X is known as the independent variable, and Y as the dependent variable. The values of Y are controlled by the values of X. Y is therefore a *response variable* whose values are a result of changes in the independent variable X. The point of view expressed by these terms is particularly important when the regression equation is used for prediction purposes. For example, the value of total stream length—$(\Sigma L)_3$ in Fig. 3.1—may be known, but not that of basin area (A_3). The value of A_3 could be estimated by substituting the measured value of $(\Sigma L)_3$ in the equation $A_3 = -0.75 + 0.525(\Sigma L)_3$ (already derived from analysis of the data in Table 3.1). Quite apart from the accuracy of such a prediction, which will be considered below, the geomorphological assumption, which may

incorrectly be made from the use of such an equation, is that changes in total stream length *cause* changes in basin area. This has not been said in so many words, but by using the mathematical terms 'dependent' and 'independent' variables, there is a danger of thinking that the two variables are in fact geomorphologically dependent and independent. This is seldom the case. All geomorphological variables are dependent variables in the genetic sense of the word. Stream lengths, within a drainage basin, may depend on the area of the basin. It is also possible that the area of the basin may be dependent upon the lengths of its streams. Headward extension by streams may cause a retreat of the basin divide and an increase in its area. In this case area is related to (dependent upon) increases in stream length. On the other hand two adjacent basins may carry streams which are no longer actively extending headwards, yet because one basin is much larger than the other it is almost bound to have a longer stream network. In this case total stream length is dependent on area, i.e. the roles of these two variables have now been reversed.

This serves to illustrate the caution which has to be employed in using the terms 'dependent' and 'independent' for the variables when geomorphological characteristics are being measured. For the purposes of numerical analysis, however, it is both useful and necessary to assume, in testing a particular hypothesis, that one variable is dependent and then to see how its values behave when compared with those of one (or more) so called independent variables.

Finding the 'best-fit' straight line
It has just been suggested that the regression equation can be used to *predict* values of the dependent variable from a known value of the independent variable. It is more difficult to define the accuracy with which that prediction can be made. The confidence which can be placed in a predicted value of Y increases if the data points are never far from the line, especially if there is also an increase in the number of points for which the regression equation is calculated. Precise values of a and b in the regression equation can be calculated. This is done by setting up a table of the type shown in Table 3.6. The coefficients a and b in the linear regression equation $[A_3 = a + b(\Sigma L)_3]$, for this example, can be derived from the following equations:

$$b = \frac{\Sigma (XY) - n\bar{X}\bar{Y}}{\Sigma (X)^2 - n(\bar{X})^2}$$

$$a = \bar{Y} - b\bar{X}$$

where, in this example, n is the number of points (10); $\bar{Y}$ is the mean value of basin area (7·16 square miles); $\bar{X}$ is the mean value of total stream length (14·898 miles). By substituting these values and those of Table 3.6:

$$b = \frac{1177\cdot8 - 10 \times 14\cdot898 \times 7\cdot16}{2401\cdot4 - 10 \times 14\cdot898^2} = 0\cdot223$$

$$a = 7\cdot16 - 0\cdot223 \times 14\cdot898 = 3\cdot84$$

By substituting for a and b the best-fit regression line for this example is:

$$A_3 = 3 \cdot 84 + 0 \cdot 223 (\Sigma \, L)_3$$

The position of this line through the scatter of points can be found by

Table 3.6 Computation of values for determining the best-fit regression line for basin area (A_3) and total stream length [$(\Sigma \, L)_3$], by the method of least squares. (Data same as Table 3.1)

Y (A_3)	X [$(\Sigma \, L)_3$]	X² [$(\Sigma \, L)_3$]²	XY [$A_3 . (\Sigma \, L)_3$]
4·08	9·40	8·4	38·4
4·53	10·85	117·6	49·2
4·92	10·69	114·2	52·5
6·18	10·97	120·5	67·7
6·22	11·94	142·5	74·4
6·54	18·69	349·3	122·0
7·55	17·21	296·3	130·0
8·92	18·41	339·0	164·1
10·16	20·14	405·8	221·0
12·50	20·68	427·8	258·5
Σ 71·59	148·98	2401·4	1177·8

giving $(\Sigma \, L)_3$ any two values, say zero and ten, and finding the associated value of A_3. This 'least-squares' regression line, as it is known, has been plotted in Figure 3.2 together with that originally drawn by eye for the same data. Clearly the latter is grossly in error, and this is brought out by comparing the coefficients of each of the two equations.

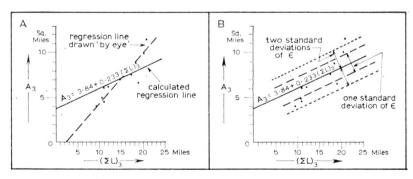

Fig. 3.2 Regression between basin area (A_3) and total stream length $(\Sigma \, L)_3$ in 10 third-order basins, Kyogya Valley, Uganda. **A**: Least-squares 'best-fit' line compared with that obtained 'by eye'; **B**; Confidence bands of the regression line.

Prediction and residuals

In geomorphological studies it is rare for all data points to fall exactly on the regression line. The prediction of a value of A_3' from $(\Sigma L)_3$ by the formula:

$$A_3' = 3 \cdot 84 + 0 \cdot 223(\Sigma L)_3$$

will give the value which A_3 would have if it actually lies *on* the regression line. In reality this may not be so, and the actual value of $(A_3)_i$ (that is to say the value of A_3 at a point i) will be:

$$(A_3)_i = 3 \cdot 84 + 0 \cdot 223[(\Sigma L)_3]_i + \varepsilon_i$$

where ε_i, the *residual* value, is the distance between the actual point on the graph and the value $(A_3)_i$ derived from the regression equation. If the point does fall on the regression line then $\varepsilon_i = 0$. ε_i will vary for each point. In calculating the least-squares linear regression equation the sum of the ε^2 values has been reduced to a minimum. In determining the accuracy of a regression line it is necessary to tabulate the ε^2 values for each data point. These values are then used to find the *standard error of estimate*. The concept of a standard error of estimate for a regression line is similar to that of standard deviation and may be illustrated by referring again to the data of Table 3.1 which are used in Table 3.7.

Table 3.7 Calculation of the standard error of estimate for the equation
$A_3 = 3 \cdot 84 + 0 \cdot 223(\Sigma L)_3$

A_3	$(\Sigma L)_3$	$(A_3)'$	ε
4·08	9·40	5·93	−1·85
4·53	10·85	6·26	−1·73
4·92	10·69	6·22	−1·30
6·18	10·97	6·28	−0·10
6·22	11·94	6·50	−0·28
6·54	18·69	8·00	−1·46
7·55	17·21	7·68	−0·13
8·92	18·41	7·94	+0·98
10·16	20·14	8·32	+1·84
12·50	20·68	8·47	+4·03

Where $(A_3)'$ is the estimated value of A_3 from the regression equation and found by substituting the value of $(\Sigma L)_3$ in the second column. $\varepsilon = A_3 - A_3'$.

The values of the residuals (ε) vary from $+4 \cdot 03$ to $-1 \cdot 85$. The standard deviation of the residuals (s) is known as the *standard error of estimate*:

$$s = \sqrt{\frac{\Sigma \, (\varepsilon_i - \bar{\varepsilon})^2}{n}}$$

where $\bar{\varepsilon}$ is the mean value of the residuals, and ε_i is the value of the residual for each of the ten drainage basins in the example.

Thus:

$$s = \sqrt{\frac{\Sigma\,(\varepsilon_i - 1{\cdot}37)^2}{10}} = 1{\cdot}1$$

This value may be marked alongside a regression line (Fig. 3.2B) by two lines drawn one on either side of the regression line and equal to one standard deviation from it. This measurement is made parallel to the Y-axis. If the residuals from an analysis of the Kyogya Valley data are normally distributed two thirds of the residual values will lie between $+1{\cdot}1$ and $-1{\cdot}1$. Once the standard deviation of the residual values has been obtained it becomes possible to comment on the degree of accuracy of the regression line ($A_3 = 3{\cdot}84 + 0{\cdot}233(\Sigma\,L)_3$). In this case only once in ten times is the area likely to be more than $2{\cdot}2$ square miles in error as an estimated value.

The standard error of estimate derived from a sample does not usually coincide with the value which would have been obtained by considering the total population. In general the values obtained from a sample are smaller than those which would occur with the whole population. To allow for this the sample standard error of estimate can be adjusted for the linear regression example by:

$$\sigma_\varepsilon{}^2 = s^2\left(\frac{n}{n-2}\right)$$

where σ_ε is the standard error of estimate for the total population; s is the standard deviation (standard error of estimate) of residuals for the sample; n is the number of sample data points. For the Kyogya Valley drainage basins, therefore:

$$\sigma_\varepsilon{}^2 = 1{\cdot}217\left(\frac{10}{10-2}\right) = 1{\cdot}52$$

The variance of the residual values for the population (σ_ε) is, therefore, $\sqrt{1{\cdot}52}$ or $1{\cdot}23$ square miles. No statement of accuracy can be made beyond the range of the sample data values. If it becomes necessary for values of X to be estimated from Y (instead of Y from X, as illustrated above) then a new regression equation has to be calculated which makes the sum of squares of the distances (measured parallel to the X-axis), between each observation and the new regression line, a minimum.

The purpose of the least-squares linear regression equation is to obtain the statement which will provide the best estimate of the dependent variable from values of an independent variable. In order to do this it is necessary always to make the sum of the squares of the residuals a minimum parallel to the axis of the dependent variable, regardless of whether it is plotted along the X- or the Y-axis.

Confidence in a *and* b

The coefficients a and b of the regression equation are also established with varying degrees of confidence according to the level of correlation which exists in the sample and the number of sample points. The accuracy with which the slope of the line b is established may be judged as follows. The variance of b is:

$$V(b) = \frac{\sigma_\varepsilon^2}{\Sigma\,(X_i - \bar{X})^2}$$

where σ_ε^2 is the variance of the residual values; X_i is each value of X between 1 and i; $\bar{X}$ is the mean of all the X values. The standard error of b is the square root of its variance:

$$\text{Standard error } (b) = \frac{\sigma_\varepsilon}{[\Sigma\,(X_i - \bar{X})^2]^{\frac{1}{2}}}$$

Table 3.8   Values required to calculate the estimated standard errors of a and b, for the Kyogya Valley example

X_i	$(X_i - \bar{X})$	$(X_i - \bar{X})^2$	X_i^2
9·40	−5·50	30·26	88·4
10·85	−4·05	16·04	117·6
10·69	−3·21	10·30	114·2
10·97	−3·93	15·45	120·5
11·94	−2·96	8·76	142·5
18·69	3·79	14·36	349·3
17·21	2·31	5·34	296·3
18·41	3·51	12·25	339·0
20·14	5·24	27·45	405·8
20·68	5·78	33·41	427·8
$\Sigma 148\cdot98$		173·62	2401·4

$\bar{X} \simeq 14\cdot90$ (where X represents $(\Sigma\,L)_3$ in this example)

An estimate (s^2) of σ_ε^2 will normally have to be used. An analysis of a sample thus gives a value for the slope of the regression line which is a best estimate of the slope that would apply to the population as a whole. Thus the estimated standard error of b is given by:

$$\text{estimated standard error } (b) = \frac{s}{[\Sigma\,(X_i - \bar{X})^2]^{\frac{1}{2}}}$$

$$= \frac{1\cdot1}{13\cdot28} = 0\cdot0835$$

On the assumption that the residual values (ε_i) follow a normal distribution, it can be shown that $100(1 - \alpha)\%$ confidence limits may be established for b by multiplying its standard error by the t-value for $(n - 2, 1 - \frac{1}{2}\alpha)$, where this is the $(1 - \frac{1}{2}\alpha)$ percentage point of a t-distribution, with $(n - 2)$ degrees of freedom. (Note $(n - 2)$ is the number of degrees of freedom on which the estimate s^2 is based.) The probability

level of $\frac{1}{2}\alpha$ is taken since the test is used to examine whether the coefficient differs from zero; it is not concerned whether b is greater than or less than zero. Thus, in the case of

$$A_3 = 3\!\cdot\!84 + 0\!\cdot\!233\,(\Sigma\,L)_3,$$

the estimated standard error of b is $0\!\cdot\!0835$ and the t-value for eight degrees of freedom at the 95% confidence level is $2\!\cdot\!306$:

$$0\!\cdot\!0835 \times 2\!\cdot\!306 = 0\!\cdot\!1925$$

That is to say:

$$b = 0\!\cdot\!223 \pm 0\!\cdot\!1925$$

at the 95% confidence level.

It is also possible to test the null hypothesis that the true value of b is zero, which would mean that in this sample there is no clear relationship between A_3 and $(\Sigma\,L)_3$. This is done from the equation:

$$|\,t\,| = b \div \text{standard error of } b$$

$$= \frac{0\!\cdot\!223}{0\!\cdot\!0835} = 2\!\cdot\!671$$

Since $|\,t\,| = 2\!\cdot\!671$ exceeds the critical value of $2\!\cdot\!306$ for $t\,(8,\,0\!\cdot\!975)$, the null hypothesis that $b = $ zero can be rejected at the 95% confidence level. Note that $t\,(8,\,0\!\cdot\!975)$ is a short-hand way of writing 'the value of t with eight degrees of freedom at the $0\!\cdot\!975$ level of significance'.

The standard error for the a parameter of a straight-line regression equation can be defined:

$$\text{standard error }(a) = \left[\frac{\Sigma\,X_i{}^2}{n\,\Sigma\,(X_i - \bar{X})^2}\right]^{\frac{1}{2}}\sigma_\varepsilon$$

and, as for b, since σ_ε is not known, although the variance of the residual values in the sample (s^2) is known:

$$\text{estimated standard error }(a) = \left[\frac{\Sigma\,X_i{}^2}{n\,\Sigma\,(X_i - \bar{X})^2}\right]^{\frac{1}{2}}s$$

As in the case of b, a confidence level can be established such that the limits of a are given by:

$$a \pm t(n - 2,\,1 - \tfrac{1}{2}\alpha)\left[\frac{\Sigma\,X_i{}^2}{n\,\Sigma\,(X_i - \bar{X})^2}\right]^{\frac{1}{2}}s$$

For the drainage basin example, therefore, let $\alpha = 0\!\cdot\!05$ so that the t-value for eight degrees of freedom and $(1 - \tfrac{1}{2}\alpha) = 0\!\cdot\!975$ is required, which is $2\!\cdot\!306$ as before. For the above equation it is also necessary to know $\Sigma\,X_i{}^2$ (Table 3.8). The value of a together with its limits is, therefore,

$$3\!\cdot\!84 \pm 2\!\cdot\!306\left[\frac{2401\!\cdot\!4}{10 \times 173\!\cdot\!62}\right]^{\frac{1}{2}}1\!\cdot\!1$$

i.e. $3\!\cdot\!84 \pm 2\!\cdot\!66$

3.6 Uganda case study—linear regression

The relevance of linear regression analysis to drainage basin studies may be illustrated by reference to the Uganda data. Significant correlation coefficients have been found to exist between many of these variables. The relationships may be further qualified by calculating regression equations for selected pairs of variables.

Basin geometry
There is a significant linear relationship between the values of the area of third- and fourth-order drainage basins and their respective breadths (Br) and lengths (Lb). Likewise there is a high level of correlation between basin breadth (Br) and the total distance around the basin watershed, or perimeter (P). These relationships are illustrated for the third-order basins by the linear regression lines of Figure 3.3, and the significant

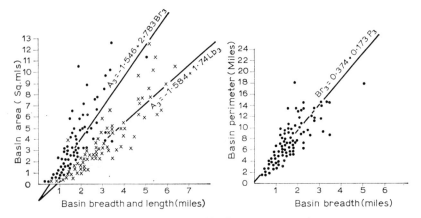

Fig. 3.3 Linear regression for aspects of basin geometry.

statistical values associated with these lines are presented in Table 3.9. A description of one of these groups of values will help in their interpretation. In the comparison of basin area in the case of the third-order basins (A_3) with basin length (Lb_3) (the subscript three denotes third-order basin) the linear regression line has the equation:

$$A_3 = -1 \cdot 584 + 1 \cdot 74 Lb_3$$

Compared with the general form $Y = a + bX$, the constant a is $-1 \cdot 584$ and the slope of the regression line b is $1 \cdot 74$. These numerical values of a and b, ignoring the difference in their signs, are of the same order, but there is a real difference between the variance and standard errors at the origin of these two values (Table 3.9). Both the variance ($0 \cdot 039$) and the standard error at the origin ($0 \cdot 197$) are much greater for a than they are for b ($0 \cdot 00486$ and $0 \cdot 0697$ respectively). Within the regression equation,

Table 3.9 Coefficients for linear regression equations between some morphometric properties of 130 third-order basins in southern Uganda

| Variables | | Constant of regression eqt. (a) | Regression coefficient (b) | Variance at origin | | Standard error at origin | | t-value | | Correlation coefficient r | Coefficient of explanation R^2 (%) |
Dependent (Y)	Independent (X)			of a	of b	of a	of b	for a	for b		
A_3	Lb_3	−1·584	1·74	0·039	0·00486	0·197	0·0697	**−8·02**	**24·93**	0·91	82·9
Lb_3	A_3	1·18	0·477	0·005	0·00036	0·0719	0·0191	**16·43**	**24·93**	0·91	82·9
A_3	Br_3	−1·546	2·783	0·0811	0·0273	0·2859	0·1659	**−5·43**	**16·84**	0·83	68·9
Br_3	P_3	0·374	0·173	0·007	0·0002	0·084	0·0109	**4·46**	**15·84**	0·81	66·3
A_3	$(\Sigma L)_3$	−0·852	0·466	0·04	0·0005	0·20	0·022	**−4·25**	**21·2**	0·88	77·8
A_3	L_1	−0·512	0·725	0·056	0·0019	0·236	0·044	**−2·17**	**16·49**	0·83	68·0
A_3	L_2	0·086	1·53	0·056	0·012	0·237	0·109	0·36	**14·02**	0·78	60·5
A_3	L_3	−0·199	1·99	0·051	0·0155	0·226	0·125	−0·877	**15·97**	0·82	66·6
$(\Sigma L)_3$	L_1	0·708	1·56	0·075	0·0026	0·275	0·051	2·58	**30·51**	0·94	87·9
$(\Sigma L)_3$	L_2	2·366	3·083	0·161	0·0341	0·402	0·185	**5·89**	**16·70**	0·83	68·5
$(\Sigma L)_3$	L_3	1·92	3·93	0·153	0·0464	0·391	0·0215	**4·91**	**18·22**	0·85	72·0
$(\Sigma N)_3$	L_3	9·181	1·14	1·143	0·014	1·069	0·118	**8·59**	**9·64**	0·65	42·1
$(\Sigma N)_3$	A_3	14·79	1·14	1·046	0·074	1·023	0·27	**14·45**	**4·20**	0·35	12·1
F_3	$D_3{}^2$	4·763	0·533	6·32	0·0097	2·52	0·098	1·89	**5·42**	0·43	18·7
$\log F_3$	$D_3{}^2$	1·528	0·0386	0·004	0·000007	0·0665	0·0026	**22·98**	**14·86**	0·79	63·3

Bold numbers are for t-values significant at better than the 0·0005 level. The t-values for r are the same as those for b, they are significant at better than the 0·001 level.

therefore, the slope of the regression line is much more precisely established than is the constant a. Nevertheless, although the t-value of b (24·93) is much greater than that of a (−8·019) both are sufficiently high to accept the actual values of a and b with more than 99·9% confidence; indeed, both values are better than that of 3·373 at the 0·0005 level of significance. The t-value for a is negative only because a is itself negative.

The correlation coefficient between A_3 and Lb_3 is 0·91. As important, however, is the t-value for this correlation coefficient. In this case it is 24·9 which is very much better than 3·373 at the 0·001 significance level indicated by t-tables. Thus, there is a 99·9% confidence that the actual correlation between these two variables is not zero. The cautious note suggested by this last phrase signifies the persistent reality that statistical tests only *indicate* probable relationships, they do not prove them. In every instance in geomorphological studies it is up to the investigator to reason the validity of a statistical relationship. It is not surprising that an increase in basin length should be associated with an increase in basin area. It is no more surprising that this is as true for fourth-order basins as it is for third-order basins (Table 3.10). The advantage in having defined the relationship between these two variables is that the equation can now be used for prediction purposes. Basin length is much easier to measure than basin area, and when new basins are examined, a rapid, and fairly precise estimate of basin area can be obtained by inserting the value of Lb in the regression equation, *assuming* that the new basin comes from a similar environment and falls within the range of values of Lb from which the regression line was calculated.

Although any pair of variables can supply the data for two possible regression lines there can only be one correlation coefficient for the relationship between the two variables. For example let

$$A = a_1 + b_1 L$$

and

$$L = a_2 + b_2 A$$

Lines can be drawn for each of these equations. It has been found, however, that there is a definable relationship between the relative positions of these two graphs and the correlation coefficient. The higher the value of r the smaller the angle between the two lines. Indeed the correlation coefficient can be calculated from the slope of the two lines:

$$r = b_1 . b_2$$

When a pair of variables is selected for regression analysis a decision has to be made as to which of them shall be considered as the dependent and which the independent variable. In the analysis of A_3 and Lb_3, just described, A_3 was taken as the dependent variable, on the assumption that an increase in A_3 would result from an increase in Lb_3. The increase in Lb_3 might result, for example, from the headward extension of valleys most distant from the basin mouth. It is also possible, therefore, to make

Table 3.10 Coefficients for linear regression equations between some morphometric properties of 34 fourth-order basins in southern Uganda

Variables		Constant of regression eqt. (a)	Regression coefficient (b)	Variance at origin		Standard error at origin		t-value		Correlation coefficient r	Coefficient of explanation R^2(%)
Dependent (Y)	Independent (X)			of a	of b	of a	of b	for a	for b		
A_4	Lb_4	−5·314	3·33	0·623	0·0356	0·789	0·189	**−6·73**	**17·64**	0·95	90·6
$\log A_4$	Br_4	−0·29	0·72	0·044	0·0073	0·21	0·0855	−1·36	**8·41**	0·83	69·6
Br_4	P_4	0·236	0·19	0·05	0·00039	0·00039	0·0198	1·05	**9·63**	0·86	74·4
A_4	$(\Sigma L)_4$	−2·63	0·348	1·705	0·00165	1·306	0·0407	−2·02	**8·57**	0·83	69·6
A_4	L_1	−1·43	0·54	2·36	0·0072	1·54	0·085	−0·93	**6·38**	0·75	56·0
A_4	L_2	−2·60	1·42	1·59	0·025	1·26	0·159	−2·06	**8·90**	0·84	71·2
A_4	L_3	−0·82	2·51	2·09	0·15	1·45	0·39	−0·57	**6·47**	0·75	56·7
A_4	L_4	−1·89	3·45	0·67	0·064	0·82	0·25	−2·32	**13·59**	0·92	85·2
$(\Sigma L)_4$	L_1	1·509	1·965	1·422	0·0044	1·193	0·066	1·27	**25·7**	0·98	95·4
$(\Sigma L)_4$	L_2	1·637	3·825	3·104	0·0497	1·762	0·23	0·93	**17·15**	0·95	90·2
$(\Sigma L)_4$	L_3	5·798	6·996	6·442	0·463	2·538	0·6806	2·28	**10·28**	0·88	76·8
$(\Sigma L)_4$	L_4	6·586	8·205	6·662	0·696	2·581	0·834	2·55	**9·83**	0·87	75·1
$(\Sigma N)_4$	L_4	21·53	2·41	36·52	0·0354	6·04	0·188	3·56	**12·79**	0·92	83·7
$(\Sigma N)_4$	A_4	60·65	3·549	89·98	0·85	9·49	0·92	**6·394**	3·85	0·56	31·7
$(\Sigma N)_4$	L_1	15·20	3·428	17·35	0·053	4·17	0·23	3·65	**14·88**	0·94	87·4

The t-values for r are the same as those for b, they are significant at better than the 0·001 level. Bold numbers are for t-values significant at better than the 0·005 level.

Lb_3 the dependent variable and A_3 the independent variable, in which case the regression equation has the form:

$$Lb_3 = 1.18 + 0.477A_3$$

The correlation coefficient and its t-value are the same as before. There is a difference, however, in the variance and standard errors at the origin of both a and b when compared with the values calculated from the first equation, when A_3 is the dependent variable. This time for b an absolute value of 0.477 has a variance of 0.000366 (compared with 1.74 and 0.00486 when A_3 was the dependent variable) and a standard error of 0.0191 (compared with 0.0697). In this second equation, the value of b is known within narrower limits than was previously the case. The main difference between the two equations concerns the constant term a. In the new equation a is 1.18, which is of a similar order of magnitude to the $(-)1.584$ obtained in the first equation. Here, however, its variance (0.005 compared with 0.039) and standard error at the origin (0.0719 compared with 0.197) are smaller, and the estimate of a is thus better than it was with A_3 as the dependent variable. The t-value for b is unchanged from that of the first equation, but for a it has improved to 16.43, justifying an increased confidence in the calculated value of a. The new equation is, therefore, the better for prediction purposes. In every analysis of the Uganda third-order basins it has been found that the best prediction equations are obtained by placing A_3 as the independent variable.

Whichever way the variables are treated it makes no difference either to the correlation coefficient or to its derivative, the R^2 value. The R^2 value is a measure of the percentage explanation of the variation in the dependent variable which is accounted for by the behaviour of the independent variable. It is known as the *coefficient of determination*. In the comparisons of basin area and basin length the R^2 values exceed 80% in both the third- and the fourth-order basins (Tables 3.9 and 3.10). Basin breadth (Br) is less well explained by basin perimeter (P), but even so R^2 is nearly 70% in both the third- and fourth-order basins. When basin area is compared with basin breadth, again nearly 70% of the variation in the one is explained by the other. It is interesting to note, however, that if the log values of A_4 are plotted against Br_4 there is a better level of explanation than if the actual values of A_4 are used. Breadth and length are not so closely related to each other. Despite the fact that the correlation coefficient between them is 0.692 (for 130 third-order basins) and that it has a t-value of 10.8, indicating the high likelihood that the two are correlated, the variation in the one is only explained to a level of 47.8% by the other.

The confidence which can be placed in a and b, of the regression equation, varies from one of the pair-wise relationships to the next. In all instances the slope of the line b has a t-value which exceeds the critical value for the 0.001 level of significance. (The critical values for this level are 3.373 for the 130 third-order basins, and 3.646 for the 34 fourth-order

basins.) This means that considerable confidence can be placed, in each case, in the slope of the line. As a result the regression lines indicate with some precision the rate at which the values of one variable are changing with respect to the other. Less confidence can be placed in the value of the constant a in some of the equations, especially those derived for the fourth-order basins. Neither the t-value of a ($-1\cdot36$) in the comparison of log A_4 with Br_4, nor that for the comparison of Br_4 and P_4 ($1\cdot05$), reach the critical t-value of $3\cdot646$. In fact only the former reaches the critical t-value of $1\cdot310$ for the $0\cdot2$ level of significance. Thus, in some cases the value of a (the point at which the linear regression line crosses the Y-axis) is less precisely determined than is the slope of the line.

An examination of the slope values, b, in Table 3.9 shows that in the third-order basins the rate of increase of A_3 is greater with a unit increase in Br_3 than it is with a unit increase in Lb_3. The rate of change in A_4 with a unit change in Lb_4 is considerably greater, however, than is the case with its third-order counterpart.

Basin area and stream lengths
The correlation coefficients between items of stream length and drainage basin area are all high and they all have a t-value significant at better than the $0\cdot001$ level (Table 3.9), suggesting that there is likely to be a close relationship between these variables. The highest percentage explanation of the variation in the values of basin area, in the case of the third-order basins, is achieved by total stream length. In the linear regression equation

$$A_3 = -0\cdot852 + 0\cdot466(\Sigma \text{ L})_3$$

the t-values for both a and b are significant at better than the $0\cdot0005$ level, indicating that this is also a good prediction equation. In the case of the fourth-order basins the rate of change of A_4 with $(\Sigma \text{ L})_4$ is well established as $0\cdot348$ square miles per mile of total stream length. Once more however, the point at which the regression line cuts the A_4 axis is not so confidently determined. Indeed, the standard error at the origin is half of the value of a, indicating that its calculated position on the A_4 axis could be as much as 50% in error.

The slope of the regression line ($b = 0\cdot348$) is of some value if a comparison is to be made between this pair-wise relationship and that which exists in other areas. It is also of some use in investigating the deviations of the individual basins of southern Uganda from this general trend. As a prediction equation, however, this regression line for the fourth-order basins is not very valuable. Even if the value of the constant a had been determined to a high level of significance the equation would still have little practical prediction value, for it is just as much labour to measure total stream length as it is to measure basin area. On the other hand, once the stream network has been ordered the measurement of the length

of the fourth-order stream is very little trouble, and at the same time is of some use for predicting values of A_4. The equation

$$A_4 = -1.89 + 3.45L_4$$

has its b value established with considerable confidence, while its a value, though not as confidently established, is significant at better than the 0.025 level.

The rate of increase of A_3 and A_4 with the values of stream length increases as the order of the streams increases. In other words, an increase in the length of the third-order stream (or fourth-order stream as the case may be) is associated with a larger increase in basin area than is an increase in the total length of the first-order streams. This does not mean, for example, that an increase in the length of first-order streams, through their headward extension, does not bring about an increase in basin area, for it will. If a new first-order stream is established as a tributary to another first-order stream within a third-order basin, then downstream of their junction a new second-order stream is created by the logic of the Strahler classification procedure (see chapter 1). As a result a complete re-ordering of the basin may be necessary, and in fact a new, but smaller, third-order basin may thus be created. As Schumm (1956) suggests, it is logical to assume that for any one order there is an upper limit to both basin area and the mean length of its streams. Above these limiting values new bifurcations will occur to form new basins. Schumm's analysis of some first- and second-order basins near Perth Amboy indicated that the transformation from first- to second-order takes place within a wide range of values. Field examination showed, however, that the first-order basins with high area values suggested imminent development into second-order basins. It was also found that the lower area values among the second-order basins were associated with basins that had only recently achieved second-order status and were capable of extension and enlargement.

Schumm (1956) also found a close relationship between mean basin area ($\bar{A}$) and mean stream length ($\bar{L}$). The ratio $\bar{A}/\bar{L}$ is in fact the area required to maintain one unit of drainage channel, and is called the *constant of channel maintenance*. This constant is a measure of texture similar to drainage density and is proportional to $1/D$. *Length of overland flow*, on the other hand, is the distance over which surface run-off will flow before concentrating into permanent channels, and is equal to $1/D^2$. Chorley (1957) sums this up by referring to *Schumm's Law of Contributing Areas* which states that the relationship between basin area and total stream length at a given order is a direct log function, the regression coefficient of which is unity. The value of the area when total stream length is unity equals Schumm's 'constant of channel maintenance (C)'. In other words:

$$\log A_u = \log C + 1 . \log (\Sigma L)_u$$

Stream lengths

There appears to be a high correlation between the total length of streams within a drainage basin and each of the lengths at the different orders (Table 3.9). This implies that all sections of the drainage net contribute to greater total stream lengths. That is to say, where one basin has a much longer total drainage net this will be associated with greater total lengths within each of the different orders. In the case of the fourth-order basins (Table 3.10) the contribution provided by the length of the first-order streams in explaining the variation in total stream length from one basin to another, is greater ($R^2 = 95.38\%$), than is the contribution made by any other order in the network. In fact there is a progressive decrease in the R^2 value with an increase in stream order. The reason why the various R^2 values do not sum to 100%, in either the third- or fourth-order basins, is that the independent variables within each of the regression equations are correlated amongst themselves.

In the case of the relationship between $(\Sigma L)_3$ and L_1, the *t*-value for the slope of the regression line is highly significant and implies that the rate of increase of $(\Sigma L)_3$ with an increase in L_1 is confidently established. In the third-order basins a unit increase in the total length of first-order streams produces an increase of 1.56 miles in the total stream length within the basin, while in the fourth-order basins this increase is fractionally greater at 1.965 miles. The rate of increase in $(\Sigma L)_u$ increases with each increase in stream order, especially within the fourth-order basins. A unit increase in L_4 within the fourth-order basin produces a much greater increase in $(\Sigma L)_4$ than a unit increase in L_3 within a third-order basin (i.e. 8.205 miles compared with 3.93 miles).

Stream numbers

There is not as close a relationship between the number of streams and the area of a drainage basin as might at first have been supposed. Variation in the value of $(\Sigma N)_3$ is accounted for by A_3 only to a level of 12.1%, and by A_4 to 31.7%. In other words, most of the variation in stream number is accounted for by variables other than basin area in southern Uganda. Field examination shows that stream number is closely related to bedrock. Higher levels of correlation may occur when the analysis is confined to one rock type, rather than to the wide range of lithologies which occur within the 130 Uganda basins. The total number of streams does, however, give a better level of correlation with total stream length in both the third- and fourth-order basins, and is very high for the comparison of $(\Sigma N)_4$ and L_4. This is interesting because, as has been shown, A_3 and L_3, and A_4 and L_4 are themselves highly correlated. It would have been likely, therefore, that they would both have shown similar levels of correlation with $(\Sigma N)_3$ and $(\Sigma N)_4$. In a case such as this, the independent relationships of basin area and total stream length to the total number of streams can best be assessed by means of partial correlation methods (section 4.2).

Drainage density and stream frequency

An analysis can be made of the relationship between drainage density (D) and stream frequency (F). In Uganda a close relationship was found to exist between $\log F_3$ and D_3^2 (Table 3.9) Elsewhere it has been found that in maturely dissected regions the relationship between F and D can be expressed as $F = 0.694\ D^2$, giving a straight line on a plot of D v F on double log paper (Melton, 1958). Melton (1957) sees the relationship as a mathematical model where $F = kD^2$, implying that with increasingly fine-textured topography F increases proportionally with D^2. Since $F = \Sigma N/A$ and $D = \Sigma L/A$ this shows that an increase in ΣL is due to increases in the number of channels and not to longer segments in each order.

Regional differences

So far this chapter has been concerned with the pair-wise relationships between variables for the 130 third-order basins measured in southern Uganda. In each analysis a regression statement has been derived that describes the general relationships and which defines the limits of the accuracy of those relationships. In many cases the correlation between the two variables has been close, and there is a strong linear relationship between them. It is possible to continue the task of adding more and more values into the data so that the regression line comes continually closer to defining the general relationships of the total population in the universe. Beyond a certain sample size this would, however, become a pointless exercise as each additional point contributes less and less in terms of its ability to alter the established trend of the line. It is more useful to consider different environments and to compare the regression line between variables as measured, by similar methods, in those different environments. It is important, however, that when two areas are compared the data are obtained by the same methods and with respect to the same criteria in each case. Comparisons can seldom be made between the absolute results obtained by different research workers simply because these conditions of comparability cannot be met.

Where a mass of data, such as that obtained in Uganda, has been drawn from a variety of morphological regions, it is also interesting to break the data down into their original components and to see how they compare with the general trend established from the total mass of data. This means that instead of just being concerned with the general relationship, as defined by the regression line, interest is now focused on the scatter of data points about that line. In Figure 3.4A, for example, each point on the graph represents one third-order drainage basin. There is a wide scatter among the data points despite the fact that nearly 70% of the variation in A_3 is defined by L_1 (see Table 3.9). It is also clear from the distribution of the data points that there is a general increase in the values of A_3 with L_1.

An additional refinement can be introduced if the data points are

divided into two sets, namely the mountain basins (marked with a dot) and all the other basins (marked with a cross). The rate at which A_3 increases with a unit increase in L_1 is not as great in the case of the mountain basins as it is in the case of the basins measured in the lowland and intermediate regions. This can be enumerated by defining separate regression equations for these two different environments (Fig. 3.4A). Thus, within the mountains a unit increase in L_1 produces an increase of 0·233 square miles in A_3, while within the rest of the basins it produces an increase of 0·863 square miles in A_3.

This difference may be a reflection of the fact that within the steep-sided mountain basins there is a higher density of first-order streams than in the non-mountain environments. This results in a high total length of

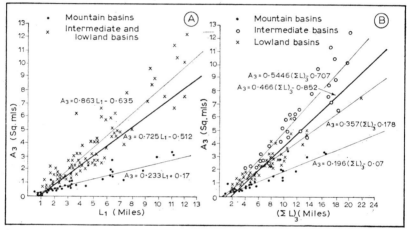

Fig. 3.4 Linear regression **A**: between area of third-order basin (A_3) and length of first-order streams (L_1) ; **B**; for area and total stream length $(\Sigma L)_3$.

first-order streams within fairly small drainage basins. Within the low-land and intermediate regions much larger third-order basins are required if they are to contain the same total length of first-order streams. It is hard to say whether this is a function of rock type or slope steepness, the point being that only the mountain basins are at one and the same time both steep-sided and cut into phyllites and quartzites.

When total stream length is compared with basin area (Table 3.9), it is found that a significant correlation exists between these two variables. In addition, the regression line $A_3 = -0·852 + 0·466(\Sigma L)_3$ is well established both in terms of its slope and in terms of the constant a. Even so, when the scatter of data points is considered (Fig. 3.4B) there is some divergence from the general trend especially amongst the high values of $(\Sigma L)_3$. A more refined description of the behaviour of the scatter of data points is obtained if the basins are grouped into three sets showing which basins were derived from the mountain, lowland and intermediate

morphological regions respectively. When this is done, and new regression lines calculated, it is apparent that some of the original scatter is reduced to three separate alignments. In each case the regression line provides a closer estimate of the behaviour of the data than the line through the total mass of data. From Figure 3.4B it is apparent that it is within the intermediate regions that a unit increase in $(\Sigma L)_3$ produces the greatest increase in A_3, while this increase is lowest within the mountain regions. The main contrast here is between the relationship derived for the mountain basins when compared with the other two. The same explanation is applicable as for the relationship between L_1 and A_3 (Fig. 3.4A). The greater density of streams on the steep mountain slopes has the effect of bringing about tributary junctions within smaller distances than are found in the lowland and intermediate regions. As a result a third-order basin is developed within a smaller area, despite the fact that total stream length may have reached significant proportions. The data also suggest that greater basin areas are obtained for the same total stream length within the intermediate regions than in the lowland regions. The intermediate regions tend to have the drainage characteristics of the lowland basins but, because their watersheds still carry remnants of the highest planation surface of southern Uganda, they tend to have larger basins.

3.7 Some complications in the analysis of pair-wise relationships

It has been shown that the relationship between drainage basin variables, taken two at a time, may be examined by correlation and linear regression methods. This type of morphometric data presents problems, however, when the nature of the process–response relationships is not precisely known. Thus in some instances it is extremely difficult to decide which should be the 'response' and which the 'process' variable. This stems from the fact that most geomorphological relationships are part of a complex system of interrelationships. In other words, most geomorphological problems are problems of multivariate analysis.

The fact that a numerically significant correlation coefficient has been found between two variables does not imply that they must be genetically related. For example, it was found that there is a significant correlation (at the 99·99% confidence level) between the values of L_1 and L_3 within the fourth-order basins of southern Uganda. There is no obvious reason why the total length of first-order streams should have any influence on the total length of third-order streams within a drainage basin. It is much more likely that both of these variables are behaving in a similar manner in response to another variable, for example basin area, than in response to each other.

In terms of process–response interaction, therefore, the correlation coefficient may indicate either that there is a genetic link between two variables, or that two (response) variables are responding in a similar

manner to a third (process) variable which may or may not be known. This illustrates the dangers involved in interpreting correlation co-efficients, and indeed the same can be said of a study of regression lines. Because a statistical interrelationship is found it does not of necessity mean that a genetic link, or process–response relationship, exists in nature. The correlation and regression analysis serves only as a search procedure by which possible geomorphological relationships may be discovered. It remains for the geomorphologist to consider and examine the implications of each analysis. In the case of all of the intercorrelations between the measures of total stream length at each order, it is likely that these are only 'apparent' correlations in a process–response sense. In reality they are all genetically linked to a common process variable. The correlation coefficients between the length of streams at each order and total stream length within the drainage basin fall into a different category. The coefficients in this case express the numerical contribution made by each order to the length of the total drainage network.

Likewise Table 3.3 showed several negative correlations between drainage density and other variables. Many of these may fall into the category of 'apparent' correlations. In Table 3.3, D_3 is seen to have a significant negative correlation with $\log A_3$, $\log L_1$, $\log L_2$, $\log L_3$, $\log (\Sigma L)_3$, $\log \bar{L}_1$, $\log \bar{L}_2$, and $\log R_1$ (1 and 2). All of these variables have a significant positive correlation with basin area. It is therefore possible that each, or at least some, of these variables have an 'apparent' rather than a genetic-ally significant relationship to D_3. There is thus no point in looking for geomorphological implications in these relationships. This example serves once again to illustrate the caution which has to be exercised in the inter-pretation of correlation coefficients. The greatest danger lies in the case when an 'apparent' correlation is mistaken for one that has genetic significance.

Process–response relationships need not be thought of only in the sense of pair-wise relationships. Indeed, the multivariate nature of geomor-phological problems demands that their analysis should involve multi-variate statistical methods. These are described, for drainage basin studies, in the rest of part I.

References

CHORLEY, R. J. 1957: Illustrating the laws of morphometry. *Geol. Mag.* **94,** 140–50.

COATES, D. R. 1958: Quantitative geomorphology of small drainage basins of Southern Indiana. *Columbia Univ., Dept. of Geol., Tech. Rept.* **10.**

EZEKIEL, M. and FOX, K. A. 1959: *Methods of correlation and regression analysis.* New York: John Wiley. (3rd edn.)

MELTON, M. A. 1957: An analysis of the relations among elements of climate, surface properties, and geomorphology. *Columbia Univ. Dept. of Geol., Tech. Rept.* **11.** (102 pp.)

1958: Geometric properties of mature drainage systems and their representation in an E_4 phase space. *J. Geol.* **66,** 35–54.

SCHUMM, S. A. 1956: Evolution of drainage systems and slopes in badlands at Perth Amboy, New Jersey. *Bull. Geol. Soc. Am.* **67,** 597–646.

SIEGEL, S. 1956: *Nonparametric statistics for the behavioral sciences.* New York: McGraw-Hill. (312 pp.)

4 Multiple relations among morphometric properties

In many geomorphological studies more than two characteristics can be measured for a particular land form or area. Table 1.1, for example, lists 28 different variables connected with drainage basin morphometry. It becomes necessary to look beyond the simple analysis of pair-wise relationships, among such variables, and to seek the numerical relationships which may exist between several variables taken together. This may be done:

1 by finding sets of intercorrelated variables within a correlation matrix
2 through *multiple correlation*
3 through *multiple regression* techniques.

When it is known that more than one independent variable is influencing the dependent variable it is useful to be able to isolate the influence of one variable by 'controlling' the influence of others. This control may be applied through the technique of *partial correlation*. The most efficient, though complex, procedure for examining a large mass of data is known

as factor analysis. Through this technique a great deal of numerical information can be reduced and meaningfully organized.

4.1 Sets of intercorrelated variables

Correlation matrices, such as Table 3.3, can be used to assess the presence, or otherwise, of groups of intercorrelated variables. This is done by re-sorting the position of the variables, around the margin of the table, in such a way that the highest positive correlation coefficients fall nearest to the central diagonal line. This is most easily achieved in the following way. The negative correlation coefficients are removed from the table, and the positive coefficients are recorded twice for each pair of variables; that is to say once on each side of the central diagonal line. A greater visual impact is made by representing the level of correlation by a density

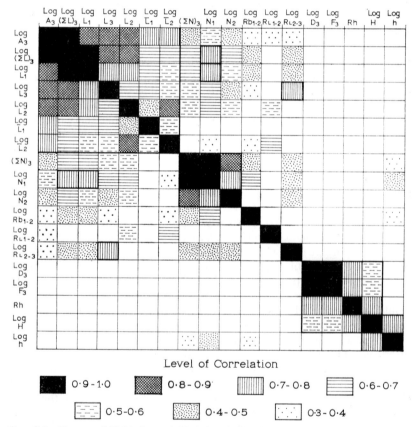

Fig. 4.1   Groups of highly intercorrelated variables found by re-sorting the data of Table 3.3. (*Mather and Doornkamp, 1970*)

shading. Thus the numerical information of Table 3.3 is replaced in Figure 4.1 by a density shading. At the same time the variables around the margin of the diagram have been re-ordered so that the densest shading, and therefore the highest correlation coefficients, fall closest to the diagonal line. Melton (1958) adopted a similar approach. This method is also described by Sokal and Sneath (1963).

The result is that groups of highly intercorrelated variables can be easily distinguished. The Uganda data for third-order basins (Fig. 4.1) suggest that four main groups exist:

Group 1 A_3, $(\Sigma L)_3$, L_1, L_2, L_3, $\bar{L}_1$, $\bar{L}_2$
Group 2 $(\Sigma N)_3$, N_1, N_2
Group 3   D, F, Rh, H
Group 4   H, h

Some significant correlations exist between individual members of two groups. For example, N_1 correlates with both $(\Sigma L)_3$ and L_1.

The re-sorting procedure, referred to above, is undertaken without reference to the nature of the individual variables, and is dependent solely on the level of correlation. It is relevant to note, therefore, that each of the groups of highly intercorrelated variables is distinct in type. Group 1 may be defined as a *basin size–stream length* group; group 2 is a *stream number* group; group 3 contains the variables which measure *intensity of dissection*, and group 4 is composed of measures of *relative relief*. Reference back to the correlation matrix of Table 3.3 shows that the variables of groups 1 and 3 are usually negatively correlated at a highly significant level. That is to say, as the members of group 1 increase in magnitude so there is a decrease in the value of the variables in group 3.

This technique of finding groups of intercorrelated variables is relatively easy to apply. It will become apparent, in section 4.7, that there is a close parallel between the results of this technique and those obtained through the much more complex multivariate procedure of factor analysis. A practical application of the technique for finding groups of intercorrelated variables lies in the time which can be saved in future studies. The measurement of many variables for a large number of drainage basins is a time-consuming task. In primary hydrological surveys, for example, it would be better if only measurements of a few variables were needed. This would be particularly useful if it were known that these few variables were likely to have a significant correlation with other, but unmeasured, variables. The difficulty frequently lies in the choice of variables to be measured. The grouping technique outlined above may help to make a decision. The maximum amount of information about the variation between drainage basins and the character of individual basins would be obtained by measuring at least one of the variables from each of the main groups. Thus in the largest group, group 1, having measured the area of the drainage basin there may be little point in also measuring total stream length for, where A_3 is large, $(\Sigma L)_3$ is also likely to be large. Likewise

as far as group 2 is concerned a knowledge of the values of $(\Sigma N)_3$ is sufficient to represent the variations in the values of N_1 and N_2. It is, however, necessary to know the values of N_1 and N_2 to find the value of $(\Sigma N)_3$, and in this case there is little time saved in the calculation of the one variable that may be chosen to represent the others. The value of $(\Sigma N)_3$ is also required for the calculation of F_3 which may be taken to represent group 3. In the case of group 3 F_3 is to be preferred to D_3 for the latter would involve the measurement of $(\Sigma L)_3$ as an additional variable (as well as A_3 in group 1). Group 4 is most easily represented by H.

The measurement of the three variables A_3, $(\Sigma N)_3$ and H and the calculation of F_3 ($= (\Sigma N)_3 \div A_3$) will supply the most information about differences in the characteristics of the third-order drainage basins of southern Uganda with the least amount of measurement effort and time. A similar analysis was made on the data supplied by Melton (1957) for some third-order basins in Arizona, Colorado, New Mexico and Utah. Four main groups of intercorrelated variables occurred, composed of the following variables.

Group 1 $A_3, L_1, L_2, L_3, (\Sigma L)_3$ (basin size–stream lengths)
Group 2 $N_1, N_2, (\Sigma N)_3$ (stream number)
Group 3 D_3, F_3 (intensity of dissection)
Group 4 H, Rh   (relative relief)

These groups are directly comparable with those obtained for southern Uganda. This suggests the possibility of consistent results of universal application to third-order drainage basins.

4.2 Multiple regression

In its simplest form multiple regression concerns the relationship between one dependent and two independent variables. It has already been said that in geomorphological studies it is very difficult, if not impossible, to state which is the 'cause' and which the 'effect' variable. Hence it is often misleading to refer to a 'dependent' variable and even more so to an 'independent' variable. Within the groups of highly correlated variables referred to in section 4.1 there is no one independent variable: all of the variables within a group are dependent on each other. Multiple regression is essentially useful for predicting one variable from a number of other variables. Whether the variables concerned are genetically related to each other may be quite another matter (Carson, 1966).

In chapter 3 close relationships were found to exist between selected morphometric variables in the third-order drainage basins of southern Uganda. The purpose of multiple regression is to discover if the variation in the values of a particular variable is accounted for to a greater extent by considering its relationship to two, or more, other variables taken together than when any of these are taken on their own. Any multivariate

analysis requires careful planning, and it is always useful to compile a flow-chart of the programme to be carried out. The organization of a multiple regression analysis, for example, is illustrated in Figure 4.2. So far, in simple linear regression, it has been assumed that the significant variation in the dependent variable (Y) can be defined by relating it to one independent variable (X) through an equation of the form:

$$Y = a + bX$$

In many instances the variation in a dependent variable can be explained only partially by reference to *one* independent variable. It is

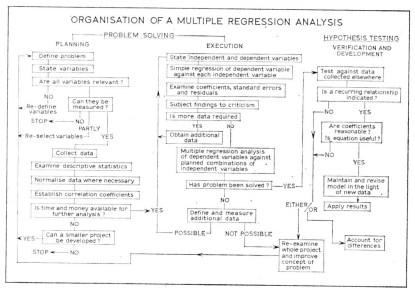

Fig. 4.2 Organization of a multiple regression analysis for linear relationships.

therefore necessary to look for some other independent variables to which it is also related. Only when all the influencing variables are taken into account can the total variance in the dependent variable be explained. It is often possible to explain a large proportion of the total variance in a dependent variable by reference to a limited number of other variables. In the case of a drainage basin, for example, it may be found that the area of a basin is related not only to total stream length, but also to the shape of the basin and the number of streams within the basin. In fact there may be several variables all of which influence the size of a particular basin. Where such a case exists the relationship between the dependent and the several independent variables can be represented by a *multiple regression equation* of the form:

$$Y = a + bX_1 + cX_2 + \ldots nX_n$$

where X_1, X_2 ... X_n represent the several independent variables.

The Kyogya Valley drainage basin example of chapter 3 (Table 3.6) can be expanded to include the additional variable of total number of streams in order to show how a multiple regression equation is obtained for the case where there is one dependent variable—basin area, A_3— and two independent variables—stream length, $(\Sigma L)_3$ and number of streams, $(\Sigma N)_3$. The multiple regression equation has the form:

$$Y = a + bX_1 + cX_2$$

or, for this example:

$$A_3 = a + b(\Sigma L)_3 + c(\Sigma N)_3$$

To calculate the regression equation by the method of least squares the data are set out as in Table 4.1.

Table 4.1 The data for the calculation of a multiple regression equation

Y (A_3)	X_1 $(\Sigma L)_3$	X_2 $(\Sigma N)_3$	YX_1	YX_2	X_1X_2	$X_1{}^2$	$X_2{}^2$
4·08	9·40	19	38·35	77·52	178·60	88·36	361
4·53	10·85	19	49·15	86·09	206·15	117·72	361
4·92	10·69	17	52·59	83·64	181·73	114·28	289
6·18	10·97	15	67·79	92·70	164·55	120·34	225
6·22	11·94	16	74·27	99·52	191·04	142·56	256
6·54	18·69	21	122·23	127·34	392·49	349·32	441
7·55	17·21	22	129·94	166·10	378·62	296·62	484
8·92	18·41	23	164·22	205·16	423·43	338·93	529
10·16	20·14	26	204·62	264·16	523·64	405·62	676
12·50	20·68	26	258·50	325·00	537·68	427·66	676
Σ 71·59	148·98	204	1161·66	1537·21	3177·93	2400·97	4298
$\bar{A}_3$ $= 7·16$	$(\Sigma L)_3$ $= 14·898$	$(\Sigma N)_3$ $= 20·4$					

The process of minimizing the square of the distance of the data point from the regression line yields equations of the following types:

$$\Sigma Y = an + b \Sigma X_1 + c \Sigma X_2$$
$$\Sigma X_1 Y = a \Sigma X_1 + b \Sigma X_1{}^2 + c \Sigma (X_1 . X_2)$$
$$\Sigma X_2 Y = a \Sigma X_2 + b \Sigma (X_1 . X_2) + c \Sigma X_2{}^2$$

By solving these equations each of the coefficients a, b and c can be calculated. Thus

$$71·70 = 10a + 148·98b + 204c \tag{1}$$
$$1161·66 = 148·98a + 2400·97b + 3177·93c \tag{2}$$
$$1537·21 = 204a + 3177·93b + 4298c \tag{3}$$

by taking equations (1) and (2), and multiplying (1) all through by 14·898

subtracting

$$1161·66 = 148·98a + 2400·970b + 3177·93c \tag{2}$$
$$1068·18 = 148·98a + 2219·504b + 3039·19c \tag{1}$$
$$\overline{\,\,93·48 = \phantom{148·98a + {}}381·466b + 38·74c} \tag{4}$$

by taking equations (1) and (3), and multiplying (1) all through by 20·4

subtracting

$$1537·21 = 204a + 3177·93b + 4298·0c \tag{3}$$
$$1462·68 = 204a + 3039·19b + 4161·6c \tag{1}$$
$$\overline{74·53 = \phantom{204a + {}}38.74b + 136.4c} \tag{5}$$

multiplying (4) by 136·4 and (5) by 38·74

subtracting

$$12750·67 = 52031·96b + 5284·14c$$
$$2887·29 = 5374·79b + 5284·14c$$
$$\overline{9863·38 = 46657·17b}$$

$$b = \frac{9863·38}{46657·17} = 0·2114$$

substituting for b in (4)

$$93·48 = 381·466 \times 0·2114 + 38·74c$$
$$c = \frac{93·48 - 80·642}{38·74} = \frac{12·838}{38·74}$$
$$c = 0·3314$$

substituting for b and c in (1)

$$71·70 = 10a + 148·98 \times 0·2114 + 204 \times 0·3314$$
$$10a = 71·70 - 31·49 - 67·61$$
$$a = -\frac{27·40}{10} = -2·74$$

by substituting the values obtained above for a, b and c in the general equation $A = a + b(\Sigma\, L)_3 + c(\Sigma\, N)_3$:

$$A_3 = -2·74 + 0·2145(\Sigma\, L)_3 + 0·339(\Sigma\, N)_3 \text{ square miles}$$

For a two-dimensional relationship a line representing the equation $Y = a + bX$ can be drawn, while a three-dimensional surface has to be used to represent $Y = a + bX_1 + cX_2$ as in Figure 4.3.

When the multiple regression involves more than two independent variables the relationship cannot be drawn on a graph or shown on a block diagram, for more than three dimensions become involved. The number of dimensions depends on the total number of variables in the

equation (i.e. dependent plus independent variables). Each data point for the value of n variables has to be thought of as a point in n-dimensional space. This concept forms the basis of many techniques of multivariate

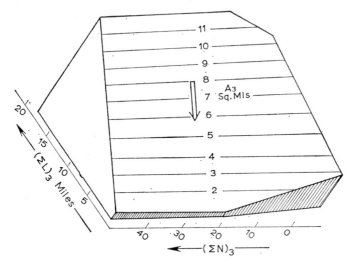

Fig. 4.3 To illustrate the relationship between changes in drainage basin area and those in total stream length and total number of streams.

analysis of the factor analysis, type, including principle components analysis discussed in section 4.7.

The t-test can be used to test the statistical significance of the parameters of the multiple regression equation. The general principles behind this are the same as those employed for the simple linear regression equation.

4.3 Uganda case study—multiple regression

The Kyogya Valley example, given above, indicates the relationship between basin area, total stream length, and number of streams in a sample of only ten basins. If the larger sample of 130 third-order basins drawn from various parts of southern Uganda is analysed, the multiple regression equation (Table 4.2) is:

$$A_3 = 0.329 - 0.14(\Sigma N)_3 + 0.639(\Sigma L)_3 \text{ square miles}$$

This equation is different from that obtained for the Kyogya Valley, but since a larger sample is being used the new equation will be more representative of the population of third-order basins in the whole of southern Uganda.

In this multiple regression relationship interest must centre on the difference in the signs preceding the two regression coefficients relating

Table 4.2 Multiple regression relationships between some selected morphometric properties of third-order drainage basins in Uganda

Variables			Constant of multiple regression eqt.	Regression coefficients			Variance at origin of		
Y	X_1	X_2	a	a	b	c	a	b	c
A_3	N_1	L_1	0·510	0·510	−0·181	1·046	0·0456	0·00038	0·0024
A_3	$(\Sigma N)_3$	$(\Sigma L)_3$	0·329	0·329	−0·140	0·639	0·026	0·00012	0·00037
D_3	Rh	H	2·16	2·16	0·003	0·001	0·041	0·0000002	0·0000002
D_3	Rh	Z	163·68	163·68	0·110	−0·037	4·036	0·0006	0·0002

Variables			Standard error at origin of			t-value for			Multiple correlation coefficient	Coefficient of explanation
Y	X_1	X_2	a	b	c	a	b	c	r	R^2 (%)
A_3	N_1	L_1	0·2138	0·0196	0·487	2·38	−9·24	21·49	0·90	81·0
A_3	$(\Sigma N)_3$	$(\Sigma L)_3$	0·1625	0·0111	0·0193	2·31	−12·88	33·06	0·95	91·6
D_3	Rh	H	0·202	0·0004	0·0004	10·66	8·26	2·88	0·76	58·0
D_3	Rh	Z	63·53	0·025	0·0136	2·58	4·36	−2·73	0·36	13·15

to $(\Sigma \; N)_3$ and $(\Sigma \; L)_3$. The negative sign associated with $(\Sigma \; N)_3$ implies that, in terms of this relationship, increases in $(\Sigma \; N)_3$ lead to a decrease in A_3; conversely the positive sign associated with $(\Sigma \; L)_3$ shows that increases in $(\Sigma \; L)_3$ give rise to an increase in A_3. These relationships can be illustrated by a block diagram (Fig. 4.3).

Basin area can also be related to the number and lengths of the first-order streams (Table 4.2). The multiple regression equation:

$$A_3 = 0 \cdot 51 - 0 \cdot 181 N_1 + 1 \cdot 05 L_1$$

accounts for 81% of the variation in A_3. Stream number again has a negative regression coefficient.

Some 58% of the variation in the values of drainage density is accounted for by the relief ratio and maximum basin relief (Table 4.2). In this Uganda example the relief ratio and the relative relief are highest in the mountain areas. Here also, as was shown in chapter 2, the greatest values of drainage density may be expected to occur. If, however, relative relief is replaced in the multiple regression equation by the height of the highest point on the watershed (Z), then the relationship with D_3 is much more tenuous. In other words, the association of high relief ratio values with high relative relief is much more important in giving rise to high drainage density values than is the association of high relief ratio values with high absolute altitudes.

Multiple regression can be applied to more than three variables at a time. For example, Morisawa (1962) in using multiple regression analysis found that the peak intensity run-off (Q_{max}) from a drainage basin was a function of the form:

$$\log Q_{max} = 17 \cdot 044 - 2 \cdot 177 \log A - 2 \cdot 302 \log P \\ + 6 \cdot 984 \log S - 6 \cdot 682 \log F_1$$

(Multiple correlation coefficient 0·9378, which has a standard error of estimate of 0·099.) Where A is the basin area; P is the P-ratio, measuring rainfall intensity; S is the S-ratio, measuring frequency of rainfall; and F_1 is the frequency of first-order streams. In situations such as this, when the relationship between five variables is being assessed at one time, it becomes difficult to visualize what the final multiple regression equation means, and so it is necessary to use techniques which can consider the behaviour of many variables at the same time (multivariate analysis), and yet produce a meaningful, but not a complex, result. To this end the most useful multivariate technique available to geomorphologists is factor analysis (section 4.7).

4.4 Multiple correlation coefficients

A *multiple correlation coefficient* (r) can be obtained to assess the significance or otherwise of a relationship between one dependent variable and two

or more independent variables. Likewise, a *multiple coefficient of determination* (R^2) can be obtained. The value of r is obtained from:

$$r = \sqrt{\frac{\Sigma\,(\hat{Y} - \bar{\hat{Y}})^2}{\Sigma\,(Y - \bar{Y})^2}}$$

where $\hat{Y}$ is a predicted value of the dependent variable obtained from the multiple regression equation, and $\bar{\hat{Y}}$ is the mean of all the predicted values. Y and $\bar{Y}$ respectively denote the actual values and the mean of these values for the dependent variable.

The data of Table 4.1 can be used to provide a worked example. For the calculation of the multiple correlation coefficient another table (Table 4.3) has to be compiled. Where $Y = A_3$, values of $\hat{Y}$ can be obtained from the multiple regression equation:

$$A_3 = -2\cdot74 + 0\cdot2145(\Sigma\,L)_3 + 0\cdot339(\Sigma\,N)_3$$

which was calculated above for the ten basins of the Kyogya Valley. Thus the predicted value of A_3 for the first basin listed in Table 4.1 is:

$$\hat{A}_3\ (\text{or } \hat{Y}) = -2\cdot74 + 0\cdot2145 \times 9\cdot4 + 0\cdot330 \times 19$$
$$= 5\cdot72 \text{ square miles}$$

Other values needed are listed in Table 4.3. It should be noted that $\sqrt{0\cdot6899} = \pm0\cdot83$ (Table 4.3). r may be taken to be positive in this case because of the results of the simple correlation analysis shown in Table 3.3.

Table 4.3 Calculations for a multiple correlation coefficient

Y (A)$_3$	$\hat{Y}$	$\hat{Y} - \bar{\hat{Y}}$	$(\hat{Y} - \bar{\hat{Y}})^2$	$Y - \bar{Y}$	$(Y - \bar{Y})^2$
4·08	5·72	(−)1·64	2·69	(−)3·08	9·49
4·53	6·02	(−)1·34	1·80	(−)2·63	6·92
4·92	5·31	(−)2·05	4·20	(−)2·24	5·02
6·18	4·69	(−)2·67	7·13	(−)0·98	0·94
6·22	5·24	(−)2·12	4·49	(−)0·94	0·88
6·54	8·37	1·01	1·02	(−)0·62	0·38
7·55	8·40	1·04	1·08	0·39	0·15
8·92	8·99	1·63	2·66	1·70	2·55
10·16	10·38	3·02	9·12	3·00	9·00
12·50	10·50	3·14	9·86	5·34	28·52
	Σ 73·62		Σ 44·05		Σ 63·85

$$\bar{Y} = 7\cdot16 \text{ square miles}; \bar{\hat{Y}} = 7\cdot36 \text{ square miles}$$

$$r = \sqrt{\frac{\Sigma\,(\hat{Y} - \bar{\hat{Y}})^2}{\Sigma\,(Y - \bar{Y})^2}} = \sqrt{\frac{44\cdot05}{68\cdot85}} = \sqrt{0\cdot6899}$$

$$r = \pm0\cdot83$$

4.5 Partial correlation coefficients

In the morphometric study of drainage basins it is seldom possible, in a genetic sense, to distinguish between an independent and a dependent variable. Indeed, as has been noted, it is possible in a correlation analysis for them to appear to have a significantly high correlation coefficient: the only reason for this must be that they are both genetically related to a third variable. In some instances this third variable may have been measured and it is possible to discern the apparent correlation situation which may have arisen. On the other hand, the third variable may not have been measured and the true nature of the relationship between two variables may remain unnoticed. One way in which true and apparent relationships may be distinguished is the technique of *partial correlation*. The partial correlation coefficient for one dependent and two independent variables is given by:

$$r_{ij \cdot k} = \frac{r_{ij} - (r_{ik} \cdot r_{jk})}{\sqrt{(1 - r_{ik}{}^2)(1 - r_{jk}{}^2)}}$$

where r_{ij} is the correlation coefficient between i and j; r_{ik}, the correlation coefficient between i and k; r_{jk}, the correlation coefficient between j and k; and $r_{ij \cdot k}$ is the partial correlation coefficient between i and j when the influence of k is controlled. For example, in Table 3.3 the following correlation coefficients are listed:

$$r_{(ij)} = 0 \cdot 92 \text{ between } A_3 \text{ and } (\Sigma L)_3$$
$$r_{(ik)} = 0 \cdot 44 \text{ between } A_3 \text{ and } (\Sigma N)_3$$
$$r_{(jk)} = 0 \cdot 68 \text{ between } (\Sigma L)_3 \text{ and } (\Sigma N)_3$$

From these values the partial correlation coefficient between A_3 and $(\Sigma L)_3$, when the influence of $(\Sigma N)_3$ is controlled, can be found from:

$$r_{ij \cdot k} = \frac{0 \cdot 92 - (0 \cdot 44 \times 0 \cdot 68)}{\sqrt{(1 - 0 \cdot 44^2)(1 - 0 \cdot 68^2)}}$$

$$= 0 \cdot 943$$

The control of the influence of stream number has led to an increase in the correlation coefficient between basin area and total stream length.

4.6 Uganda case study—partial and multiple correlation coefficients

In Table 4.4 three sets of correlation coefficients have been listed between basin area (A_3), basin length (Lb) and perimeter (P) for the Uganda basins. The first group of correlation coefficients are the simple coefficients stated in the form of a matrix. The second and third matrices give the partial and multiple correlation coefficients respectively. The matrix of simple correlation coefficients indicates that there is a high positive

correlation between the three variables. The correlation between Lb and P is stronger than that between A_3 and either Lb or P.

Table 4.4 Simple and partial correlation coefficients between A_3, P and Lb for some third-order basins in Uganda

Simple correlation coefficients

	A_3	P	Lb
A_3	1·0000	0·8392	0·7515
P	0·8392	1·0000	0·9603
Lb	0·7515	0·9603	1·0000

Partial correlation coefficients

1 Between A_3 and the variable listed along the row, holding the variable listed in the column constant.

	A_3	P	Lb
A_3	—	—	—
P	—	—	0·6387
Lb	—	−0·3585	—

2 Between P and the variable listed along the row, holding the variable listed in the column constant.

	A_3	P	Lb
A_3	—	—	0·6387
P	—	—	—
Lb	0·9187	—	—

3 Between Lb and the variable listed along the row, holding the variable listed in the column constant.

	A_3	P	Lb
A_3	—	−0·3585	—
P	0·9187	—	—
Lb	—	—	—

The matrix of partial correlation coefficients defines the correlation coefficient between A_3 and P, when Lb is held constant, as 0·6387. Thus when Lb is held constant, there is still a significant correlation between A_3 and P, but the level of correlation is lower than when Lb is allowed to exert an influence. That is to say, when the three variables A_3, P and

Lb are allowed to change together there is a closer relationship between A_3 and P than would be the case if Lb were to be held constant.

What is much more interesting, however, is the partial correlation coefficient between A_3 and Lb when P is held constant. In this case the coefficient is -0.3585. This negative relationship between basin area and basin length is not suggested by the simple correlation coefficient. The apparent correlation between A_3 and Lb, according to the correlation coefficient of 0.7515 given in Table 4.4, is a significantly high positive one. However, from the partial correlation analysis it appears that if the effect of P on Lb is controlled the relationship between A_3 and Lb is negative. In other words, if the length of the watersheds around these third-order drainage basins in Uganda could be held constant, then there would be a tendency for the area of those basins with the longest long-axis to be less than the area of those basins with a short long-axis. This is related to pure geometry, for as the form of any body with a constant circumference approaches nearer to a circle so its area will increase. The positive correlation between A_3 and Lb which appears in the simple correlation matrix indicates that the much stronger influence of P on A_3 masks the actual relationship between A_3 and Lb; and so not until the effect of P is controlled can the true relationship between A_3 and Lb be detected.

From the second part of the partial correlation table (Table 4.4) it is apparent that when the effects of A_3 are controlled there still remains a highly significant positive partial correlation coefficient (0.9187) between P and Lb. Thus the basins with the longest basin axes also have the longest perimeters.

Through the multiple correlation coefficients it is possible to see what the combined influence of two (or more) variables is on another variable. In Table 4.5 the multiple correlation coefficients are given expressing the combined influence of Lb and P on A_3 (0.8616), of Lb and A_3 on P (0.9767), and of P and A_3 on Lb (0.9655). It will be recalled that the total variation in the behaviour of the values of one variable accounted for by the other is the correlation coefficient squared. In the same way the variation in one variable accounted for by the combined effect of the other two is the multiple correlation coefficient squared. Thus the variation in A_3 explained by the combined effects of P and Lb is $0.8616^2 = 0.7423$, which may be multiplied by 100 to convert it to a percentage value. Thus:

1 74.23% of the variation in A_3 is accounted for by the combined effects of P and Lb

2 95.4% of the variation in P is accounted for by the combined effects of A_3 and Lb

3 93.22% of the variation in Lb is accounted for by the combined effects of A_3 and P.

In order to see if this level of explanation has any real significance it is necessary to convert the simple correlations between each of the variables

into the same type of percentage figure. From the simple correlation given in the upper matrix of Table 4.4 the following values are obtained:

1   70·43% of the variation in A_3 is accounted for by P
2   56·48% of the variation in A_3 is accounted for by Lb
3   92·15% of the variation in P is accounted for by Lb

By comparison with the multiple correlation values, therefore, there is an increase of only 3·8% in the explanation of the variation in A_3 accounted for by the joint effects of P and Lb over and above that achieved

Table 4.5 Multiple correlation coefficients between A_3, P and Lb for some third-order basins in Uganda

1 Between A_3 and the two column and row variables.

	A_3	P	Lb
A_3	—	—	—
P	—	—	0·8616
Lb	—	0·8616	—

2 Between P and the two column and row variables.

	A_3	P	Lb
A_3	—	—	0·9767
P	—	—	—
Lb	0·9767	—	—

3 Between Lb and the two column and row variables.

	A_3	P	Lb
A_3	—	0·9655	—
P	0·9655	—	—
Lb	—	—	—

by P alone. Thus, if it is necessary to predict the value of A_3 in a given situation it can be done almost as effectively from P alone as it can from P and Lb combined. It may not be worth measuring Lb at all for this particular purpose. On the other hand since only 56·48% of the variation in A_3 is accounted for by Lb alone it is well worth combining it with P, thus increasing the level of explanation to 74·23% rather than relying on Lb alone in predicting values of A_3. Since 92·15% of the variation in P is explained by variations in Lb alone, it certainly would not be worth measuring A_3 as well in order to estimate values of P. By adding the influence of A_3 to that of Lb, on the values of P, only 1·07% is added to the level of explanation achieved.

Table 4.6 A comparison between the correlation coefficient (upper value) and the partial correlation coefficient holding area constant· (lower value), for the morphometric properties of some Uganda drainage basins (*cf.* Table 3.3)

	$\log L_1$	$\log L_2$	$\log L_3$	$\log (\Sigma L)_3$	$\log \bar{L}_1$	$\log \bar{L}_2$	$\log N_1$	$\log N_2$	$\log Rl$ 1&2	$\log Rl$ 2&3	$\log Rb$ 1&2	Rh	$\log H$
$(\Sigma N)_3$	74 / 14	79 / 26			5 / −50								
$\log L_1$		66 / −29		85 / 33	65 / 4	56 / −24						−52 / 35	1 / 55
$\log L_2$				86 / 33	48 / −50	52 / −51	61 / 32			12 / −43		−66 / 0	−27 / 3
$\log L_3$					61 / −20	67 / −16	14 / 52		18 / −28			−69 / 4	
$\log (\Sigma L)_3$					65 / −20	56 / −8						−64 / 29	−10 / 57
$\log \bar{L}_1$							31 / −21	6 / −49	3 / −43			−52 / 19	
$\log \bar{L}_2$								12 / −39	−12 / −73			−67 / −18	
$\log N_1$													21 / 50

The correlation coefficient between basin area and total number of streams is 0·35 (when neither of these variables is log-normalized). The partial correlation coefficient between them is −0·66 when total stream length is held constant. In other words, what at first appeared to be a suggestion that increases in basin area are associated with increases in the number of streams does not now appear to hold good. Controlling the influence of total stream length indicates that if the total stream length in each basin could be unity, there would be a significant decrease in the number of streams with increases in basin area. Multiple regression between the same three variables reflects the importance of considering total stream length as well as the number of streams in explaining variations in the value of basin area. Only 12·1% of the variation in A_3 is explained by $(\Sigma N)_3$ alone. This increases to 88% when the combined influence of $(\Sigma N)_3$ and $(\Sigma L)_3$ is considered.

Basin area is often highly correlated with many other variables (Fig. 4.1), so much so that it has been called the Devil's own variable. The problem of disentangling the influence of area on other apparent relationships frequently has to be faced. This can, in part, be done through partial correlation analysis, and has been undertaken for the variables listed in Table 3.3. A new table has been compiled (Table 4.6) which lists both the correlation coefficient between variables and also the partial correlation coefficient between them when basin area is controlled. For simplification purposes those cases where this involves little change in the coefficient have been omitted from Table 4.6. Since drainage density and stream frequency are both ratios involving basin area these too have been omitted.

The most interesting differences between the correlation coefficient and the partial correlation coefficient are those when there is a change in the sign. For example, the correlation coefficient between L_3 and L_2 (log-normalized in each case) is 0·66. The partial correlation coefficient is −0·29. Thus when A_3 is controlled L_3 does not increase with increases in L_2, but on the contrary tends to decrease. The relief ratio (Rh) has a significant negative correlation coefficient with the measures of stream length at each order and with total stream length. In other words, high values of L_1, L_2, L_3 or $(\Sigma L)_3$ tend to be associated with low values of R_h. When A_3 is controlled, however, the partial correlation coefficients become positive, and those between R_h and L_1, and R_h and $(\Sigma L)_3$, are significant at the 99% level. Thus, if basin area could be controlled the value of both L_1 and $(\Sigma L)_3$ would increase with the relief ratio, and not decrease as they actually do. Basin area can be seen, therefore, to be an important influence on the relationship between these variables.

4.7 Factor analysis

In multiple regression analysis, as has been shown, the numerical relationship between more than two variables at a time can be examined. As the

number of variables increases, however, so the multiple regression equation becomes increasingly unmanageable. Through *factor analysis* many variables may be reduced to a few factors, and this procedure provides a summary of the original data. The closer the amount of intercorrelation among the variables the fewer will be the number of factors produced. Factor analysis is only a feasible undertaking when a computer is available. The purpose of this section, therefore, is not to provide a worked example of the mathematics of the technique, but to illustrate its concepts and results by reference to the Uganda case study. As the following example will show, there is such a close link between the results of factor analysis and the pattern of intercorrelated variables derived in Figure 4.1 that the latter can be used effectively without a computer being available. One of the procedures in factor analysis is to replace the correlation matrix (Table 3.3) with a *factor loadings matrix*. The latter expresses the relationships contained in the correlation matrix in terms of the degree of association between the variables and a lesser number of 'abstract' variables.

It is possible to plot a graph, in two dimensions, of one variable against another, and to construct a model, in three dimensions, of the relationship between three variables. With a computer it is possible to 'think' algebraically in terms of as many dimensions as there are variables. In this *n*-dimensional space (where *n* is the number of variables) it is possible to consider each drainage basin, for example, as having a location related to the values of each of its *n* variables. In very simple terms it is possible to think of factor analysis by imagining a hyper-ellipsoid around and enclosing all these 'location' points. This hyper-ellipsoidal body will have *n* dimensions and thus *n* axes. The longest of these axes follows the general trend, or elongation, of the majority of 'locational' points. It will, therefore, be most closely related to the dominant trend in the data and will go further than any other axis in 'describing' the data values. Some of the variables will fall close to this longest axis (which is known as *factor axis* I, or simply *factor* I) and these are said to have a high *loading* on the first factor. The variables most distant from factor I, in the *n*-dimensional space, will have a low loading on this factor. They will be more closely related to one of the other axes of the hyper-ellipsoid. The axes of this geometrical body are labelled I, II, III. . . in decreasing order of their lengths. Thus the variables lying closest to the second longest axis are said to load highly on factor II, and so on.

Since the variables in most multivariate geomorphological studies show some degree of intercorrelation, more than one variable will lie close to each of the longer factor axes. This is because in a group of highly intercorrelated variables (see Fig. 4.1) one variable tends to change in response to changes in the other variables in that group. These variables tend, therefore, to have similar trends in *n*-dimensional space, and this will be reflected by the factor axes. The algebraic 'distance' between the scatter of values for a variable and the factor axis can be calculated on a computer,

and from these the factor loadings matrix is compiled. This is in essence a summary of the correlation matrix as calculated through the groups of intercorrelated variables. It is not surprising that the variables which load together on different factors are also those which group together on the correlation matrix (Fig. 4.1). This will be illustrated below for the Uganda data.

Factor analysis provides other information which is of great value in analysing a large mass of data. For example, in addition to calculating which variables load highly on each factor the computer can also be programmed to calculate how much of the variability in the original data is accounted for, or 'explained', by each factor. If the length of the first factor axis is P units, then factor I is said to explain $P/n \times 100\%$ of the total variance. The importance of the factor (i.e. its length) is proportional to the corresponding *eigenvalue* of the correlation matrix. (Eigenvalues are the characteristic or latent roots of the correlation matrix, which are somewhat analogous to the roots of a linear equation.) In practice the analysis of the factors can be restricted to all those factors that have an eigenvalue greater than one.

When the hyper-ellipsoid is established its axes are assumed to be at right-angles to each other. Thus, once the longest axis has been found the second axis must be measured at right-angles (or orthogonal) to the first, and so on. This places a constraint on the position of the axes. There are a number of methods by which this can be overcome. One of these allows the factor axes to 'rotate' so that their positions are determined by the pattern of the data in n-dimensional space. The rotational scheme in fact leads to the position where only a few variables have a high loading on each factor and the remaining variables have near-zero loadings. This makes the interpretation of the results much easier. In the example which follows the rotated scheme has been used.

It is also useful to consider the relationship between the units (e.g. drainage basins), for which the variables were measured, and the factor axes. A new matrix, known as the *factor scores matrix*, can be compiled which defines the contribution of each factor to the variation within each unit. This becomes particularly important when it is necessary to see if there are groups of similar and dissimilar units (see chapter 5). For a further discussion, and references, on factor analysis see Mather and Doornkamp (1970).

4.8 Uganda case study—factor analysis

Factor analysis was applied to the 18 morphometric variables measured for the 130 third-order Uganda drainage basins. This involved 2,340 individual observations. Through factor analysis it was found that these could be very largely summarized by the information contained in the first six factors. Together these take into account 94·9% of the variation in

the original data. The separate contribution of each rotated factor is listed in Table 4.7, for example factor I explains 48·14% of the variation in the data. In this particular case the factor axes were rotated so as to provide the most efficient 'fit' to the 18 variables. That is to say, the

Table 4.7   The first six eigenvalues of the correlation matrix of Table 3.3

Factor	I	II	III	IV	V	VI
Eigenvalue	8·666	3·854	1·706	1·465	0·836	0·559
% of total explanation	48·14	21·41	9·48	8·14	4·65	3·11
Cumulative % of total explanation	48·14	69·55	79·03	87·17	91·82	94·93

distribution of the individual loadings of each variable is as simple as possible. As a result each of the variables is closely associated with at least one of these first six factors (Table 4.8). From Table 4.8 it is seen that factor I is predominantly related to basin area, stream lengths, and is negatively related to drainage density, stream frequency and the relief ratio. Reference to Figure 4.1 shows that this conforms to the first set of highly intercorrelated variables. These variables are all negatively correlated with D_3, F_3 and R_h (Table 3.3). Factor I is essentially composed

Table 4.8   Rotated factor loading matrix (expressed as the percentage contributed to the total variance of each variable by each rotated factor)

	Factor					
	I	II	III	IV	V	VI
A_3	**72·30**	17·13	3·87	(−)0·36	1·56	3·52
$(\Sigma L)_3$	45·08	39·92	2·19	0·62	5·46	3·46
L_1	34·38	39·38	0·15	1·02	8·39	4·59
L_3	40·28	18·60	1·13	(−)0·05	2·73	32·42
L_2	35·67	38·10	22·79	(−)0·52	0·08	(−)0·61
L_1	**88·05**	0·01	(−)1·92	0·45	0·10	0·36
L_2	**49·72**	3·24	33·44	(−)1·46	3·97	(−)3·83
$(\Sigma N)_3$	0·00	**84·64**	0·04	2·19	6·49	2·03
N_1	1·26	**75·63**	0·90	2·98	15·59	2·35
N_2	0·34	**87·62**	0·10	0·30	(−)7·70	2·82
R_b 1 & 2	1·11	4·05	1·23	4·02	**89·20**	0·05
R_l 1 & 2	2·81	0·51	**90·25**	(−)1·63	0·56	(−)0·91
R_l 2 & 3	2·60	8·49	(−)3·91	0·45	(−)0·00	**82·58**
D_3	(−)**72·65**	0·51	(−)4·70	4·55	0·32	(−)0·40
F_3	(−)**90·23**	(−)0·01	(−)3·61	2·41	0·02	(−)1·99
R_h	(−)**38·75**	(−)4·33	(−)7·11	15·55	(−)0·79	(−)4·00
H	(−)11·38	1·24	(−)2·76	**61·47**	1·68	(−)0·04
h	0·02	4·24	(−)0·45	**90·28**	2·19	0·71

of variables relating to basin size and may be referred to as a *size factor*, with the implication of a negative association with intensity of dissection and the relief ratio. Factor II, on the other hand, is essentially related to the variables of *stream number*, although stream lengths appear also to be closely grouped with respect to this factor as well as to factor I. Reference to the correlation diagram (Fig. 4.1) shows that this has come about because, although the stream number variables form a highly inter-correlated set, there are also individual significant correlations between these and some of the variables of stream length.

Together factors I and II account for almost 70% of the explanation of the variation in the original data (Table 4.7). The other factors contribute progressively less, and they are related in sequence to the following variables:

Factor III   stream length ratio (between first- and second-order streams)

Factor IV   relative relief (H and h)

Factor V   bifurcation ratio (between first- and second-order streams)

Factor VI   stream length ratio (between second- and third-order streams)

In this way factor analysis indicates the relative importance of the variables in accounting for the behaviour of all of the data. The variables of basin size (and intensity of dissection) are the most important, but adding those of stream number is significant in raising the level of explanation to almost 75%. Factors III–VI are each associated either with single variables, or in the case of factor IV, with two similar measures of relative relief. Since the eigenvalues for factors V and VI do not reach one, the analysis of the results should concentrate on the first four factors. The variables R_b 1 & 2 and R_l 2 & 3 play little significant part in 'explaining' the variation in the rest of the data.

The position has now been reached where a large mass of data may be reduced either to groups of significantly intercorrelated variables (Fig. 4.1) or to a few summary factors (Table 4.8). This has brought about a certain amount of organization in the data which can be used as a basis for making regional comparisons between the basins originally measured. In chapter 2 it was shown that an analysis of individual variables was relatively unsuccessful in discriminating between the morphological regions from which the data was originally obtained. The next chapter shows how, through cluster analysis, the summary of the data in the form of factors may be used to discriminate between groups of basins which are significantly different in character. At the same time this information is used to establish a model of the relationship between the variables and the areas from which they were obtained.

References

CARSON, M. 1966: Some problems with the use of correlation techniques in morphometric studies. In Slaymaker, H. O., editor, Morphometric analysis of maps, *British Geomorph. Research Group, Occ. Paper* **4**, 49–67.

MATHER, P. M. and DOORNKAMP, J. C. 1970: Multivariate analysis in geography with particular reference to drainage-basin morphometry. *Trans. Inst. Br. Geogr.* **51**, 163–87.

MELTON, M. A. 1957: An analysis of the relations among elements of climate, surface properties and geomorphology. *Columbia Univ.: Dept. of Geol., Tech. Rept.* **11.** (102 pp.)

1958: Correlation structure of morphometric properties of drainage systems and their controlling agents. *J. Geol.* **66**, 442–60.

MORISAWA, M. E. 1962: Quantitative geomorphology of some watersheds in the Appalachian Plateau. *Bull. Geol. Soc. Am.* **73**(9) 1025–46.

SOKAL, R. R. and SNEATH, P. H. A. 1963: *Principles of mumerical taxonomy.* San Francisco: W. H. Freeman (359 pp.)

5 Basin morphometry and morphological regions

5.1 Cluster analysis
  [*similarity coefficient, cluster centroid, pairing sequence, linkage tree (dendo-gram)*]
5.2 Uganda case study—cluster analysis
  [*factor weightings, linkage tree, similarity coefficient*]
5.3 Multiple discriminant analysis
  [*statistical significance of cluster groups, sources of inter-group differences, Wilks' Lambda test, F-ratio, significance levels, null hypothesis, discriminant axes, discriminant functions*]
5.4 Uganda case study—multiple discriminant analysis
  [*cluster groups, factor analysis, discriminant functions, scores on the discriminant functions, classification of uncertain cases*]; A functional model; [*discriminant functions, factors, group centroids, discriminant axes*]
5.5 Basin groupings in relation to the morphological regions
  [*cluster analysis*]

The data for the morphometric analysis of the third-order basins of southern Uganda have been drawn from eight areas which were subjectively defined as constituting different morphological regions (chapter 2). These eight regions are subsets of three larger groups, namely, the mountain, intermediate and lowland areas (Fig. 2.1). Until now the validity of this subjective classification of the regions has not been questioned. However, now that the original morphometric data have been reduced to manageable proportions by factor analysis (section 4.7) it will be possible to see if the subjective definition of different morphological regions is supported by the morphometric data drawn from those areas. This chapter draws freely on the more detailed discussion in Mather and Doornkamp (1970).

The conclusions reached in this chapter are dependent on the results of the factor analysis described in section 4.7. It is necessary to introduce two further techniques, namely *cluster analysis* and *multiple discriminant analysis*, though it is not possible in the compass of this account also to describe their mathematical derivation. Nevertheless, though more advanced fields are being entered, it is of value to see how these techniques may be used in the case of cluster analysis to provide an objective regional classification of drainage basins, and in the case of multiple discriminant analysis to test if there is a significant difference between the regions so

defined. In addition it is possible to relate these regions to the factors of the factor analysis and thus to establish a relationship with the original morphometric variables. This leads to the building of a model which most efficiently describes the relationship of the behaviour of the variables to each other.

Studies in other areas have shown that where factors I and II together account for a high proportion of the variation in the original data, a plot of the individual factor weightings of each area on factor I *v.* factor II makes possible a distinction between different regions (see Fig. 15.3 as well as Cole and King, 1968, Fig. 7.5). In the case of the Uganda morphometric data a similar analysis did not provide a discrimination between the eight morphological regions. This being the case the whole problem may be turned round by asking the question: 'Which drainage basins are most similar to each other, regardless of the morphological regions from which they were originally obtained?' In this way the subjective classification is temporarily forgotten and an attempt made to obtain an objective classification by an examination of the properties of the individual drainage basins. This is possible through cluster analysis (Mather, 1969).

5.1 Cluster analysis

If the drainage basins are very similar in all their measured attributes then they will fall very close to each other in the n-dimensional space described in section 4.7. Their similarity can be assessed by 'measuring' the distance between them, and this provides a single value known as the *similarity coefficient* which summarizes the 'alikeness' of two basins. The smaller the coefficient the more similar the two basins. When the two closest basins have been found they form a proto-cluster, and the point half-way between them becomes the *centroid* of this proto-cluster. Other proto-clusters are formed in a similar way, and early proto-clusters may, at a lower level of similarity, become incorporated with a single group. This process is repeated until all points are included. The order of combination is known as the *pairing sequence*, and may be represented by a *linkage tree* or *dendogram*. In this way cluster analysis produces a summary statement of the similarity, on the basis of the measured variables, of all of the drainage basins. This procedure has its limitations in that additional information cannot be incorporated without recalculating the entire set of data, and the validity of the cluster groups still remains to be tested. This may be done by means of multiple discriminant analysis. In the situation where there is a very large amount of information (i.e. when the data matrix is very large), the cluster analysis may be used on the factor weightings (see section 4.7) of each drainage basin rather than on the original data.

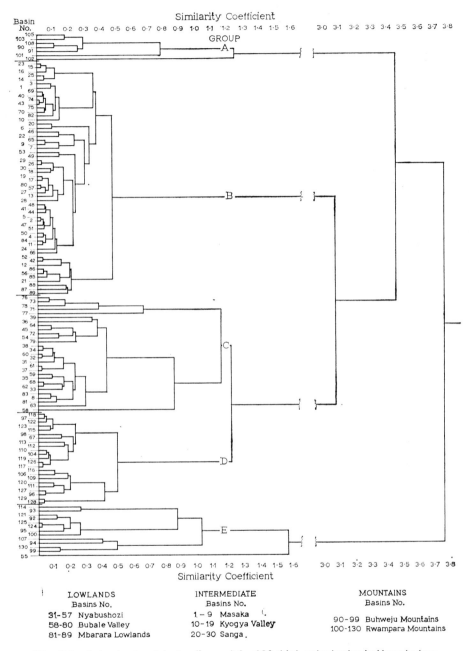

Fig. 5.1 A dendogram (cluster diagram) for 130 third-order basins in Uganda (see Fig. 2.1). (*Mather and Doornkamp, 1970*)

5.2 Uganda case study—cluster analysis

A linkage tree (Fig. 5.1) has been compiled for the 130 third-order basins of Uganda, based on a cluster analysis of the factor weightings of each drainage basin. In some instances groups or clusters of basins are established, because of their similarity to each other, at very low similarity coefficients. In other cases groups of basins are clearly much less alike for they do not join within the linkage tree until much higher values of the similarity coefficients are reached. If a limiting value of 1·2 is assigned to the similarity coefficient (a value arbitrarily selected by examining the cluster diagram) then five separate clusters of basins occur. At this value of 1·2, however, two of the drainage basins (i.e. numbers 55 and 102 in Fig. 5.1) remain without having joined any of the five clusters. Their similarity to one or other of the groups remains to be tested. It is possible, in the context of the discussion in section 1.1 concerning basin morphometry and geomorphological processes, that these two basins may be singled out as those which have not yet reached a steady state (equilibrium) between process and form.

If fewer than five groups were chosen, by setting a higher limiting value to the similarity coefficient, then there would probably be too few groups to allow a comparison to be made with the original eight morphological regions. If more groups than this had been selected the number of groups would have been fairly large (because at only a slightly lower value of the similarity coefficient the number of separate groups is much larger), and would have led to further complexities— when the aim of the analysis is to try and simplify the mass of original observations in terms of the few morphological regions.

An initial examination of the groups (labelled A–E in Fig. 5.1) provided by the cluster analysis indicates that they show some relationship to the morphological regions (see Figs. 2.1 and 2.5). Group A, for example, is composed of the smaller of the mountain drainage basins, while the largest basins of the mountain regions occur in group E. Groups B, C and D contain a mixture of basins drawn from the lowland and intermediate regions. In group B the basins are predominantly large, with an area in excess of 4·5 square miles (see Fig. 2.5). Group C consists of much smaller basins, and group D, in the main, of mountain basins which belong to neither of the extremes of size associated with either group A or group E.

5.3 Multiple discriminant analysis

There is some similarity between this technique and the *t*-test (section 2.5). In the case of *multiple discriminant analysis*, however, its purpose is to discriminate between two groups when each is characterized by several

different variables. In the context of the Uganda case study multiple discriminant analysis is seen as a method of:

1   assessing the statistical significance of the hypothesis provided by the cluster analysis
2   determining the major sources of inter-group difference
3   setting up some system whereby unidentified or previously unconsidered objects may be categorized with the smallest probability of error.

Items (2) and (3) are required only if the groups postulated during cluster analysis prove to be statistically significant.

The starting point of multiple discriminant analysis is the classificatory scheme produced by cluster analysis. The scores of the drainage basins in each cluster group are examined first of all to see whether the between-group differences are sufficiently large in comparison with the within-group variability in order to justify the non-acceptance of the null hypothesis that there is no significant difference between the groups—i.e. that the observed differences are the result of chance or of the peculiarities of the clustering scheme employed. A statistical test of this hypothesis is the *Wilks' Lambda test*; this produces an F-ratio which is compared to the tabled value of F at a selected significance level. If the null hypothesis is accepted it is pointless to proceed, but if it is not accepted then the analysis may be continued, bearing in mind the fact that a significant value of F does not mean that every pair of groups is necessarily separate and distinct (see Krumbein and Graybill, 1965, chapter 14).

To determine the nature of the inter-group differences, presuming that the null hypothesis is not accepted, a method similar to factor analysis is adopted. This involves the setting-up of a set of mutually orthogonal coordinate axes which have the property that inter-group differences are maximized on each axis, the first axis accounting for most of the difference, and so on. The maximum number of axes required to account for 100% of the difference between the groups is equal to the number of variables or one less than the number of groups, whichever is the smaller. As in factor analysis, those discriminant axes which add little to the total explanation of inter-group differences may be discounted. As the data are assumed to be multivariate-normal in their distribution a precise statistical test can be used to determine the number of significant discriminant functions. For example, consecutive analyses may be run to test the hypotheses that the explanation provided by discriminants other than first, first and second, first and second and third, and so on, is due to chance. In practice it may be simpler to use discriminant axes to which a meaning can be attached, and presume that the remaining axes are a response to some unidentified source of variability, or to error.

When the number of axes to be used has been established then unidentified or unclassified objects may be allocated to their most likely class. This is done by plotting the points (representing objects) on a

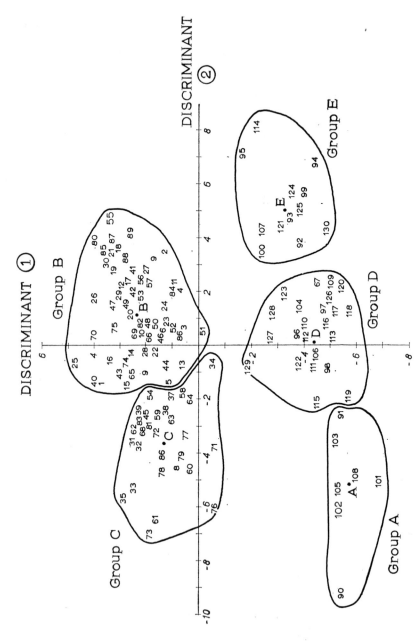

Fig. 5.2 The score of each basin with respect to discriminant functions 1 and 2. *(Mather and Doornkamp, 1970)*

scatter diagram and allocating each unidentified point to the class whose centroid lies nearest. In more complex cases a distance function, similar to that used in cluster analysis, can be calculated and the unidentified point allocated to the cluster whose centroid is closest. Interpretation of the results of multiple discriminant analysis is aided by the table of 'loadings' of variables on the discriminant functions. This enables each discriminant function to be identified with some of the original variables. Furthermore, the variables may be ranked in order of importance and those which make a negligible contribution may even be discarded in future studies.

5.4 Uganda case study—multiple discriminant analysis

Each of the cluster groups A–E (Fig. 5.1) has a centroid whose position in the n-dimensional space of the factor analysis (section 4.7) is the mean value for the score (or weighting) of each basin in the cluster group on each factor. The centroid values are used to see if there is a significant difference between each of the groups. The application of multiple discriminant analysis to the Uganda data indicated that a significant difference exists between the five groups at the 99% level. The calculations showed in this case that over 98% of the difference between the groups was accounted for by the first two discriminant functions. Each of the drainage basins can be related to these two discriminant functions through their score on these functions. In Figure 5.2 these values have been plotted and a boundary drawn around each of the groups defined through cluster analysis and found to be significantly different from each other by multiple discriminant analysis. It is also possible, within this diagram, to plot the scores of the two basins (nos. 55 and 102) which did not combine with any of the groups before the similarity coefficient of 1·2 was reached on the cluster dendogram. Figure 5.2 shows that basin 55 lies closest to group B with which it may be included, and basin 102 may now be included with group A. This illustrates the use of these techniques in classifying uncertain cases.

A functional model
From this point it is now possible to build a model which is related to the original morphometric variables measured. This has to be done, however, by relating the discriminant functions to the factors of the factor analysis. It was found (Table 5.1) that discriminant function 1 was most closely related to factors I, IV and VI, that is to say (see section 4.7) to the variables which measure the size, relative relief and stream length ratio (between second- and third-order streams). The negative sign associated with the relative relief factor (factor IV in Table 5.1) indicates that the discriminant function is responding to decreases in relative relief while responding to increases in basin size and the stream length ratio. In

Table 5.1   Relative contribution of each factor to each discriminant function

		Factor					
		I	II	III	IV	V	VI
Discriminant	1	4·305	0·564	1·909	−9·991	0·267	6·542
function	2	4·152	5·118	3·746	10·148	3·322	0·768
Character of factor		size	number	R_l 1 & 2	relative relief	R_b 1 & 2	R_l 2 & 3

addition, factor analysis (section 4.7) showed that the size factor is concerned not only with increases in basin area and stream lengths but is also related to decreases in the measure of intensity of dissection. Discriminant function 2 is particularly associated with factors II (number of streams) and IV (relative relief), and with both in a positive sense. Furthermore, it also has some relationship to factors I (size), III (stream length ratio between first- and second-order streams) and V (bifurcation ratio between first- and second-order streams).

The general situation summed up in Table 5.1 can be shown in dia-

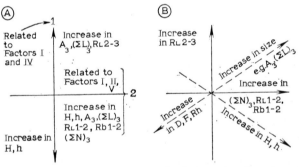

Fig. 5.3   The general relationship between discriminant functions 1 and 2 and the six factors of Table 4.8. (*Mather and Doornkamp, 1970*)

grammatic form (Fig. 5.3A). A vertical axis representing discriminant function 1 is positively related to basin size (e.g. basin area and total stream length) and to the length ratio between second- and third-order streams, and is negatively related to the intensity of dissection (drainage density and stream frequency) together with the relative relief variables. A horizontal axis representing discriminant function 2 is positively related to relative relief, size, stream numbers, stream length ratio between first- and second-order streams, and the bifurcation ratio between first- and second-order streams. The fact that basin size and relative relief

appear on both of the axes representing the discriminant functions suggests that their true direction of increase can be represented by vectors lying somewhere between the two axes, yet in the same two-dimensional plane. Increases in drainage density and stream frequency will, of course, also move onto the line of the vector representing basin size, for these two variables always act in the opposite sense to basin size. Figure 5.3A has therefore been modified (Fig. 5.3B) to take these facts into account.

Figure 5.3B represents the first stage in the building of a model that presents in summary form the general behaviour of the variables in the

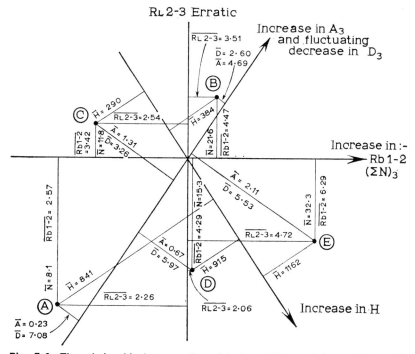

Fig. 5.4 The relationship between the original variables and the model axes for each of the cluster groups. (*Mather and Doornkamp, 1970*)

130 third-order drainage basins being analysed. The easiest way to relate the actual drainage basins to this scheme is by plotting the mean values of the five groups obtained from the cluster diagram (Fig. 5.1) on the coordinate system shown in Figure 5.3B. The vertical and horizontal axes (Fig. 5.4) remain at the same scale as they are in Figure 5.2. The centroid of each group of basins is then plotted in the same position as that shown in Figure 5.2 and marked by the group letter.

The next stage in the analysis involves returning to the raw data to find the mean value of each of the variables recorded in Figure 5.3B, for each of the groups A–E of the cluster diagram. These mean values,

Table 5.2 Data characteristics for each of the cluster analysis groups

Group	A_3 mean	A_3 range	$(\Sigma L)_3$ mean	$(\Sigma L)_3$ range	$(\Sigma N)_3$ mean	$(\Sigma N)_3$ range	R_b 1 & 2 mean	R_b 1 & 2 range	D_3 mean	D_3 range	F_3 mean	F_3 range	H mean	H range	h mean	h range
A	0·23	0·11 — 0·35	1·54	1·00 — 2·26	8·10	7·00 — 10·00	2·57	2·00 — 3·50	7·08	3·86 — 9·10	39·35	25·00 — 63·70	841	525 — 970	193	50 — 400
B	4·69	1·40 — 12·50	11·14	4·65 — 22·42	21·60	11·00 — 48·00	4·47	2·40 — 6·50	2·60	1·11 — 4·18	5·40	1·57 — 12·88	384	215 — 802	280	100 — 490
C	1·31	0·61 — 2·23	4·16	2·07 — 6·30	11·80	7·00 — 20·00	3·42	2·00 — 7·00	3·26	1·99 — 4·12	9·42	5·30 — 14·19	290	140 — 640	161	50 — 350
D	0·67	0·32 — 1·17	3·77	1·44 — 5·89	15·30	10·00 — 24·00	4·29	2·70 — 7·00	5·97	1·68 — 7·95	25·40	8·55 — 45·00	915	730 — 1175	503	300 — 900
E	2·11	1·18 — 3·30	11·00	7·83 — 16·57	32·30	18·00 — 51·00	6·29	3·30 — 10·00	5·53	3·47 — 8·37	17·50	7·50 — 38·15	1162	810 — 1450	783	540 — 1120

together with the range of values for each variable within each group, are recorded in Table 5.2. A perpendicular is dropped from each group centroid position to each of the four axes shown in Figure 5.4. Each of these axes is associated with an increase or decrease in the values of at least one variable. Alongside each of the perpendiculars drawn in Figure 5.4 are the mean values of the variables for each of the groups A–E.

To check if the model (Fig. 5.4) is working, it is necessary to determine whether the mean values of the variables increase in the right direction with respect to the set of axes. Thus, for example, group A has a mean basin area of 0·23 square miles while group D has a mean basin area of 0·67 square miles. Since area is, according to the model, increasing from bottom left to top right in Figure 5.4, the perpendicular from the centroid of group D to this axis should lie above (north-east of) the perpendicular from the centroid of group A to that axis. In other words since the perpendiculars from the centroids join the 'size' axis in the order A, D, C, E and B, moving progressively in the supposed direction of increases in basin area, the mean values of area for these groups must increase in the same order. Reference to either Table 5.2 or Figure 5.4 shows that this is the case. The same should be true for all other variables in relation to the appropriate axis. Of the variables shown in Figure 5.4 only the length ratio between second- and third-order streams, measured along the vertical axis, is really erratic. This variable is identified with factor VI and explains just over 3% of the variation in the original data. Its erratic behaviour with respect to the general model may not be greatly significant, and this variable axis may be ignored since this length ratio variable plays little part in determining basin behaviour. There is a fluctuating decrease in the mean values of drainage density in the direction in which mean basin area increases. The one exception concerns groups E and C: the value for group E should be less than that for group C, but in fact it is larger. Nevertheless this one exception does not destroy the general model. Of the variables not recorded in Figure 5.4 and shown in Table 5.2 stream frequency fluctuates in precisely the same way as drainage density. The only other variable to differ slightly from the general model is local relative relief which is greater for group B than it is for group A, whereas the model suggests that this relationship should be reversed.

These findings can be summed up by Figure 5.5. The original 130 drainage basins have been reduced to five groups which are shown to be significantly different from each other at the 99% level. These differences can be described by reference to four principal variables—basin area, the total number of streams in the basin, drainage density and the maximum relief of the basin. Additional variables can be brought in to increase the scope of the model; these include total stream length, the bifurcation ratio between first- and second-order streams, stream frequency, and the maximum local relative relief within the drainage basin. Thus the original 18 variables have been reduced to eight significant variables, four of which are fundamental in discriminating between the

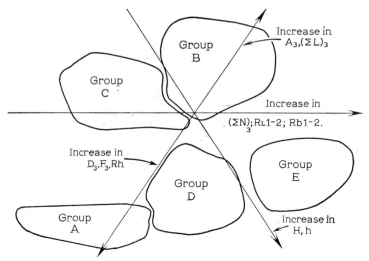

Fig. 5.5   A general model of the cluster groups in relation to changes in the values of the morphometric characteristics. (*Mather and Doornkamp, 1970*)

separate groups of basins (see also section 4.1). Furthermore, the character of each of the five groups can be described by reference to the general model (Fig. 5.5).

5.5 Basin groupings in relation to the morphological regions

The groupings suggested by cluster analysis can be conveniently compared with the original morphological regions (Fig. 2.1) from which the basins were drawn by plotting the areal distribution of the basins and by classifying them according to the groups A–E in which they occur (Fig. 5.6). A perfect match would have resulted if there had been as many groups as there are morphological regions, and if each group were composed only of the basins from one of these regions. This is not the case. Morphological regions 2 and 3 contain basins which belong to only one group, a condition almost fulfilled by region 1. All of these basins belong to group B, and there is therefore no case for distinguishing between regions 2 and 3 and indeed, for the most part, region 1, as separate areas of distinctly different character. To this end the original subjective classification of the area into morphological regions does not appear to be justified. On the other hand out of the 30 basins selected from these three regions only one does not belong to group B. There is hence a strong case for feeling more confident about including these three regions within a larger category, as in chapter 2 when they were referred to as areas transitional between mountains and lowland.

By contrast, the two mountain areas (regions 7 and 8) are composed of

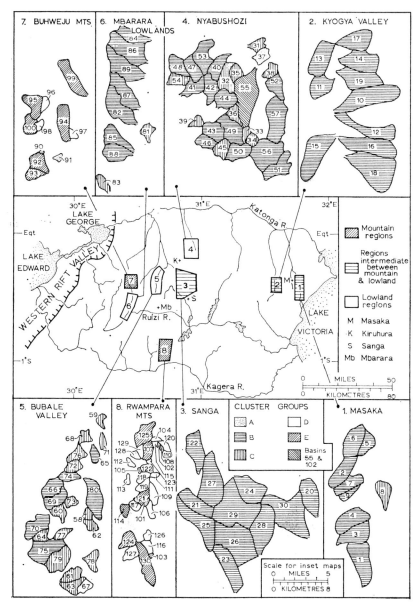

Fig. 5.6   A classification of the Uganda drainage basins into the cluster groupings of Figure 5.1. (*Mather and Doornkamp, 1970*)

basins which fall into three distinct groups (A, B and E). There are differences between the morphometric properties of the third-order basins within these two mountain areas. Nevertheless, it is also relevant to note that of the 89 basins within the non-mountain areas only one (basin 67) belongs to one of the groups (D) containing mountain basins, and even in this isolated case there are special local conditions which account for this.

The subdivision of the basins of the mountain regions into three different categories is directly related to the sites occupied by the third-order basins within the mountains. The basins of group A reach third-order status while still flowing on the steep mountain slopes and before reaching the floor of the most deeply incised valleys. The basins of group D reach to the foot of the mountain slopes (i.e. they extend over the whole distance from mountain summit to the floor of the more deeply incised valleys). The basins of group E, however, do not reach third-order status until they have flowed for some distance along the length of the more deeply incised valleys. Thus there are basically three different classes of third-order mountain drainage basins. None of these, however, is like any of the classes of lowland or intermediate region basins. The lowland basins (regions 4, 5 and 6 of Fig. 2.1) conform less regularly to the groupings suggested by cluster analysis, as they are composed of members of both groups B and C. Variations in the lithological properties of the bedrock do not appear to be responsible for creating these two separate groups. What may be significant is that all but one of the basins within the transition regions also belong to group B. It may be that those members of the lowland regions which belong to group B maintain some of the properties of the basins in the transition regions, while others (group C) have passed beyond this and have acquired a new set of characteristics more truly associated with lowland conditions.

The character of the basins within the mountain regions can be stated in the form of a table (Table 5.3) which contains a summary of the geomorphological characteristics of these regions, as known before the multivariate analysis was performed, and which lists the numerical values of some of the morphometric properties of third-order drainage basins within these regions. The terms 'low', 'medium' and 'high' as used in this table (and in Table 5.4) are relative terms based on the values found in this study only. Absolute values are also listed so that comparisons between the groups and with other areas can be made.

The observation above that while group B is typical of the transition regions, the lowland basins seem to reflect a new set of conditions (those of group C), suggests that an evolutionary sequence can possibly be established. The transition and lowland regions are described in Table 5.4 where the characteristics of their third-order basins are also summarized. Within the transition regions only the smallest portions of a pre-existing planation surface (the upland landscape) remain. In the lowland basins of group C the interfluves no longer show any traces of this planation

Table 5.3 The character of the mountain regions

Third-order drainage basin characteristics

			Range of mean values for groups A, D and E	
Relative relief:	H	high	841–1162	feet
	h	high	193– 783	feet
Drainage density:	D_3	medium–high	5·53– 7·08	miles/square mile
Stream frequency:	F_3	medium–high	17·50– 39·35	number/square mile
Relief ratio:	R_h	medium–high	484·25–1257·00 feet	
Basin area:	A_3	medium–low	0·23– 2·11	square miles
Total stream length:	$(\Sigma L)_3$	medium–low	1·54– 11·00	miles
Number of streams:	$(\Sigma N)_3$	large range	8·10– 32·30	
Bifurcation ratio:	R_b 1 & 2	large range	2·57– 6·29	

Geomorphological character: extensive planation surface remnants, deep river incision, steep slopes, high relative relief.

Table 5.4 An evolutionary sequence of basin morphometry

Transition region (group B)   Lowland basins (group C)

Geomorphological character

some remnants of planation surface	interfluves reduced below planation surface
some alluvial infill in valley floors	large amounts of alluvial infill
slopes steep only in upper part	gentle slopes, some inselbergs

Third-order drainage basin characteristics

			Mean value		Mean value
Relative relief:	H	low	384	low	290 feet
	h	low	280	low	161 feet
Drainage density:	D_3	low	2·60	medium–low	3·26 miles per square mile
Stream frequency:	F_3	low	5·40	medium–low	9·42 number per square mile
Relief ratio:	R_h	low	87·96	medium–low	140·10 feet
Basin area:	A_3	high	4·69	medium	1·31 square miles
Total stream length:	$(\Sigma L)_3$	high	11·14	medium	4·16 miles
Number of streams:	$(\Sigma N)_3$	medium–high	21·60	low	11·80
Bifurcation ratio:	R_b 1 & 2	medium–high	4·47	low	3·42

surface. In the context of the whole range of data values used in this study (and this includes the data for the mountain regions), relative relief is low in both of these groups, though in absolute terms it is higher in the case of group B. The values of drainage density, stream frequency and relief ratio increase slightly from group B to group C. Although the differences in each case are not great, the direction of this change may be contrary to what would have been expected. Basin area, total stream length, the number of streams and the bifurcation ratio between first- and second-order streams all reduce in magnitude from group B to group C (i.e. from an area still carrying the higher planation surface to one which is not). The reduction in basin area is probably associated with changes in the position of divides once the last vestiges of the pre-existing planation surface had been removed. These changes lead to a local reorganization of drainage basins and stream networks, and apparently, if the results of this study are valid, to a reduction in the size of third-order basins. The positive correlation (Fig. 4.1) between basin area and stream lengths suggests that such changes in basin area may be expected to lead to a reduction in total stream length. Conversely, the strong negative correlation (Table 3.3) between basin area and drainage density, stream frequency and the relief ratio, suggests that these are likely to increase as basin area decreases. Thus at least four other variables [$(\Sigma L)_3, D_3, F_3$, and Rh] may change in response to the changes in basin area that result from drainage reorganization following upon the removal of the remnants of the pre-existing planation surface. The reduction in the number of streams within the third-order basins as the last vestiges of the earlier surface are removed, may be due to an increase in the efficiency of the drainage network in that fewer streams are needed to meet the conditions in the areas without the old residual surface.

The increase in drainage density as the divides are lowered shows that in the ratio $(\Sigma L)_3 : A_3$ the area decreases more rapidly than total stream length. Likewise the increase in stream frequency shows that in proportion to the decrease in basin area there is a smaller proportional decrease in stream numbers. Thus although there is a decrease in the absolute values of A_3, $(\Sigma L)_3$ and $(\Sigma N)_3$ as divides are lowered below the old planation surface, there is, at the same time in this area, an increase in the intensity of the development of the drainage network.

References

COLE, J. P. and KING, C. A. M. 1968: *Quantitative geography.* London and New York: John Wiley (692 pp.)

KRUMBEIN, W. C. and GRAYBILL, F. A. 1965: *An introduction to statistical models in geology.* New York: McGraw-Hill (475 pp.)

MATHER, P. M. 1969: Cluster analysis. *Univ. of Nottingham, Dept. of Geogr., Computer Applications in the Natural and Social Sciences* **1.** (22 pp.)

MATHER, P. M. and DOORNKAMP, J. C. 1970: Multivariate analysis in geography with particular reference to drainage-basin morphometry. *Trans. Inst. Br. Geogr.* **51,** 163–87.

Part II   Slopes

6 The measurement and organization of slope data

6.1 Introduction

Many slope studies have been concerned with producing order out of the multitude of combinations of slope forms which can exist in an area. Not until it is actually known which forms are present is it possible to explain those forms. Just as an area can be broken down into its constituent drainage basins, thus reducing a large area to a limited number of manageable smaller ones, so it is possible to consider in detail the slope forms of which a drainage basin is composed. One possible subdivision of the slopes of an area is into *morphological units*. A morphological unit may be rectilinear in profile and a plane surface in plan, in which case it is a slope *facet*; or it may be a smoothly curved surface in which case it is a slope *element* (Savigear, 1960). Morphological units can be given numerical values. Thus a slope facet can be defined by its angle of slope, while an element can be defined in terms of the rate of change of slope between a known maximum and minimum slope angle. In each case the morphological unit can also be defined by its position either in plan within a drainage basin or on a hillside, or in profile down the hillside of which it is a part. To this end the two most useful techniques for measuring slopes in the field are:

1   *morphological mapping*, which defines the steepness of a slope and its relationship in plan to adjacent slopes
2   *slope profiling*, by which the steepness of a slope can be measured as well as both its vertical extent and its position down a hillside along an established profile line.

Through these two techniques a great deal of quantitative information can be collected and this can be subjected to numerical analysis. It is

An incising stream is creating new slopes, and their location and character is at least in part determined by the activity of the stream. Likewise the earlier development of the stream may have left a legacy in terms of slope form within the landscape, but these slopes now lie higher up the hillside than the newly created slope forms. Information concerning the hydrological characteristics of the drainage net is therefore relevant in terms of the efficiency of the removal of waste from the lower valley sides. Although the earlier history of the development of the drainage may have had a large effect in determining present slope forms it is very difficult to show that this was the case. The reason for this is that a circular argument often arises. Evidence of past drainage activity is usually sought in the evidence of the slopes themselves. How then is it possible to argue back from the deduced drainage history to the slope forms? The result must inevitably be a high correlation between slope form and drainage history.

Similar problems occur in comparing slope form with the climatic history of the area in which they occur. There may be some records of climate going back for many years, but the time period is usually very short when compared with the time available for the development of the slope. Some indications of changes in the micro-climate may be deduced from evidence, for example, that deforestation occurred within the area being examined. This will have led to a greater intensity of rainfall on the deforested slopes, and thus an increase in erosional activity. In addition the soil will lose the binding effect of tree roots, and again there is a greater likelihood of erosion. It is seldom possible, however, to translate such historical evidence into numerical terms.

Bedrock also provides an environmental control on slope form. Resistant elements in the bedrock are revealed through the presence of locally steeper slopes, or even the formation of a cliff. Useful information may, therefore, result from a comparison of a slope map with a geological map. But this is only a helpful exercise if it is known that the geology was not mapped from surface form in the first instance. 'Feature mapping' of geology is bound to lead to the conclusion that there is a close relationship between geology and slope form. Where there has been sufficient confidence in this relationship for feature mapping to have taken place the correlation between form and geology may indeed be close. Feature mapping of the geology tends, however, to overstate a relationship which may not always be so precise in reality.

There is a variety of bedrock characteristics which might influence slope form. These include its dip, both in amount and direction, the spacing and degree of development of joints, and its resistance to weathering and erosion. Dip can be measured in absolute terms, as can the density of joints. Absolute values cannot be given to rock resistance to weathering, though it is possible to rank the various outcrops in an area into an order of increasing resistance (Gregory and Brown, 1966, 254).

The position of a morphological unit may be significant in influencing

the purpose of this chapter to describe both of them and to discuss some of their limitations as well as their advantages. The type of quantitative information which they provide will also be described. First, however, there are some major difficulties about slope studies which need to be considered. It is important too that the role of numerial analysis as an aid in the solving of these problems is appreciated.

A major difficulty arises through the element of subjectivity which is present in looking at slope form. Each geomorphologist tends to look at a landscape in terms of the training and experience which he has had (Clayton, 1970). Any methods of geomorphological analysis which allow a non-subjective approach to land form studies is, therefore, to be welcomed. Unfortunately there is a certain amount of subjectivity present in the construction of a morphological map, though less so in the case of the measurement of a slope profile. Nevertheless both go a long way towards a definition of what *is* present in the landscape as opposed to the more subjective view of what the observer thinks is there.

Numerical analysis allows the interrelationships between one slope characteristic and another to be examined, and this provides the theme for chapter 8. For example, it is possible to test such hypotheses as 'the longer a slope the gentler on average it will be', or 'the steeper the slope the higher, in general, it will rise above its base', or yet again, 'the steepest parts of a slope occur at a particular position on its profile'. In addition, the relationships between slopes and measurable environmental properties can be assessed. In a particular area it may be thought that aspect plays a part in determining slope steepness, or, on the other hand, distance from the mouth of the drainage basin may be considered to have a significant relationship to slope steepness. In each of these cases correlation and regression techniques can be used to test such hypotheses. On the other hand, although a relationship between angle of slope and bedrock lithology may be suspected, regression techniques will be of no value for it is not possible to define lithology on the interval or ratio scales. It is true that various properties of the strength of bedrock can be measured, but no single number can be used as a measure of its resistance to weathering and erosion. Likewise a numerical value cannot be assigned to a particular stage in the history of the development of a landscape, and yet it may be that particular slope values are associated with a clearly recognized stage in landscape evolution.

In addition to the measurement and analysis of slope form the geomorphologist is concerned with slope processes. All too little is known, however, about the precise nature and intensity of processes currently at work on different slopes. The development of this side of geomorphology awaits the development and perfection of techniques of slope instrumentation. General observations can be made about the location of areas of rock fall from the accumulation of boulder screes, or concerning the presence of slope creep from the presence of terracettes or the accumulation of slope material behind a large tree. It is true that limiting factors

for the development, for example, of landslides can be defined, and these may include various aspects of slope form, such as a minimum angle of slope. In the case of less dramatic geomorphological events such a definition is not always possible.

If it is difficult to measure and to define the extent and intensity of processes active today, how much more difficult it is to do so for the processes active in the past. Without such knowledge, however, the analysis of slope development is speculative to say the least. Slope development, reduced to its simplest terms (Fig. 6.1), is the result of processes in the past working on an initial form giving rise to the slopes which are seen and measured today. Processes actively at work today give rise to new land forms which can be measured tomorrow. In many cases we do not know yesterday's processes, and we cannot know tomorrow's slope forms. Usually it is only possible to measure the form properties of a slope as it exists, and to look for relationships among those properties. It is sometimes possible, over a limited period of time and in isolated areas, to measure

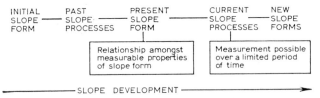

Fig. 6.1 The course of slope development.

some of the active slope processes. (Rapp, 1960a and b). Only in a very few cases can the whole sequence from measured early form through process to later form be analysed. This is possible only where the process is a catastrophic one, such as a landslide, or when a process has been measured over a very long period of time or at a micro-scale. Methods of effective instrumentation are too recently developed for there to have been many case studies, as yet, of either slow processes or micro-scale effects.

Numerical analysis of the relationship between slope form and environment may go some way, however, towards giving clues about slope development. For example it may be, as Pallister (1956) found in Uganda, that a particularly resistant rock capping hill crests gives rise to very steep slopes around their summit margins. Where no such cap rock exists this steep unit of the hillside is absent and the relative relief from the bottom to the top of the hillside is less. It may be inferred, therefore, that when the cap rock is finally removed from a hill crest its summit will decrease in elevation and the steepest portion of the slope profile will disappear. In other words one aspect of the development of slopes may be deduced from a comparison of the measured form of neighbouring hillsides.

There are many slope studies in which differences observed among neighbouring hillsides are accepted as indicators of changes which have

taken place or will take place on one hillside through time. This is not a completely logical step for it assumes that all of the slopes studied on an aerial basis are developing in the same way as each other. This idea may be represented diagrammatically as in Figure 6.2A: the two slopes have a similar form to begin with; slope 2, however, develops new forms much more quickly than does slope 1. Thus, when time B is reached the two forms are different. They may be sufficiently similar, however, for there to be a subjective impression on viewing them that slope 2 may once have had the form now displayed by slope 1. This was indeed the case at time unit A. If at time B there is a significant change in climate then the rate of slope development will change for both slopes 1 and 2. Since their forms are now different, the processes active upon them may also differ at least in degree if not in kind, and their rates of development to a new form, Z,

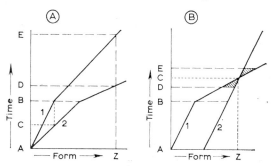

Fig. 6.2 A comparison of the development of two slopes: **A**: slopes begin with same form, **B**: initial forms are unlike each other.

will continue to be different (Fig. 6.2A) and will prevent them from being alike at any one moment in time.

An infinite variety of possibilities of slope development can be represented in this way. Another one is shown in Figure 6.2B. In this case slopes 1 and 2 do not have the same form at time A. Time A may be their moment of inception or it can be some moment in time during their development. Up to time B the rate of development to new forms is the same for both slopes, but at B climatic change, for example, speeds up the development of new forms in the case of slope 1, though it does not do this to the same extent in the case of slope 2. The result may be that at moment C both slopes will have the form Z. It is most unlikely that a geomorphologist would happen to study and compare both slopes exactly at moment C. These slopes could, however, be studied at some time between moments D and E when their forms may be very similar. This similarity may also suggest that slope 2 once had the form now shown by slope 1. This will only be the case at C; it would be a mistake to conclude that both slopes have had an identical or even similar history.

Numerical analysis is not yet sufficiently developed to enable it to be generally used in deciphering the nature of slope evolution. In can

however, make possible the classification of slopes into separate groups within each of which the slopes are alike in their present form. In this sense the classification of slopes, based on their form, presents problems similar to those encountered in zoological taxonomy (Sokal and Sneath, 1963; Cole, 1969). Methods of slope classification will be examined again in chapter 9.

If numerical analysis is to be applied to slopes by different people in different areas then their methods both of field measurement and of subsequent analysis must be similar in order for their results to be comparable. A great deal depends, for example, on the way in which the measured sections of a slope profile are grouped together. Certain precise rules for slope profile analysis have been developed, especially by Young and Savigear. These are described in later sections of this chapter and form the basis for the case studies analysed in chapter 8.

6.2 The measurement of slope data

Morphological maps
Slope form can be mapped by delimiting the position of breaks and changes of slope. A break of slope occurs where there is a sharp junction line between two slopes, or morphological units, of differing steepness. A change of slope occurs where the junction is more gradual and occupies a zone on the ground. The symbols used for the mapping of breaks and changes of slope are illustrated in Figure 6.3. The mapping of the junctions between morphological units leads to a delimitation of the morphological units. As was suggested at the beginning of this chapter, these units may be plane or curved. The plane surface, or facet, may be horizontal (in which case it is a *flat*) or it may be inclined (in which case it is a *slope*). On the other hand a curved surface, or element, may be either curved in a concave manner (upwards), in which case it is a *concave element*, or it may be convex upwards (a *convex element*). The range of slope values embraced by the element can be shown on a morphological map by recording the steepness of the slope at its upper and lower extremities. The amount of detail which can be recorded on a morphological map varies with the map scale. Problems of scale are discussed in detail by Savigear (1965). Most landscapes contain slope boundaries which are of varying magnitude. The visual impression of a morphological map is enhanced if different thicknesses of line are employed, for the breaks and changes of slope, to correspond with their relative magnitudes. Likewise slopes of different steepness occur within the landscape, and if slope classes are defined then the morphological units can be shaded or coloured with respect to their steepness category.

Morphological maps do not of themselves show heights. For this reason they are made more comprehensive by superimposing the mapping symbols on a contour base map. The resulting map gives a much more

detailed and precise statement of surface form than is possible from the contours alone. An example of a morphological map superimposed on a contour base is shown in Figure 6.4. This map is of a part of a composite cuesta consisting of two scarps developed in carboniferous sandstones to

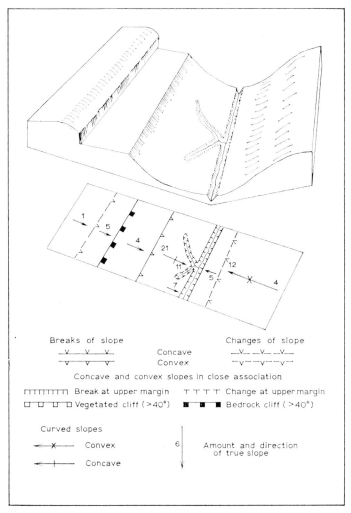

Fig. 6.3 Morphological mapping symbols.

the west of Sheffield. These scarps are separated from each other by a vale formed in shales. The slopes are shaded to indicate their steepness. Variations in slope angle are more easily appreciated from such shading than they are simply by assessing changes in the spacing of contours. In addition morphological boundaries can be shown on the morphological

map which would remain undetected on the contour map. It is possible to compile extremely detailed contour maps, but their construction is much more time-consuming than morphological mapping and would in any case require further categorization to be of use for numerical analysis. An additional, and valuable, advantage of morphological maps is that they indicate both the continuity and lateral variations in the nature of

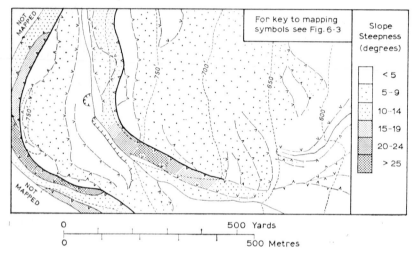

Fig. 6.4 A morphological map superimposed on a contour base map and shaded according to slope steepness.

the morphological boundaries. A break of slope can merge laterally with a change of slope, or it may divide into two separate breaks between which a distinct morphological unit occurs. It is also possible for a morphological boundary to fade out and to become indistinct along a slope. Other published examples of morphological maps may be found in Bridges and Doornkamp (1963), Curtis, Doornkamp and Gregory (1965), Doornkamp (1964), Gregory and Brown (1966), Savigear (1960; 1965) and Waters (1958).

Slope profiles
A statement of the form of a hillside in profile can be arrived at in a number of ways. The least detailed of these may be compiled from the contours of published relief maps. The numerical data which result are of little value, and at times may be quite misleading, depending on the scale of the map and the vertical interval between the contours. Slope profiles can also be compiled from morphological maps by taking account of the uphill succession of morphological units, their up-slope extent and their steepness. The detail of the resulting profile is commensurate with the scale and detail shown on the morphological map. More frequently, however, slope profiles are obtained by direct measurement in the field.

The field techniques involve the measurement of slope angle over known lengths of slope. The shorter the measured lengths the more detailed the slope profile will be. Three types of equipment have been used in the past for slope profiling. These are:

1   *Ranging rods and Abney level:* The distances between the rod positions are measured with a tape. The angle of slope between successive ranging rods is measured with the Abney level to 10 minutes of arc.

2   *Level and staff:* Distances and height differences are recorded between stations by viewing through the eye-piece of the level.

3   *Slope pantometer:* The measured length is fixed (usually at five feet,) and the angle of slope measured on a protractor attached to the vertical leg, and registered by the slope of a cross-bar.

4   *Automatic recording devices* for measuring slope profiles are being developed commercially. In essence they consist of a rotating drum on which the angle of slope and distance traversed are recorded as the instrument is pulled up or down the slope. These automatic recorders have not yet been widely used for geomorphological studies.

The first two of these methods are borrowed from surveying practice, and are documented in surveying manuals. The slope pantometer, on the other hand, was constructed specifically for the task of providing slope profile information for geomorphological purposes and is described by Pitty (1968). It provides the most rapid method for collecting numerical information about the slope in profile, especially where both the topography is highly accidented and where many readings would be required with either of the other two methods. In addition the use of a constant five-feet measured length between stations is in some ways a great advantage when making a statistical analysis of the data (Pitty, 1967).

Slope studies using the Abney level or a level and staff are open to a degree of subjectivity in that there is a choice between using fixed interval stations or of measuring only from one change, or break, of slope to the next. In the former case there is the possibility of successive station positions standing astride a significant break of slope and not recording its full magnitude or its correct position (Fig. 6.5). In the latter case great care has to be taken that all of the breaks and changes of slope are spotted as profiling proceeds. Even when this is achieved some of the measured lengths will be extremely small while others may be too large to record the slope angle with a high degree of accuracy. A compromise is therefore more successful and measuring from one slope boundary to the next may be adopted as a general principle, though where these are very widely spaced the intervening ground should be measured at constant station intervals.

Slope profiling should never be undertaken in isolation. It should always be tied to morphological mapping in such a way that the slope boundaries crossed by the profile line can be located on the morphological map.

Indeed, the morphological map may be the best form of base map for recording slope profile positions. The morphological map also provides a ready means of assessing the lateral continuity of a morphological

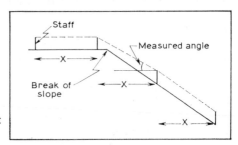

Fig. 6.5 Slope profile measurement with a fixed station interval.

boundary detected on adjacent slope profiles. Slope profiles may be used either to check the accuracy of the location of symbols on the morphological map, or in their own right for slope form analysis. In both cases they provide numerical data which can be subjected to detailed analysis.

The data supplied by morphological maps
Morphological maps provide a statement of the areal distribution of morphological units, their steepness, and the space which they occupy. Thus for any hillside it is possible to obtain a complete assessment of the range of slope steepnesses, the area of the slopes at each slope angle or in each slope category, and the relative positions of the slopes at each slope angle. This type of data is highly amenable to numerical analysis.

The analysis of drainage basins in part I included a measure of the area of a drainage basin. This was in fact a measure of its map area and not its ground area. Ground area will always exceed map area. The steeper the slopes within a drainage basin the greater its ground area will differ from its map area. When a morphological map is compiled it is possible to convert the measured map area into ground area. This is achieved by measuring the area of each morphological unit, and dividing this value by the cosine of the angle of slope. The total ground area of a drainage basin is then obtained by summing the ground areas of all of its morphological units. This is an extremely laborious procedure, but a necessary one if slope studies are to be scientifically refined. Gregory and Brown (1966) found, for example, that a simple count of the number of morphological units at a particular slope angle did not provide the same frequency distribution as a measure of the area of the units at each slope angle. This was especially true for slopes of 2–8°.

The data supplied by slope profiles
Slope profiles provide information about the angle of slope along a measured length within a complete profile whose position has been

accurately located on a base map. The detail shown on the measured profile depends on the measurement interval, but once obtained the profile can be generalized to any required degree. Certain principles for doing this have been established and these are described below.

Other slope properties
Although a measurement of surface form is the fundamental prerequisite for slope studies it is not the only item which may usefully be measured. Most slopes carry a mantle of regolith. This is usually of variable depth on any one hillside, and the implication is, therefore, that the form of the solid rock surface beneath the regolith cover is not the same as that of the ground slope. The nature of the bedrock surface beneath a measured slope profile may be examined by measuring the thickness of the regolith at selected points along the slope profile. From this a bedrock profile may be constructed. An extremely detailed example may be quoted to illustrate the type of difference that may occur between the slope and the bedrock profiles. Construction of both the slope and bedrock profiles on the easternmost of the two scarp faces shown in Figure 6.4 showed that

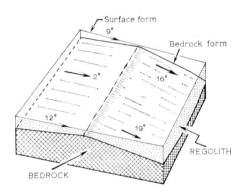

Fig. 6.6 A comparison of a surface and bedrock-profile on the scarp face of a cuesta west of Sheffield.

while the surface profile was an extremely smooth feature, it masked a highly irregular bedrock profile. The bedrock profile consisted of a series of steps (Fig. 6.6), with slopes of between 2° and 19°, and there was a considerable variation in the thickness of regolith. Differences between the surface and bedrock profiles are most marked where an area has been subjected to deep weathering. The maps and cross-sections presented by Thomas (1966) indicate how different the surface and bedrock forms may be within a tropical area. These differences are of considerable geomorphological significance.

It is difficult to obtain precise information about the relationship between slope form and climate. In terms of present-day processes it is the micro-climate of the slope which is relevant. In only a few studies has there been detailed measurement on an intensive scale of this important factor. Melton (1957) was able to gather enough information to use

Thornthwaite's precipitation-effectiveness index (or P-E index) (Thornthwaite, 1931) as a general measure of climate. The P-E index is derived from:

$$\hat{I} = 10 \sum_{12} \frac{P}{E}$$

where P is the mean precipitation for each month; E is the average evaporation for each month. When no evaporation data are available a formula which provides a reasonable estimate of $\hat{I}$ is:

$$I = 115 \sum_{12} \left(\frac{P}{T-10}\right)^{1 \cdot 11}$$

where T is the monthly average daily temperature. It has often been suggested that slope aspect is an important determinant of micro-climate and since aspect can be easily measured it has sometimes been used as a numerical substitute for a statement concerning micro-climate. Slope aspect may also be used as a measured variable quite independent of any relationship it may have to climate. This may be done by measuring the orientation of the true slope direction of a morphological unit with respect to grid north. These aspect values may then be grouped into classes of equal orientation interval. The study by Gregory and Brown (1966) employed 36 categories, each occupying 10°.

Several studies have been concerned with the relationship between slope form and the regolith cover. The physical properties of the regolith which may be measured include the size and shape characteristics of the individual regolith particles. Other physical properties of the regolith which may be related to slope form, and even to slope processes, include its shear strength, bulk density, pore pressure and cohesion. More and more numerical studies of slope development are taking these measurable characteristics into account. The relationships which are beginning to emerge between some of these and the form of the slope itself are discussed in chapter 8.

Direct measurements of slope processes have been predominantly concerned with measuring the rate at which a surface is degrading or aggrading, as well as attempting to measure the rate at which regolith material is moving across a surface. A great deal of work remains to be done in this field, but its more important conclusions will result from numerical analysis of the type being discussed in this book.

6.3 Environmental influences

Slope form is dependent on many environmental conditions which may not be entirely measurable on the slope itself. Thus, for example, slope form is related to the erosional history of the stream flowing at its base.

its form. Gregory and Brown (1966) suggest that morphological units may be located with respect to three major slope types:

1   *stream-side slopes*, extending up from the stream to a free face low down the hillside
2   *valley side slopes*, extending from the stream-side slopes to the top of the valley side (which may be represented by another free face)
3   *summit slopes*, which reach from the top of the valley-side slopes to the interfluve summit.

Greater precision is possible if these definitions are replaced by those of Savigear (1960), who uses and defines the terms footslope, backslope (which may or may not be cliffed) and crestslope (see chapter 8). Nevertheless, the method of measuring the position of a morphological unit is similar in either case. This is achieved by calculating its percentage position, within the slope type of which it is a part, according to the equation:

$$P = \frac{H_b}{H_b + H + H_t} \times 100\%$$

where H_b is the difference in height between the base of the major slope type and the lowest part of the morphological unit; H_t is the difference in height between the highest part of the unit and the top of the slope type; and H is the height range of the morphological unit. In addition, the height of a morphological unit above the stream to which it is supplying debris can be measured.

6.4 The application of numerical analysis to slope form data

The usefulness of morphological mapping and slope profiling depends on the nature of the investigation being undertaken. This book is concerned with methods of numerical analysis, and for this purpose there are no other techniques which will supply the same amount or the same kind of data. Such data may be used in order to assess the interrelationship between slope properties, and also to investigate the relationship between slope forms and other aspects of the environment. Both of these provide the subject matter of chapter 7. The data supplied by the techniques of morphological mapping and slope profiling may in addition be used to compare slope form with that suggested by theoretical models. Chapter 9 deals with this subject, and it stresses the importance of working from the measured slope form to the slope model, rather than the other way as has been the case so often in the past. Before these things can be done, however, it is necessary to consider how the field data can be assembled into a manageable form for numerical analysis.

6.5 The organization of slope data

The morphological map

A morphological map presents, at the map scale, a summary statement of the morphological units of the ground. Figure 6.7A shows the nature of a third-order drainage basin in the lowland area to the west of Weymouth (Dorset, England). As it stands this map shows the distribution and steepness of each morphological unit, and as such, it provides a statement of the *pattern* of the surface. For numerical analysis, however, this information needs to be organized into a tabular form. In addition to listing the steepness and area of each morphological unit the table can also include information about its shape and position. In areas of high relative relief there will also be large differences in the altitudes of the

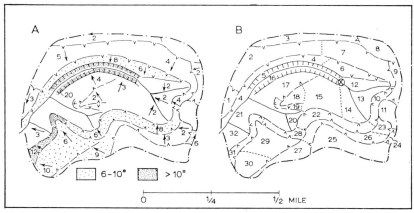

Fig. 6.7   A morphological map of a third-order basin near Weymouth : **A**: the map ; **B**: the slope unit numbers used in Table 6.1.

morphological units: these too may be recorded. Morphological units also vary in their relationship to the units above and below. In particular, some of the boundaries between units are very distinct because there is a large difference between their inclinations. Thus the magnitude of this *discontinuity* may be recorded as the difference in angle between the slopes of the two units.

Table 6.1 has been compiled for the drainage basin shown in Figure 6.7. The angles of slope were measured with an Abney level. Ground area has been obtained by measuring the map area and dividing by the cosine of the angle of slope. This area may also be expressed as a percentage of the total area of the basin, for this gives a measure of the magnitude of the unit within its local setting. The shape of the morphological unit can be measured in a similar manner to that of a drainage basin, but can also be measured in relation to the way it lies within the valley. Thus it is possible to measure the maximum length of the unit parallel to, and perpendicular

to, the trend of the valley floor. The height of the unit above the valley floor can be measured for both its upper and lower boundaries, and a mean value calculated for the vertical centre of the unit. If the unit boundaries vary in height then an average value will have to be calculated. This can be done by marking the boundary into ten equally spaced intercepts and by calculating the mean height for these ten points. Another useful measure of the position of the morphological unit can be obtained by measuring the distance of the centre of the lower boundary of the unit from the basin mouth as well as this distance for the upstream and downstream limits of the unit. To be meaningful the distance must be measured along the length of the valley floor, including its twists and turns. A line is drawn from each of the appropriate points on the lower edge of the morphological unit, down the true slope to the valley floor. The distance up the valley is measured to the junction of these lines with the valley

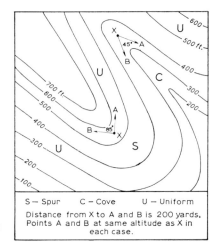

S — Spur   C — Cove   U — Uniform
Distance from X to A and B is 200 yards.
Points A and B at same altitude as X in
each case.

Fig. 6.8 The measurement of plan curvature on a spur or a cove.

floor. Discontinuities may be either positive or negative (i.e. related to convex and concave slope boundaries respectively). For the preliminary analysis of an area the sign may be ignored and just the magnitude of the discontinuity recorded for both the upper and lower boundaries of the unit.

Morphological units are not always straight in plan. When viewed from above they may be either concave (termed a *cove* by Aandahl, 1948), convex (a *spur*), or straight (uniform). The amount of curvature of the unit can be measured by taking two points (A and B in Fig. 6.8) on either side of the site (X in Fig. 6.8) and along the contour passing through X. Bearings to A and B from X can be measured and the smallest difference between them provides the angle of the curvature (α).

The compilation of a table for a whole drainage basin is dependent on the assumption that all of the slopes are facets—that is to say, they are not curved in profile. In this example it was possible to define the whole

Table 6.1 Slope properties of a selected third-order drainage basin near Weymouth (see Fig 6.7)

Slope unit	θ	G	g%	L_p	L_r	Z_b	Z_t	Z_b-Z_t	Z_c	B_u	B_d	B	S_u	S_l	α	F_c
1	3	34	1·3	12	4	31	51	20	31	15	4	10	3	5	180	4
2	5	216	8·1	38	7	50	84	34	67	45	15	30	3	3	123	9
3	2	94	3·5	40	3	84	87	3	88	56	27	42	2	3	170	10
4	8	104	3·9	45	2	36	50	14	43	45	0	22	3	12	150	3
5	20	68	2·6	39	2	0	36	36	18	56	17	37	12	20	140	1
6	6	72	2·7	24	5	25	60	35	42	70	45	58	2	14	180	4
7	4	200	7·5	30	8	60	116	56	88	74	45	60	4	2	175	12
8	0	78	2·9	16	5	65	65	0	65	70	56	63	0	4	180	18
9	2	42	1·6	3	14	35	39	4	37	80	78	79	0	2	147	21
10	4	60	2·3	4	16	20	34	14	27	72	69	71	4	4	120	14
11	0	52	2·0	6	8	34	34	0	34	78	72	75	0	2	180	20
12	2	70	2·6	18	4	0	8	8	4	74	56	65	4	2	146	8
13	2	72	2·7	12	12	7	20	13	13	69	60	64	2	2	85	10
14	2	70	2·6	10	14	0	31	31	15	56	72	64	6	17	152	8
15	3	160	6·0	16	12	20	46	26	33	60	42	51	5	20	130	8
16	20	68	2·6	39	2	0	30	30	15	56	17	37	16	16	140	1
17	4	172	6·5	24	8	36	64	28	50	46	15	30	2	2	175	6
18	2	40	1·5	10	6	64	71	7	67	33	23	28	1	2	180	10
19	1	16	0·6	10	2	54	59	5	56	35	21	28	1	4	70	12
20	3	20	0·8	4	6	49	63	14	56	30	24	27	4	4	120	18
21	3	110	4·1	11	20	0	28	28	14	74	23	48	6	5	75	8
22	8	100	3·8	30	4	31	73	42	52	70	28	49	5	2	108	13
23	2	30	1·1	5	6	41	48	7	44	82	76	79	2	6	120	22
24	6	18	0·7	6	2	34	38	44	56	84	82	83	6	3	170	27
25	0	160	6·0	26	10	78	78	0	78	82	54	68	0	5	180	20
26	3	110	4·1	36	4	73	78	5	75	79	21	47	3	3	92	18
27	9	50	1·9	23	8	98	145	47	121	25	0	12	9	4	180	17
28	6	40	1·5	18	4	64	70	6	67	28	20	24	2	4	120	17
29	6	140	5·3	20	16	45	98	53	71	24	24	10	4	2	180	16
30	10	80	3·0	10	10	45	133	88	99	12	0	6	5		150	15
31	12	48	1·8	22	4	28	45	17	36	16	0	8	3	10	150	8
32	3	70	2·6	8	8	0	28	28	14	8	0	4	10	3	180	4

basin in terms of slope facets. This is not always the case. When both facets and elements occur separate tables must be compiled for each. This is because the elements have measurable properties which do not apply to facets. In particular curved units (elements) have different angles of slope at their upper and lower extremities (θ_U and θ_L respectively), and they are either steeply or gently curved according to their radius of curvature (R). A table of slope properties for slope elements must, therefore, also include three columns for θ_U, θ_L and R. Such a table can be made comparable with that for slope facets only by finding a mean slope

$$\left(\frac{\theta_U + \theta_L}{2} = \theta \right)$$

for the element and by ignoring its radius of curvature.

The slope profile
Scale is an important influence upon the amount of detail that is recorded by both morphological maps and slope profiles. There are different scales of data recording and there are also different scales of data generalization. These may be illustrated with respect to slope profile measurement. In addition two schemes of slope profile classification, those of Young (1963; 1964) and Savigear (1967), will be examined to show how they can provide data for numerical analysis.

The initial slope profile is always the *measured profile* (Fig. 6.9A). It consists of a record of the stations between which slope angles were measured.

where:

θ	degrees	angle of slope
G	thousandths of square miles	area of unit
g	%	area of unit as percentage of total area of basin
L_p	hundredths of a mile	length of unit parallel to the long-axis of the valley
L_r	,,	length of unit at right-angles to the long-axis of the valley
Z_b	feet	height, above the valley floor, of the lower boundary of the unit
Z_t	,,	height, above the valley floor, of the upper boundary of the unit
Z_c	,,	height, above the valley floor, of the centre of the unit
B_u	hundredths of a mile	distance, from the basin mouth, of the upstream end of the unit
B_d	,,	distance, from the basin mouth, of the downstream end of the unit
B	,,	mean distance of unit from the basin mouth
S_u	degrees	strength of upper discontinuity (boundary)
S_1	,,	strength of lower discontinuity (boundary)
α	,,	curvature of spur or cove
F_c	hundredths of a mile	distance of the centre of the unit from the valley floor

These stations are marked by short ticks, spaced according to the distance measured between them, and joined by a line drawn at the angle of slope. The profile can be annotated with both the slope angles and the difference in angle between one measured length and the next. A positive change occurs when the angle is increasing downslope and negative when the angle decreases. These differences are *survey discontinuities* since they arise directly from the position of the survey stations (Savigear, 1967).

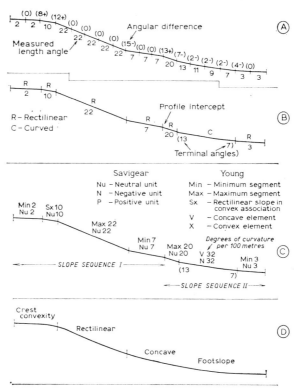

Fig. 6.9 Stages in the generalization of a slope profile (*after Savigear, 1967*): A: Measured profile; B: Intercept profile; C: Unit profile; D: Component profile.

Some of these survey discontinuities will be large, as when the survey station coincides with a distinct break or change of slope; at other places they may be small.

A slightly generalized profile, known as an *intercept profile* (Fig. 6.9B), may be derived from the measured profile (Savigear, 1967). The generalization involves the recognition of the curved and rectilinear portions of the profile as a whole. A curved unit occurs when each successive measured length is either steeper or gentler than the previous one. The profile intercepts occur when a particular type of succession ends. Profile intercepts

also occur on either side of a rectilinear portion of the profile. Where two successive measured lengths differ by less than $1°$, or where three successive lengths differ by less than $2°$, they may be considered to form a continuous rectilinear unit. If a measured length differs in inclination by more than this from its neighbour, it either remains as a rectilinear unit on its own and the bounding slope intercepts coincide with the original station positions, or it forms a part of a larger curved portion of the profile. This will depend on the behaviour of the set of measured lengths above and below the section being examined. When all the rectilinear portions of the profile have been delimited the remaining parts of the profile must all be curved.

Once the intercept profile has been compiled it is possible to go one stage further and to classify each of the curved and rectilinear units of the profile. The result is the *unit profile*. Savigear (1967) and Young (1963; 1964) both illustrate the nature of unit profiles and basically their methods are the same. The difference between them lies in the terms which they use. In Figure 6.9c the unit profile has been subdivided into its curved and rectilinear units and the key shows the terms which would be used by Savigear and Young for this profile. Curved units are referred to by Young as convex elements and by Savigear as positive units, when the angle of slope is increasing downhill. Where Young would recognize two sectors of an element, each having a different radius of curvature but still continuing the sense of the downward trend of slope, Savigear recognizes two separate positive units. Curved units are defined by both the terminal angles and the radius of curvature of the unit. In the same sense Young refers to a progressive decrease in slope angle as a concave element, whereas Savigear uses the term negative unit.

A rectilinear portion of the profile is called a segment by Young and a neutral unit by Savigear. It can be defined by its angle of slope, and its length can be measured from the profile drawn to scale. Rectilinear units can hold a variety of relationships to the slopes which occur above and below. If the terminal angle of the units both above and below the rectilinear portion of the profile are steeper than the slope of the rectilinear unit itself then Young refers to it as a *minimum segment*. Conversely if both the slopes on either side of the rectilinear unit are gentler then it is known as a *maximum segment*. Segments may form the last units at either end of a slope profile which reaches from valley floor to hill crest. The segment at the hill crest is called a *crest segment* (Young, 1964); the segment at the base of the slope can equally well be called a *foot segment*. Savigear (1967) defines a comprehensive set of profile unit groups. These have not yet been applied and subjected to numerical analysis, but they may well provoke a great deal more research into slope form classification.

At a yet more generalized scale there is the *component profile* (Fig. 6.9D). These components are the flats, slopes and curved portions of a hillside normally appreciated during field observation (Savigear, 1967). They are composed of slope units, and can be derived from the unit profile by

generalizing the profile form into its main curved and rectilinear portions. This scale of generalization involves a loss of much of the measured detail, and is thus not an efficient scale at which to analyse the data laboriously collected in the field.

References

AANDAHL, A. R. 1948: The characteristics of slope positions and their influence on the total nitrogen content of a few virgin soils of Western Iowa. *Soil. Sci. Soc. Am. Proc.* **13,** 449.

BRIDGES, E. M. and DOORNKAMP, J. C. 1963: Morphological mapping and the study of soil patterns. *Geography* **48,** 175–81.

CLAYTON, K. M. 1970: The problems of field evidence in geomorphology. In Osborne, R. H., Barnes, F. A. and Doornkamp, J. C., editors, *Geographical essays in honour of K. C. Edwards*, University of Nottingham, 131–9.

COLE, A. J., editor, 1969: *Numerical taxonomy*. New York: Academic Press.

CURTIS, L. F., DOORNKAMP, J. C. and GREGORY, K. J. 1965: The description of relief in field studies of soils. *J. Soil Sci.* **16,** 16–30.

DOORNKAMP, J. C. 1964: Subaerial landform development in relation to past sea levels in a part of south Dorset. *Proc. Dorset Nat. Hist. and Arch. Soc.* **85,** 71–7.

GREGORY, K. J. and BROWN, E. H. 1966: Data processing and the study of land form. *Zeit. für Geomorph.* NF **10,** 237–63.

MELTON, M. A. 1957: An analysis of the relations among elements of climate, surface properties, and geomorphology. *Columbia Univ., Dept. of Geol., Tech. Rept.* **11,** 1–102.

PALLISTER, J. W. 1956: Slope development in Buganda. *Geogr. J.* **122,** 80–7.

PITTY, A. F. 1967: Some problems in selecting a ground-surface length for slope-angle measurements. *Rev. Géomorph. Dynamique* **17,** 66–71.

1968: A simple device for the field measurement of hillslope. *J. Geol.* **76,** 717–20.

RAPP, A. 1960a: Recent development of mountain slopes in Kärkevagge and surroundings, northern Scandinavia. *Geogr. Ann.* **42,** 71–123.

1960b: Talus slopes and mountain walls at Tempelfiorden, Spitsbergen. *Norsk Polarinstitutt, Skrifter Nr.* **119.** Oslo.

SAVIGEAR, R. A. G. 1960: Slopes and hills in West Africa. *Zeit. für Geomorph.* Suppl. **1,** 156–71.

1965: A technique of morphological mapping. *Ann. Ass. Amer. Geogr.* **55,** 514–38.

1967: The analysis and classification of slope profile forms. In L'evolution des versants, *Les Congrès et Colloques de l'Université Liège* **40,** 271–90.

SOKAL, R. R. and SNEATH, P. H. A. 1963: *Principles of numerical taxonomy*. San Francisco: W. H. Freeman.

THOMAS, M. F. 1966: Some geomorphological implications of deep weather-

ing patterns in crystalline rocks in Nigeria. *Trans. Inst. Brit. Geogr.* **40,** 173–93.

THORNTHWAITE, C. W. 1931: The climates of North America according to a new classification. *Geogr. Rev.* **21,** 633–55.

WATERS, R. S. 1958: Morphological mapping. *Geography* **43,** 10–17.

YOUNG, A. 1963: Some field observations of slope form and regolith and their relation to slope development. *Trans. Inst. Brit. Geogr.* **32,** 1–29.

1964: Slope profile analysis. *Zeit. für Geomorph.* Supp. Band **5,** 17–27.

7 Analysis of slope form

7.1 Profile analysis

Convex and concave profiles; A slope classification map; Slope profiles as mathematical curves; [*regression equation, parabolic curve, logarithmic curve, hyperbolic curve, polynomial curve, power functions, linear, quadratic, and cubic equations, residual value, differentiation, differential coefficient, roots of an equation, maximum and minimum turning points, points of inflexion*]

7.2 Descriptive statistics and slope data

[*random sampling, linear regression, histogram, point sampling, frequency distribution tables, normality of data*]; Characteristic and limiting slope angles; Comparison between basins

7.1 Profile analysis

Convex and concave profiles

The numerical analysis of profile form is best carried out on data derived from the *unit profile* (section 6.5). These units are capable of providing a measure of the amount of convexity or concavity possessed by the profile as a whole. For example, Figure 7.1 shows the form of seven profiles measured along the face of an escarpment, developed in sandstones, to the west of Sheffield (England). These are unit profiles, and they have been classified on the basis illustrated in Figure 6.9c. Against each X (convex element) and V (concave element) is a number which is the radius of curvature of the element measured in degrees per 100 feet. The lengths of these curved units and their terminal angles are known as well as the radii of curvature. For each of these profiles two separate calculations can be made in order to establish the amount of profile convexity or concavity. The first requires a measurement of the total ground length of those units which taken together tend to give an increase in the steepness of the profile. These are (Fig. 7.1) the convex elements (X), the segments (or facets) in a convex situation (Sx) and the minimum (MIN) and maximum segments (MAX) which lie above and below these convex units respectively. Conversely the flattening out of the profile is produced by a combination of the concave elements (V), the concave segments (Sv), and those maximum segments which occur above a concavity and the minimum segments which lie below it. This means that some of the maximum and minimum segments are included twice in the calculations. The total convexity (Lx %) and total concavity (Lv %) values, as defined in terms of length, may be obtained by expressing the length of those units which are tending to steepen the profile—or to flatten it as the case may be—as a percentage of the total length of the profile. Table 7.1 gives

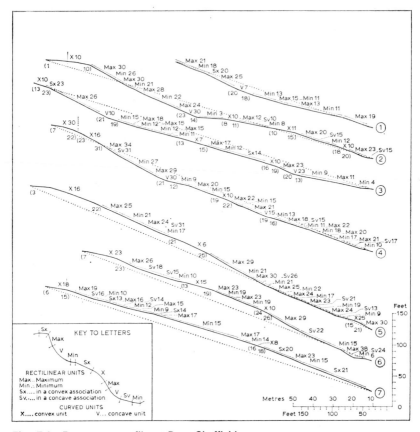

Fig. 7.1 Escarpment profiles at Dore, Sheffield.

Table 7.1 Convex and concave lengths of the slope profiles as a percentage of the total length of the profile

Profile no.	1	2	3	4	5	6	7
Lx %	78·75	76·0	88·0	81·5	86·5	66·5	91·0
Lv %	81·5	65·0	75·0	76·0	64·5	64·5	51·0

the results of this analysis for the profiles in Figure 7.1. In assessing the sense of the main curvature of a profile it is not sufficient, however, to consider only the lengths of these curved portions; account must also be taken of the changes involved in the angles of slope. This may be done by summing the angular values of the convex discontinuities (Ax) and then of the concave discontinuities (Av). From these the total amount of angular

change down the profile can be ascertained (Ax + Av) and the total for each group (Ax % and Av %) found as a percentage of the total angular change (Table 7.2).

Table 7.2 Convex and concave angular changes of the slope profiles as a percentage of the total amount of angular change

Profile no.	1	2	3	4	5	6	7
Ax %	49·0	51·8	49·0	54·4	51·0	40·2	59
Av %	51·0	48·2	51·0	45·6	49·0	59·8	41

The measure of the amount of convexity or concavity may now be found by:

$$Lx \% \div Ax \% = \text{convexity (Cx)}$$
$$Lv \% \div Av \% = \text{concavity (Cv)}$$
$$\text{General form of curve} = \frac{Cx}{Cv}$$

If $Cx/Cv = 1$, then no total curvature is involved; if $Cx/Cv > 1$, slope is generally convex; if $Cx/Cv < 1$, slope is generally concave. The values of Cx/Cv for the profiles of Figure 7.1 are listed in Table 7.3.

Table 7.3 The ratio of slope curvature in profile (Cx/Cv) for the slope profiles of Figure 7.1

Profile no.	1	2	3	4	5	6	7
Lx/Ax = Cx	3·9	0·9	1·7	1·5	1·7	1·7	1·5
Lv/Av = Cv	4·0	1·3	3·8	1·7	1·3	1·1	1·2
Cx/Cv	0·98	0·69	0·45	0·88	1·31	1·55	1·25

By this means, therefore, the visual impression gained from looking at Figure 7.1 that profiles 1–4 are concave and profiles 5–7 are convex is given numerical expression. In addition it can be seen from Table 7.3 that of the concave slopes profile 3 has the greatest amount of concavity and that profile 1 is almost without a dominance of concave changes on the one hand or convex on the other. Profile 6 has the most marked convex tendencies of all of the profiles, as is shown by its Cx/Cv ratio of 1·545, which is well in excess of $Cx/Cv = 1$.

All of these profiles lie along the same escarpment. They have been drawn, in Figure 7.1, in the same order as they occur on the ground with profile 1 lying downstream of the other profiles. The analysis shows, therefore, that it is the downstream profiles which are generally concave

in form, and the upstream which are predominantly convex. The ratio values indicate, however, that profiles 1–4 do not become progressively less concave as the convex group is approached. On the contrary profiles 2 and 3 are successively more concave than profile 1. Neither, in fact, do the profiles 5–7 become successively more convex the farther they are away from the concave group. In fact, although profile 6 is more convex than profile 5, profile 7 which is the most upstream of all of the profiles, is less convex than either of the two lower profiles. Change in the general curvature of the profile form of the escarpment in this case is not simply, therefore, a function of distance upstream.

A slope classification map
The techniques of morphological mapping and slope profiling can be combined to produce a slope classification map. The purpose of such a map is to show the lateral continuity of the slope units as classified according to the system illustrated in Figures 6.9c and 7.1. The production of

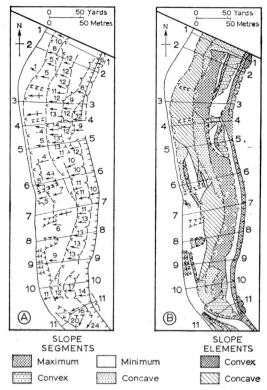

Fig. 7.2 The form of a sandstone escarpment at Totley, Sheffield : **A** : morphological map showing the position of the slope profiles of Figure 7.3 ; **B** : slope form map.

such a map requires that morphological mapping and slope profiling are carried out with the profiles measured close to each other and each of the profiles accurately located on the morphological map. The profiles, when drawn up, may be subdivided as shown in Figures 6.9c and 7.1 and the morphological units recorded along the lines of the profiles marked on the morphological base map. This task is made easier if the profiles are marked in the field to show where their position crosses a change or break of slope recorded on the morphological map. Having superimposed the slope profile information on to the morphological map along the recorded position of these profile lines, the inter-profile areas can be filled in by continuing the boundaries between the morphological units laterally across the area from one profile to the next. The result is a slope form map superimposed on a morphological base map.

This technique may be described with respect to a specific example. Figure 7.2A is a morphological map of the escarpment of a cuesta developed within the Greenmoor Rock, a sandstone of the Lower Coal Measures, in the headwater basin of the River Sheaf at Totley, west of Sheffield. The selected area was initially mapped at a scale of 1 : 2640. Profiles were measured (Fig. 7.3) along the lines marked on the morphological map, and a record kept of every occasion on which the position of a break or change of slope, recorded on the morphological map, was crossed. As each profile was measured, significant changes of slope angle became apparent. When these occurred, their position was checked against the morphological map to ensure that they coincided with a mapped break or change of slope. In this way the morphological mapping and slope profiling were checked for inaccuracies. The value of the final slope form map depends on the spacing of the profile lines. The closer the profile lines the smaller the inter-profile area will be, and the smaller the distance over which the boundary between adjacent morphological units has to be conjectured. The morphological subdivisions of the slope profiles were drawn up and then transferred from the profiles to their corresponding positions on the morphological map. The final slope form map, Figure 7.2B, was then produced by extrapolating the junctions of the morphological units across the inter-profile area. The order of generalization is directly proportional to the profile spacing.

This technique allows an expression of the lateral variations in the morphological units that are defined on adjacent slope profiles. It also makes possible a numerical examination of these units as they occur in area as well as in profile. Developing these ideas and collecting sufficient information to produce worthwhile results remains as a topic for further research. Its value lies in providing data capable of numerical analysis, and it should be possible for any two people to use this technique and obtain comparable results. It must always be remembered that measuring two like things in two different areas produces results dependent not only on environmental differences but also on differences in measurement procedure. The sooner measurement differences can be eliminated the sooner

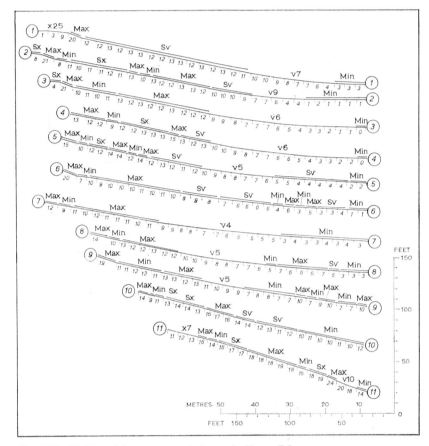

Fig. 7.3 Profiles of the escarpment shown in Figure 7.2.

valid comparisons can be made between results obtained in different areas and by different people.

Slope profiles as mathematical curves

A slope profile can be treated as a mathematical curve. That is to say an equation, similar to a regression equation, can be found for the curve which most closely fits the shape of the measured slope profile. The simplest form that can be fitted to a slope profile is the straight line:

$$Y = a + bX, \quad \text{or} \quad Y = a - bX$$

Once a curve is introduced on the profile the mathematical equations become more complex (polynominal equations), and several alternative curve forms may be compared with the slope profile. Some of these alternatives are illustrated in Figure 7.4, and are described below.

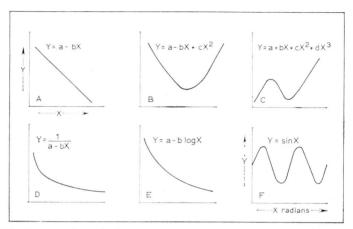

Fig. 7.4 Some mathematical curves with which slope form may be compared.

a   The parabolic curve:

$$Y = a + bX + cX^2$$

In this case an additional item, cX^2, has been added to the equation for a straight line, making it a *quadratic equation*. The effect of this is to give the regression line a curved form, and the graph has a turning point at a maximum value of Y beyond which its gradient is reversed. Up to this maximum the gradient flattens progressively. There are times when just the portion of the parabolic curve on one side or the other of the turning point comes close to defining the best-fit regression line for a set of profile data. If an equation has the form $Y = a - bX + cX^2$ (Fig. 7.4B), then the curve would have a turning point at a minimum value of Y. Variations in the form of the parabolic curve can be introduced by adding increased powers of X to the equation. Thus if the term dX^3 is added (Fig. 7.4C) the equation for the simple parabola becomes a *cubic equation* and takes the form:

$$Y = a + bX + cX^2 + dX^3$$

The curve of such an equation will display two turning points, one being a maximum and the other a minimum. These properties may be used to advantage in some geomorphological studies, as will be shown later in this chapter. The number of turning points in a parabolic curve is always one less than the highest power of X included (in this case $3 - 1 = 2$).

b   The logarithmic curve: In some instances the scatter of points can be most closely defined by a regression curve if its logarithmic values are plotted instead of its actual values. Sometimes the best results are obtained by plotting log Y against X, and at other times by plotting Y against log X

(Fig. 7.3E). Sometimes a plot of log Y against log X may be usefully employed. The equation of the best-fitting curve in each case is then:

$$\log Y = a + bX$$
$$Y = a + b \log X$$
$$\log Y = a + b \log X$$

Similar logarithmic modifications can be made to the equations of the parabolic curves. The curve, $\log Y = a + bX$ never actually passes through the origin; it flattens out as the X-axis is approached.

c   The hyperbolic curve: The equation which defines a hyperbolic curve has the form:

$$Y = \frac{1}{a + bX}$$

This curve also has the property of flattening out as the X- and Y-axes are approached. The position of the curve in relation to these axes, however, unlike the logarithmic case, depends on the values of *a* and *b* in the equation.

When measured profiles are being compared with these mathematical forms, the hillside is sometimes sufficiently smooth and regular in form to make the related best-fit curve a fairly simple one. On the other hand, the measured profile may be so complex to make it necessary to break it down into smaller portions and for mathematical curves to be found for

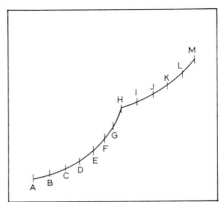

Fig. 7.5 Analysing a slope profile consisting of two concave parts.

each portion separately. The question as to whether or not there is a significant break in the profile may be decided from mathematical analysis itself. For example, a profile with the form shown in Figure 7.6 has a distinct boundary (at station H) between the upper and lower parts of the profile. One possible method of analysing the form of such a curve is to find the form of a polynomial equation which fits the profile first of all for the portion between A and D then between B and E, then between C and

F, and so on. As long as the polynomial curves of each of these portions have similar coefficients there is a possibility of fitting a 'sensible' polynominal curve to the whole of the profile. If on the other hand a significant change takes place in the form of the profile, such as would be the case in finding an equation for the portion of the profile between stations G and J, then this would be reflected in a change in the coefficients. They would start to become similar to each other again, on successive steps up the profile, once the section between stations G and I was no longer included in the analysis. The form of the polynominal equation for the section of the profile above and below the discontinuity between them might even be similar, nevertheless the discontinuity will have been detected.

Savigear (1956) compared the goodness of fit of smooth mathematical curves with some slope profiles measured on seaward-facing slopes in Devon and Cornwall. He also found the goodness of fit of a generalized profile composed of straight-line sections. In two instances the rectilinear generalization gave a better fit to the measured profile than did the smooth mathematical curves. In the third case the better fit was provided by a quadratic curve. The evidence from this study was not conclusive, therefore, that polynomial curves provide a valid means of describing the general character of a slope profile.

Hack and Goodlett (1960) were not concerned with the whole slope profile, instead they confined their attention to the upper convex portions of the slopes, in the central Appalachians. They found that these forms could be fitted to straight lines if they were plotted so that the centre of the coordinate system lay at the centre of the ridge or hilltop. A graph is drawn of the logarithm of the height below the ridge top (on the ordinate, or Y-axis) against the logarithm of slope length (abscissa, or X-axis). The graphs which result can be approximated by simple power functions of the form:

$$\log H = \log C + f \log L$$

or

$$H = CL$$

where H is the fall from the ridge centre; L is the horizontal distance or slope length; and C and f are constants. C is in effect a coefficient of steepness for it defines the fall (H) at a given slope length for a slope with a curvature (f). C is generally a very small fraction and the exponent f is usually a number lying between 1 and 2. Hack and Goodlett found that the equations for convex slopes were a good means of comparing those slopes. In addition, they noted that variations in the soil or bedrock commonly produced variations in surface form that showed up strikingly on these graphs as breaks in the slope of the line, even though they were scarcely noticeable on the ground.

The use of mathematical curves in relation to the form of a hillside may also be illustrated with reference to the profiles shown in Figure 7.6.

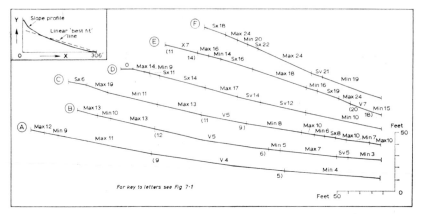

Fig. 7.6 Slope profiles of a sandstone escarpment for which equations have been derived.

The polynomial equations which most closely fit these profiles are listed, up to the third power of X, in Table 7.4. As the power of X increases so does the complexity of the mathematical equation. At the same time the curve has increasing flexibility and has the opportunity to fit the measured

Table 7.4 The mathematical curves which best fit the slope profiles of Figure 7.6

Linear equations

Profile A	$Y = 38\cdot45 - 0\cdot135X$
B	$Y = 40\cdot79 - 0\cdot149X$
C	$Y = 55\cdot35 - 0\cdot189X$
D	$Y = 72\cdot90 - 0\cdot243X$
E	$Y = 99\cdot03 - 0\cdot319X$
F	$Y = 119\cdot03 - 0\cdot396X$

Quadratic equations

Profile A	$Y = 42\cdot32 - 0\cdot23X + 0\cdot0003X^2$
B	$Y = 50\cdot02 - 0\cdot29X + 0\cdot0004X^2$
C	$Y = 62\cdot33 - 0\cdot29X + 0\cdot0003X^2$
D	$Y = 77\cdot78 - 0\cdot30X + 0\cdot0002X^2$
E	$Y = 85\cdot28 - 0\cdot18X - 0\cdot0003X^2$
F	$Y = 143\cdot92 - 0\cdot62X + 0\cdot0005X^2$

Cubic equations

Profile A	$Y = 42\cdot54 - 0\cdot24X + 0\cdot0004X^2 - 0\cdot0000002X^3$
B	$Y = 48\cdot76 - 0\cdot25X + 0\cdot0002X^2 + 0\cdot0000005X^3$
C	$Y = 62\cdot37 - 0\cdot29X + 0\cdot0003X^2 - 0\cdot00000002X^3$
D	$Y = 70\cdot35 - 0\cdot16X - 0\cdot0006X^2 + 0\cdot0000013X^3$
E	$Y = 71\cdot69 + 0\cdot03X - 0\cdot0010X^2 + 0\cdot000002X^3$
F	$Y = 168\cdot09 - 0\cdot96X + 0\cdot0020X^2 - 0\cdot000002X^3$

where Y is the height of the slope in feet above its base, and X is the horizontal distance from the origin shown in Figure 7.6.

profile more closely. Each of the mathematical curves has its uses: some of these may be listed as follows:

a Linear equation: The linear equations of Table 7.4 are similar to the linear regression lines between two variables in that, in this case, height above the foot of the slope and horizontal distance are taken as the two variables. The equation in each case describes the straight line which most nearly passes through all of the data points. The X coefficient defines the gradient of the line and therefore the general gradient of the slope, while the constant term is the height the slope has if it continues to the point where X = o (Fig. 7.6). Thus, in general, the profiles A–F form a sequence which become steeper from A to F, and which meet the Y-axis at increasingly greater heights from A to F. Profile A has a general gradient of 0·1347 feet in height per one foot horizontal distance, compared with a gradient of 0·396 feet in height per one foot horizontal distance for profile F. This is reflected in the fact that if both profiles were equally long (in this example 306 feet, see Fig. 7.6), the crest of A

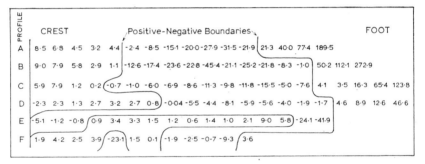

Fig. 7.7 Analysis of residual values as a percentage of observed values for the linear equations of Table 7.4.

would be only 38·447 feet above the base of the slope compared with 119·03 feet for profile F. This is, therefore, an easy method of describing the general steepness of these slopes.

The linear equation has a second use, as a search tool for distinguishing between irregularities in the profiles. For example, the equation can be used to calculate what the height of any point along the line will be, given a knowledge of the horizontal distance (i.e. X can be used to predict Y). Unless the measured profile is a straight line—and this will very rarely be the case—the calculated height at any point will not be the same as its actual height. The difference between these two values is known as the *residual value*. Residual values will be positive when the ground profile lies above the mathematical line, and negative when below. The residual values for the profiles of Figure 7.6 are plotted in schematic form in Figure 7.7. In general they show that the upper and lower portions of each slope (except E and F) lie above the calculated 'best-fit' line, while

the central portions are below it. This is characteristic of concave slope profiles. Those residuals which are excessively large are underlined, and nearly all of these occur at the base of the slope. This indicates that this concave slope flattens out very quickly towards its foot, leaving the mathematical slope far below. This sharp flattening is also responsible for the greater differences in the residual values on either side of the positive–negative boundary at the foot of the slope than is the case near the crest.

b   Quadratic equation: The quadratic equation describes a curve which has one turning point. This may either be at a maximum or a minimum value of Y. (In the former case the curve is convex upwards, and vice-versa.) An individual slope profile will not pass beyond the maximum or minimum value, as the case may be, because in the former this would imply the crossing of a divide and in the latter would mean a crossing of a valley floor. There is some benefit to be derived from calculating the minimum and maximum values—as will be seen—but first it is relevant to note another use of the quadratic equation. The algebraic procedure of *differentiation* is a simple but useful method by which the gradient at any point on the mathematical curve can be calculated. If the quadratic curve is a good fit then these gradient values may closely approximate those of the surface, and an equation may be given which can be used to compare the gradients of different slopes. The differentiation of a quadratic equation is performed as follows:

If the quadratic equation is

$$y = a + bx + cx^2$$

then the *differential coefficient* of y with respect to x (known as dy/dx) is

$$\frac{dy}{dx} = b + 2cx$$

and this is the gradient of the quadratic curve at any point whose horizontal distance from the origin is X.

Thus in profile A, where $Y = 42 \cdot 32 - 0 \cdot 23X + 0 \cdot 0003X^2$:

$$\frac{dy}{dx} = -0 \cdot 23 + 2 \times 0 \cdot 0003X$$

$$= -0 \cdot 23 + 0 \cdot 0006X$$

when $X = 100$ feet, $dy/dx = -0 \cdot 23 + 0 \cdot 06 = -0 \cdot 17$ feet/one foot horizontal distance. For profile F:

$$\frac{dy}{dx} = -0 \cdot 62 + 0 \cdot 0010X$$

when $X = 100$ feet, $dy/dx = -0 \cdot 52$ feet/one foot horizontal distance. Thus the large difference in the general gradient of profiles A and F, suggested by the linear equations, is also apparent in the difference in the gradients of the quadratic 'best fit' curves when $X = 100$ feet.

It may be argued that much of this information could have been assessed

from an inspection of the measured profiles. This is in part true, but the information of gradients would have related to *actual* ground slope at a certain point, rather than the *general* slope of the whole curve as it passes through that point. The latter may, in fact, be more valuable for the purpose of comparing slope profiles with each other.

c   Cubic equation: The cubic equation is capable of providing information which is not obvious from the measured profile. For example, the cubic equation has both a maximum and a minimum value of Y coinciding with changes in the direction of slope of the mathematic curve (Fig. 7.4c). The determination of each of these can be useful in certain geomorphological situations. For example, a bedrock slope may pass beneath a spread of alluvium in a valley floor. If a cubic equation is calculated for the bedrock portion of the slope only, then a determination of the minimum value of the curve (Fig. 7.8) may be the most accurate means of

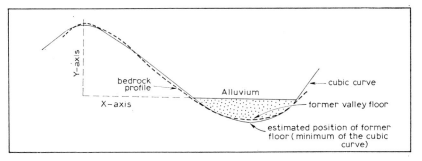

Fig. 7.8   To illustrate the use of the cubic curve in estimating the depth of the old valley floor below an infill of alluvium.

estimating the depth of the alluvium, i.e. of finding the height of the former valley floor with respect to the X-axis of the measured profile. In similar manner the former position and depth of an earlier valley in a landscape could be calculated from the cubic curves obtained for profiles that have been measured on the slopes above a steepening related to the incision which has cut into and removed the former valley floor. Both of these analyses would have to assume:

1   that no significant change had taken place in the form of the remaining portions of the old hillsides
2   that there is no major change from the smooth form of the cubic curve in the slope below the alluvium or of the slope cut away by recent rejuvenation.

In some cases these assumptions may not be valid.

In a similar sense it would be possible, by calculating the maximum of the cubic curve, to estimate the height of a divide before reduction to its present level. However, in such a case it would be dangerous to assume

that no significant change had taken place in the form of the hillside on which the calculations are based. Any equation of the profile form of the hillside would probably show little relationship to the slope form prior to the reduction of the hill crest.

Calculating the minimum of a cubic equation is achieved by differentiating twice. Thus, for profile D of Figure 7.6:

$$Y = 70 \cdot 35 - 0 \cdot 16X - 0 \cdot 0006X^2 + 0 \cdot 0000013X^3$$

$$\frac{dy}{dx} = \quad\quad - 0 \cdot 16 \quad - 0 \cdot 0012X + 0 \cdot 0000039X^2$$

$$\frac{d^2y}{dx^2} = \quad\quad\quad\quad - 0 \cdot 0012 \quad + 0 \cdot 0000078X$$

Since dy/dx defines the gradient of the curve at any selected point, the minimum will occur when $dy/dx = 0$. The gradient will also be zero at a maximum and tests need to be made to distinguish between them (see below). Putting $dy/dx = 0$:

$$0 \cdot 0000039X^2 - 0 \cdot 0012X - 0 \cdot 16 = 0$$

The roots of this equation (i.e. the two possible values of X) can be found by applying the standard formula:

$$x = \frac{-b \pm \sqrt{b^2 - 4ac}}{2a}$$

Thus

$$x = \frac{0 \cdot 0012 \pm \sqrt{(0 \cdot 0012)^2 - 4(0 \cdot 0000039)(-0 \cdot 16)}}{2 \times 0 \cdot 0000039}$$

$$= \frac{0 \cdot 0012 + 0 \cdot 001984}{0 \cdot 0000078} \quad \text{or} \quad \frac{0 \cdot 0012 - 0 \cdot 001984}{0 \cdot 0000078}$$

$$\therefore X = 408 \cdot 2 \quad \text{or} \quad X = -100 \cdot 5$$

One of these values will define the position of the maximum, the other the minimum of the mathematical curve. To see which is which it has to be appreciated that, at a maximum, as X increases dy/dx changes from positive to negative. At a minimum the slope of the curve changes gradient from negative to positive. If in the equation:

$$\frac{dy}{dx} = -0 \cdot 16 - 0 \cdot 0012X + 0 \cdot 0000039X^2$$

X is made a little less than 408·2 in the first instance, and then a little more than 408·2, the change in the sign of dy/dx will indicate whether 408·2 is associated with a maximum or a minimum turning point of the cubic curve: for example let X = 400 (i.e. a little less than 408·2) then:

$$-0 \cdot 16 - 0 \cdot 0012(400) + 0 \cdot 0000039(400)^2 = -0 \cdot 02$$

Let X = 420 (i.e. a little more than 408·2) then:

$$-0 \cdot 16 - 0 \cdot 0012(420) + 0 \cdot 0000039(420)^2 = +0 \cdot 03$$

As X increases dy/dx changes from negative to positive through the point where X = 408·2, which is therefore a minimum. The coincidence of the maximum with the value X = −100·5 can be tested in a similar manner.

Another use for the equation of the cubic curve is in finding the point of inflexion on a convex–concave curve. On many convex–concave slope profiles it is extremely difficult to decide at which point the upper convexity meets the lower concavity. This is possible, however, through the differentiation procedure, because the point of inflexion occurs when $d^2y/dx^2 = 0$ and changes with increasing values of X from negative to positive through that point. Thus for profile D:

$$\frac{d^2y}{dx^2} = -0\cdot0012 + 0\cdot0000078X = 0$$

$$X = 168\cdot7 \text{ feet}$$

When X = 168·7 feet there is a point of inflexion. To test that this occurs between a convex upper portion and a concave lower portion, two values on either side of 168·7 feet are selected: when X = 100, d^2y/dx^2 is negative; when X = 200, d^2y/dx^2 is positive. This shows that, as X increases through 168·7 feet, d^2y/dx^2 changes from negative to positive and the inflexion occurs between an upper convexity and a lower concavity.

In profile E (Fig. 7.6) the point of inflexion occurs at 166·6 feet: it has moved nearer the origin of X and thus slightly up-slope between the two profiles. This means that the concave portion of the profile extends higher up-slope on profile E than on profile D. By these methods, therefore, lateral changes in the position of the point of inflexion can be studied from profile to profile. Thus yet another characteristic of profile form can be defined with precision by numerical analysis.

7.2 Descriptive statistics and slope data

Slopes form a continuous feature across each continent. They may differ in angle and direction, but they run into each other to form a complete and continuous surface. Slope analysis demands that this surface be sub-divided into units that are convenient for definition and about which numerical data can be obtained. In the main this chapter has been concerned to show how slope forms may be organized and classified to make numerical analysis possible.

Statistical analysis requires certain types of data about which it is possible to be specific. For example, how close do the data values come to representing the whole area of study? Were the data values obtained by methods of random sampling? Is there enough information available to come to valid conclusions about the whole population from the sample? It has become apparent that the most successful statistical studies of slope information are those which relate to slopes confined to a particular drainage basin, or group of drainage basins. For example, Gregory and

Brown (1966) compared angle of slope and area of occurrence within a drainage basin and found a correlation coefficient of 0·916 (significant at better than the 0·05% level) to occur between them. The linear regression relationship which best described the data was:

$$\log Y = 0·848 - 0·067\,X$$

where Y is area in square miles and X is the angle of slope. However, when an area, bounded by rectangular grid lines and covering seven square miles, was analysed in the same way the simple relationship was no longer apparent. In addition, Gregory and Brown found that a histogram to show the relationship between angle of slope and area of occurrence was unimodal when the data was collected within a drainage basin. For the rectangular grid area the distribution was no longer unimodal. Clearer results are obtained when the drainage basin forms the unit for slope study than when other units are chosen. Another good reason for analysing slopes with reference to drainage basins is that hydrological data, to which slopes may be related, are only available for drainage basin units.

It has also been found that general relationships may not become apparent from the study of single slopes or small areas. What is emerging from numerical studies in geomorphology is that a large body of data may suggest a general rule of development or a general relationship, even when individual cases may not conform to this rule. For example, Carter and Chorley (1961) found that a plot of individual observations of maximum angle of slope against stream gradient showed no straightforward relationship. However, when they re-examined the data taking *the area as a whole* they found a positive correlation existed such that:

$$\log Sg = 0·6 + 0·8 \log Sc$$

where Sg = mean valley side slope, and Sc = mean channel gradient.

When a group of drainage basins have been selected for slope form analysis there are at least five separate ways available in which numerical information may be collected. On the assumption that these basins have been mapped at a detailed scale by the system of morphological mapping described in section 6.2, then data may be collected by:

1 point sampling
2 map area of a sample number of morphological units
3 ground area of a sample number of morphological units
4 map area of the whole population of units
5 ground area of the whole population of units.

These five methods have been listed in order, from those requiring the least amount of physical labour to those which demand most time and effort.

To test the differences between the results obtained by some of these methods data were collected by methods (1), (4) and (5), for seven third-order drainage basins in the Jurassic lowlands near Weymouth, Dorset.

Point sampling (1) involved placing a transparent grid across the morphological map that included these seven basins, and noting the angle of slope at each grid intersection, giving 647 observations. This method differs from that of a simple count of morphological units at each angle which Gregory and Brown (1966) found to be unsatisfactory, for the grid intersect method makes it possible to give extra weight by including the same unit more than once in the count if it has a very large area. The map area of the whole population (4) was obtained by measuring the area of each unit separately from the morphological map. The ground area (5) was calculated for each unit by dividing the map area

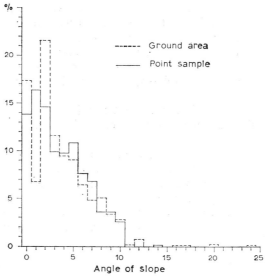

Fig. 7.9 Frequency histogram of slope values obtained by point sampling (647 observations) and ground area measurements for the total population of seven basins in the Weymouth lowland.

by the cosine of the angle of slope. The total map and ground areas were found by summing the individual values in each case.

Frequency distribution tables were compiled and histograms drawn for each of the three sets of data. Table 7.5 shows only the cumulative frequency as a percentage of the total frequency for each method of obtaining the data, and this allows a comparison to be made between them. A striking difference is that the point sampling method failed to identify the few high angle slope units, for the highest angle recorded by this method is 14° when, in fact, slopes up to 24° occur. There is also a considerable difference between the cumulative frequency obtained by point sampling as compared with the other two methods, especially for the lowest slope angles of 0° and 1°, and a 3% difference at 3° and 7°. There is no great difference between the cumulative frequencies for the two

Table 7.5 Comparison of the cumulative frequency (as a percentage of the total) for slope data collected by three different methods

Angle of slope	0	1	2	3	4	5	6	7	8	9
Point sampling	13·8	30·2	44·8	54·7	64·5	75·3	82·9	89·7	93·4	97·0
Map area	17·4	24·3	45·9	57·4	66·9	75·5	82·0	86·9	91·8	96·1
Ground area	17·3	24·1	45·7	57·3	66·8	75·8	82·3	87·2	92·3	95·7

Angle of slope	10	11	12	13	14	15	16	17	20	24
Point sampling	99·4	99·4	99·8	99·8	100·0	—	—	—	—	—
Map area	98·7	98·8	99·5	99·5	99·5	99·5	99·7	99·7	99·9	100·00
Ground area	98·6	98·7	99·4	99·4	99·5	99·5	99·6	99·7	99·9	100·00

methods of area measurement for the whole population. This being the case it is not worth the labour of converting the data to ground area values for the purposes of frequency distribution analysis.

In the histogram (Fig. 7.9) only the point sample and ground area data are shown, as the map area histogram is almost identical to the one for ground area. For the seven basins in question the ground area histogram represents the total population. The discrepancy between the two histograms represents the error arising from point sampling. The greatest difference lies in the first four slope categories, namely 0–4°. The amount of slope at 1° is grossly overstated by the point sample method, with a considerable loss of information concerning the 2° slopes. This is significant considering that 647 points form a large sample, and the possibility of such a difference must be recognized in future studies based on point sampling. The normality of the distribution represented by these histograms may be tested by plotting the cumulative frequency values (Table 7.5) on arithmetic probability paper (Fig. 7.10). The point sample data have a generally good straight-line plot, especially in the range 1–9°. Divergence from the straight line is more prominent at the extremities in the case of the ground area data, that is to say outside the range of values 2–12°. Few of the higher values were recorded by the point sampling method and they do not, therefore, appear on the arithmetic probability graph. Thus, when data are collected by point sampling methods this tends to give a stronger impression of normality than is in fact the case. Strahler (1950) also examined the normality of slope data in this way, but his data was related entirely to *maximum* valley side slope angle, and not to the total population of all slopes as is the case in the analysis of ground area data (Fig. 7.10). The frequency distribution of the data can also be tested for normality by the computation of χ^2. This is illustrated by Krumbein and Graybill (1965, chapter 8).

Another method of comparing the slope data obtained by the different methods is to compare the mean angle of slope obtained in each case.

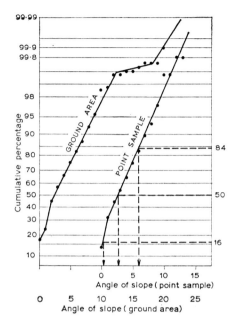

Fig. 7.10 Values of angle of slope (cumulative frequency) plotted on arithmetic probability paper.

For the point sample data the mean angle of slope is calculated by dividing the cumulative total of all slope angles by the number of observations (3·6°). For the ground area data the mean angle of slope is obtained by multiplying each unit area by its angle of slope, finding the total, and dividing this total by the total area of the seven drainage basins from which the data was obtained (3·3°). The difference between these two mean values (3·6 − 3·3 = 0·3°) is small, and suggests that if a mean angle of slope is required for an area, then point sampling will give a result close to that which would have been obtained by considering the ground area of all slope units in the population. A rapid method of estimating the mean angle of slope may be obtained by adopting the method of Folk and Ward (1957)—originally devised for finding the mean value of the size of sediments—to the arithmetic probability graph of the point sample data (Fig. 7.10). The mean angle of slope can be estimated from:

$$\bar{\theta} = \frac{\theta_{16} + \theta_{50} + \theta_{84}}{3} = \frac{0\cdot4 + 2\cdot6 + 5\cdot9}{3} \simeq 3°$$

where θ_{16} is the angle of slope at the 16% cumulative frequency level etc.

Characteristic and limiting slope angles
It is often the case that certain angles of slope are frequently repeated within an area, and these may be said to be characteristic of the area.

The concept of characteristic slope angles has been discussed by Young (1961) and by Gregory and Brown (1966). Young defines the term as:

characteristic angles of slope are those which most frequently occur, either on all slopes, under particular conditions of rock type or climate, or in a local region.

Gregory and Brown adopt a more restricted definition:

characteristic angles are . . . those angles which occur on a specific type of morphological unit under controlled conditions, such as geology or orientation, within a drainage basin.

In either case characteristic angles are detected through an examination of the slope histograms. The differences between the histograms that result from point sampling and from an analysis of ground area (Fig. 7.9) are now seen to be important. The point sampling data suggest that, in the Weymouth lowland, characteristic angles occur at 1°, 5° and 12°. The ground area histogram shows, however, that this is not the case, for the peaks on the ground area histogram occur at 0°, 2°, 12° and 20°. Thus, a point sample may not represent the characteristic slope angles very precisely and a ground area (or map area) analysis may be necessary. On the other hand, more meaningful results arise from relating angle of slope to the rock type on which it occurs, and by this means breaking the data down into subsets related to geology. This is described in section 8.6.

Limiting angles of slope are defined by Young (1961) as those slopes that define the range within which particular types of ground surface occur, or particular denudational processes operate.

It has already been shown that point sampling failed to detect the higher angles of slope within the area and that this method will lead to an incorrect assessment of the limiting angle of slope in the area. A closer examination is made of limiting angles of slope in section 8.6, where again it is found to be of greater value to consider the slope data in terms of the geological outcrops on which they occur. In addition both characteristic and limiting slope angles may be considered in terms of the individual drainage basins which constitute the area represented by the histogram of Figure 7.9. For example, Figure 7.11 shows the histograms for two of the third-order basins in the Weymouth lowland. In general terms (see chapter 9) the slopes of these basins can be classified into three types— the crestslope, backslope and footslope (see Figs. 9.1 and 9.3). In basin A, 0°, 2° and 5° are characteristic of the crestslope, 8° and 12° characteristic of the backslope, and 2° characteristic of the footslope. In basin B the characteristic angles are:

crestslope: 0°, 2°
backslope: 5°, 8° and 12°
footslope: 2°

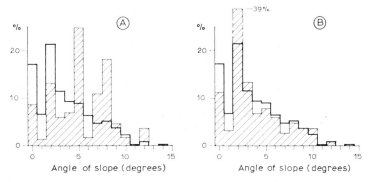

Fig. 7.11   A comparison between slope histograms for two individual drainage basins and the total data for seven basins near Weymouth.

Limiting slope angles tend to overlap between these units: they are in basin A, 0° and 5° for the crestslope, 2° and 12° for the backslope, 0° and 3° for the footslope. In basin B the limiting angles are similar, though not quite the same, being 0° and 6° for the crestlope, 2° and 12° for the backslope and 1° and 5° for the footslope. These two examples are discussed again in the next paragraph.

Comparison between basins
The slope histograms for two of the seven third-order drainage basins measured in the Weymouth lowland are shown in Figure 7.11. Superimposed in each case is the histogram for all of the basins combined. In this way the differences can be quickly spotted between an individual basin and the general behaviour of all the data combined. In a sense this is equivalent to an examination of individual cases with respect to a general trend established by regression analysis.

For example, in Figure 7.10A the basin has a very high proportion of slopes at 5°, 7°, 8° and 12° compared with the total population; it has relatively few low angle slopes, and none over 14°. The latter is the case because there has been no undercutting of the river banks which elsewhere, in the seven basins, has led to the development of slopes up to 24°. The relatively small amount of low angle slopes is due to the fact that, compared with the rest of the Weymouth lowland, the interfluves carry much smaller remnants of the hill and ridge top planation surfaces than is usual. In addition, some of the crestslopes are inclined at up to 8°: so although the crestslopes occupy 34% of the area of the basin, some of their slopes are as steep as units found within the backslope (see chapter 9). The backslope with units ranging from 2° to 12° in steepness occupies 53% of the area of the basin and accounts for the slope maxima at 7–8° on the histogram.

In the example shown in Figure 7.10B the data apply to a basin a portion of which is illustrated in Figure 9.7. In this basin 39% of the

slopes are at 2°, which is the main anomaly in the data compared with the general character of the total population. Re-examining the form of the basin shows not only that there are some large remnants of planation surfaces, with slopes of 2°, on the ridge crests, but also that this one basin has a very high proportion of low angle footslopes. Indeed, most of the main valley floor is composed of long slopes at 2°. Over 44% of the total area of the basin is contained within the footslopes.

Histograms are of value, therefore, as both a descriptive and as a search device. The histograms summarize the frequency distribution and the range of the data. They also allow a comparison to be made between one drainage basin and the whole population of available data. In this way it is possible to search for discrepancies between the single basin and the general frequency distribution. This search procedure helps to indicate any local peculiarities which a single basin may display.

Throughout this chapter methods of organizing slope form data have been considered, and the quantitative information which they provide has been described. In the next chapter examples are provided of the use of other methods of numerical analysis on this and other types of information.

References

AANDAHL, A. R. 1948: The characteristics of slope positions and their influence on the total nitrogen content of a few virgin soils of western Iowa. *Proc. Soil Sci. Soc. Am.* **13,** 449.

CARTER, C. S. and CHORLEY, R. J. 1961: Early slope development in an expanding stream system. *Geol. Mag.* **98,** 117–30.

GREGORY, K. J. and BROWN, E. H. 1966: Data processing and the study of land form. *Zeit. für Geomorph.* NF **10,** 237–63.

HACK, J. J. and GOODLETT, J. C. 1960: Geomorphology and forest ecology of a mountain region in the Central Appalachians *U.S. Geol. Surv. Prof. Paper* **347,** 1–66.

KRUMBEIN, W. C. and GRAYBILL, F. A. 1965: *An introduction to statistical models in geology.* New York: McGraw-Hill.

SAVIGEAR, R. A. G. 1956: Technique and terminology in the investigation of slope forms. *Premier rapport de la commission pour l'étude des versants* (IGU, Rio de Janeiro), Amsterdam.

   1967: The analysis and classification of slope profile forms. In L'evolution des versants, *Les Congrès et Colloques de L'Université de Liège*, **40,** 271–90.

STRAHLER, A. N. 1950: Equilibrium theory of erosional slopes approached by frequency distribution analysis. *Am. J. Sci.* **248,** 673–96; 800–14.

YOUNG, A. 1961: Characteristic and limiting slope angles. *Zeit. für Geomorph.* **5,** 126–31.

   1963: Some field observations of slope form and regolith and their relation to slope development. *Trans. Inst. Brit. Geogr.* **32,** 1–29.

8 Slope and environment

In this chapter numerical analysis is used to examine the relationship between slope form and definable characteristics of the environment. The relationships which exist among the properties of slope form need to be examined before dealing with the relationships between form and environment.

8.1 Relationships among properties of slope form

The data presented in Table 6.1 have been used for a Pearson product-moment correlation analysis, except that six of the slope units were omitted. These six units are all remnants of hill crest planation surfaces. They are thought to be unrelated to the morphology of the drainage basin as developed by the incision of the present stream network. The results of the correlation analysis are presented in matrix form in Table 8.1, from which the lower levels of correlation have been omitted. High angles of slope appear to be associated with morphological units that are long in a direction parallel to the valley axis, but short at right-angles to this axis. Most of the units are elongated parallel to the long-axis of the valley in this basin, and the steeper units (Fig. 6.7) tend also to be short ones at right-angles to this axis. This is most clearly shown, for example, by

Table 8.1 Product-moment correlation coefficients between the variables listed in Table 6.1

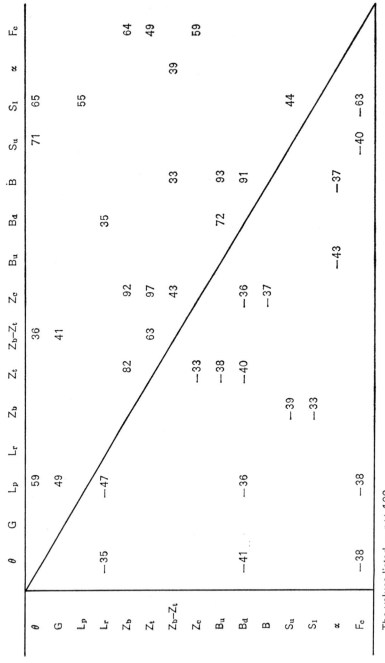

	θ	G	L_p	L_r	Z_b	Z_t	$Z_b\!-\!Z_t$	Z_c	B_u	B_d	$\dot{B}$	S_u	S_l	α	F_c
θ															
G	59														
L_p		49													
L_r	−35		−47												
Z_b															
Z_t					82										
$Z_b\!-\!Z_t$	36	41				63									
Z_c					92	97	43								
B_u	−41	−36													
B_d				35					72						
$\dot{B}$								33	93	91					
S_u	71				−39	−38									
S_l	65		55		−33	−33						44			
α								39	−43		−37				
F_c	−38	−38			64	49		59				−40	−63		

The values listed $= r \times 100$
For key to letter abbreviations see Table 6.1

units 5 and 16, each of 20° and forming the sides of a narrow, but fairly long, ravine which contains the river. The less steep slopes above this ravine, especially those on the southern side, are of only 2–4° and are relatively wide. Not until the steeper slopes higher up the valley side (e.g. units, 22, 28 and 31) are reached do the units become narrow and elongated again. Their sinuosity in plan, however, decreases the level of correlation which would otherwise have been reached between θ and L_p and L_r—where θ is the angle of slope, L_p the length of the unit parallel to the valley floor and L_r its length up the slope.

There is a close relationship between the absolute angle of slope and the strength of the discontinuities bounding the upper and lower margins of the slope unit. In general as the angle of slope increases, the junction it makes with higher and lower units becomes more abrupt. Both angle of slope and strength of discontinuity are a feature of the system of slope profile analysis suggested by Savigear (1967). The application of this system will, therefore, supply much more data for which these relationships may be tested. The relationship probably does not apply to the boundary between adjacent measured lengths, but only to the strength of the boundary between the larger subdivisions of the slope.

The correlation of −0·41 between θ (angle of slope) and B_d (the distance from the basin mouth) and of −0·38 between θ and F_c (the distance of the centre of the unit from the valley floor) indicates that there is a tendency for the angle of slope to flatten both up-valley and up-slope. The morphological map (Fig. 6.7) shows that the steepest slopes are those of 20° forming the sides of the ravine: these are close to the valley floor, and above the ravine, on the northern side, the slopes become progressively less steep. A high correlation coefficient (with a negative sign) between θ and F_c would occur in a river valley whose slopes become progressively steeper towards the stream (e.g. a polycyclic or V-in-V valley form). The correlation coefficient between θ and F_c would be higher were it not for the units of gentle slope immediately above the ravine and on its southern side. These gentle slopes lead up to steeper slopes (up to 12°) before the flat crest of the hillside. A strong positive correlation between θ and F_c will occur where the general form of the hillside is concave with a very narrow crest and without stream incision at the base of the slope. In these circumstances angle of slope will progressively decrease with distance from the valley floor. The positive or negative sign associated with the correlation coefficient between θ and F_c will, therefore, be a reflection of whether the slopes of the basin are generally inclined to flatten downslope (no progressive stream incision) or to steepen downslope (as in the case of polycyclic incision).

The negative correlation between θ and B_d indicates that the steepest slopes tend to occur near the basin mouth. This suggests that in general the basin is composed of low angle slopes at its head which, like a tilted shallow cream jug, steepen towards the basin exit. This implies that the basin is approaching a gorge. Here this is not the case, but the parallel

is important. Where an open, perhaps shallow, valley head composed of low angle slopes leads down to a narrower section composed of steeper slopes the correlation coefficient between θ and B_d will be negative. If, on the other hand, the basin has a narrowing valley head, with steep slopes V-ing upstream (in plan), this correlation coefficient will be positive. The sign of the correlation coefficient is thus a reflection of the dominant slope form in plan of the drainage basin. It will also be related to its state of recent development. In the example of Figure 6.7 the basin has not been recently influenced by an incision which has steepened the slopes at the valley head.

Both steepness of slope up the valley side and along the length of the valley may be related to bedrock as well as to the history of drainage development. The presence of a free face, or steep slope, in a comparatively resistant bed of rock high up the valley side will tend to result in a positive correlation between θ and F_c (distance from valley floor), whatever the history of drainage development. Similarly where a resistant bed of rock is dipping downstream, the steep slopes to which it may give rise will tend to funnel downstream, leaving a valley head of gentler slopes and an area of steeper slopes close to the stream lower down the valley.

Associated with the relationships mentioned above is the negative correlation coefficient between F_c and both S_u and S_l (strength of upper and lower morphological boundaries respectively). As distance from the valley floor increases so the change in slope steepness which occurs at unit boundaries tends to decrease. This is partly a function of angle of slope, for this is positively correlated with S_u and S_l, and also decreases as F_c increases. The general trend within this basin appears to be for slopes to become less steep up-slope, and for discontinuities to become less marked. This again is the relationship which could be expected in a valley whose predominant form might be described as V-in-V.

These correlation coefficients therefore provide an indication of the general slope form with respect to steepening down the valley side and down the length of the valley. A visual inspection of the ground, or the morphological map (Fig. 9.1) does not allow such a rapid appreciation of the strength of the overall tendency in either direction. It may eventually prove possible to use coefficients such as these as a means of comparing the form of one drainage basin with another, regardless of the size of the basins in question.

A factor analysis (sections 4.7 and 4.8) was performed on the data of Table 6.1 and it was found that the variables loaded on the rotated factors in the following way:

Factor   I   height above, and distance from, the valley floor
Factor   II   length perpendicular to the valley axis
Factor   III   size of morphological unit
Factor   IV   distance from basin mouth

Factor V   angle of slope and the strength of the upper boundary of the
             morphological unit
Factor VI   curvature of the unit.

These first six factors 'explained' more than 90% of the variation in the
data. The first two factors combined explained more than 50% of the variation in the data, and each successive factor contributed less to the 'explanation' than the previous one. In this context it is worthy of note that the
angle of slope is related to the fifth factor and is therefore by no means the
most important variable in accounting for the variation in the slope form
data of Table 6.1.

8.2 Relationships between slopes and the gross morphometry of drainage basins

Emphasis has been placed in this book upon the importance of studying
slopes within the context of drainage basins. It is relevant, therefore, to see
whether or not any significant relationships are to be found between slope
characteristics and the morphometric properties of drainage basins. At
the most generalized scale some relationships may be deduced between
morphometric properties of the drainage basin and the angle of slope.
For example, drainage density (D) is defined as the ratio of total stream
length to basin area. The average horizontal distance between valley
floors is the reciprocal of drainage density (Horton, 1945), and the horizontal distance from the watershed to valley floor is therefore $1/2D$. The
angle of slope of the intervening hillside (θ) is geometrically related both
to this horizontal distance and the relief (H) between the valley floor and
the watershed by the equation:

$$\tan \theta = \frac{H}{1 \div 2D} = 2HD$$

In a right-angled triangle formed with sides equal to H and $1/2D$, the
hypotenuse is the length of the slope L, which is:

$$L = \sqrt{\left(\frac{1}{2D}\right)^2 + H^2}$$

From these relationships, which are well displayed if they are constructed
in the form of a series of right-angled triangles in which H and $1/2D$ are
varied, it can be seen that the greater the drainage density in an area
of constant relative relief, the shorter and steeper will be the slopes. On
the other hand, if the relief is increased in an area of constant drainage
density the longer and steeper will be the slopes. Within any one area of
fairly constant drainage density and relative relief it might be expected
that similar valley-side slopes will be found in adjacent drainage basins.

This is the theoretical link between slopes and drainage basin morphometry which gives rise to characteristic angles (Young, 1961) for particular areas. Relationships between the gross morphometry of drainage basins and their valley-side slopes are seldom quite so simple. Variations are introduced by differences in lithology and by the different rate of response of drainage basins to external influences, such as climatic change or renewed stream incision.

In the study of a small area containing drainage basins up to the sixth-order, Carter and Chorley (1961) found a significant relationship between the order of the drainage basin and the maximum angle of slope which it contained. The mean maximum angle of slope progressively increased from the first- up to the fourth-order basins. They used *analysis of variance test* (a worked example appears on p. 165) to show that there was a significant difference between the mean value of maximum slope angle for the second- , third- and fourth-order basins at an *F*-level of 0·95. This increase is attributed, by these authors, to an increase in stream discharge with basin order. Maximum valley side slope cannot increase indefinitely with basin order, for above a certain angle the slopes would become unstable and new, lower angles of slope would again be created. No significant difference was found by Carter and Chorley between the mean maximum angles of slope for the fourth- and fifth-order basins, showing that the limit of increasing slope with order had probably been reached. Likewise a *t*-test for small samples indicated that there was no significant difference between the mean maximum angles of slope for fifth- and sixth-order basins. Field inspection showed that deposition at the base of the valley side slopes, rather than the active removal of material, introduced lower slope angles within the higher order drainage basins.

In a detailed analysis of the environmental influences on drainage basins and slopes Melton (1957) showed that angle of maximum valley side slope correlated positively with relative relief (H) and negatively with drainage density (D). This may be compared with the theoretical observations made above. In this case, under natural conditions, neither H nor D is being held constant for the convenience of demonstrating the theoretical relationship to slope angle. Instead H and D appear to be changing together. The relationship of the one or the other to maximum angle of slope could be seen through a partial correlation analysis (see section 4.5) by first holding one and then the other variable constant.

Melton also discovered that there was a significant positive correlation between maximum valley side slope (θ_{max}) and ruggedness number (Rg), and a negative correlation with channel frequency. He found that on a regression line plotted for *valley side slope v. ruggedness number* the points for third-order basins tended always to fall on one side of the line, while those for fifth-order basins tended to fall on the other. He therefore adopted a method of adjusting for basin order by dividing Rg by u (basin order). On this basis there is a significant correlation between θ_{max}

and Rg/u where:

$$\theta_{max} = 23 \cdot 71 \log \frac{Ru}{u} + 30 \cdot 84$$

(standard error of estimate $= 4 \cdot 6$, $r = 0 \cdot 777$, $R^2 = 60 \cdot 4\%$)

Maximum valley side slope also varies with the relative relief:

$$\theta_{max} = 18 \cdot 69 \log H + 13 \cdot 1$$

(standard error of estimate $= 5 \cdot 7$, $r = 0 \cdot 625$, $R^2 = 39\%$)

8.3 Analysis of morphological units within drainage basins

As distinct from an analysis of the relationship between angle of slope and the morphometric properties of drainage basins, it is also possible to consider the size of the morphological units at each slope angle. Within the Weymouth lowland, for example, it was found that the mean size of morphological units in the seven third-order drainage basins was much higher for the slopes of 0–10° than for the slopes above 10° (Table 8.2).

Table 8.2 Mean area of morphological units within the Weymouth Lowland

Angle of slope	0	1	2	3	4	5	6	7	8
Mean area (thousandths of a square mile)	7·7	8·1	11	9·8	10	12·7	8·6	8·1	10·8

Angle of slope	9	10	11	12	14	16	20	24
Mean area (thousandths of a square mile)	10·1	7·7	1	4	0·7	0·9	2·5	0·5

Range of area values for slopes 0–10° is 7·7 to 12·7 ; range for slopes of more than 10° is 0·5 to 2·5.

The *analysis of variance* test may be used (a worked example is given below), to see if there is a significant difference between the areas of the units above and below 10°. The analysis of variance test is a very useful one and has many applications in geomorphology. Its main purpose is to test for a significant difference between sample means. In this way an indication is obtained as to whether or not two samples could have been drawn from the same population. The null hypothesis, H_0, for this test states that the samples have been drawn from the same population, and the test is applied to see if H_0 can be rejected, and if so at what level of confidence. The test is applicable to data measured on the interval or ratio scale, so long as the data are normally distributed and the two samples have identical standard deviations. These conditions refer to the

ideal case, though small departures from them appear not to influence the validity of the results. A more important condition concerning the data is that the values in one sample should in no way influence the values which occur in the other sample. The samples should also be random. If the data are collected selectively then the results can only be applied to a population that has the specific characteristics of the sample.

In the case of the Weymouth lowland data (see Table 8·2) there appears to be a significant difference between the area occupied by each of the lower angle slopes and each of the slopes over 10°. If the slopes without inclination (i.e. those of 0°) are ignored, then the mean area (Table 8.3) for the slopes of 1–10° is 6·221 (square miles × 100) and for those of

Table 8.3 To find group means for analysis of variance test

	Slopes 1–10° Area (square miles × 100)	Slopes 11–20° Area (square miles × 100)
Basin 1	2·118	0·275
2	4·221	0·186
3	1·287	0·070
4	11·152	0·482
5	8·421	0·130
6	12·296	0·040
7	4·044	0·440
Σ A	43·550	1·623
Ā	6·221	0·232

Total number of basins = N = 14
Grand total of all areas = T = 45·173
Grand average of all areas = $\dfrac{T}{N}$ = 3·227
Total number of degrees of freedom = N − 1 = 13

11–20° is 0·232 (square miles × 100). This is a large difference. To test whether or not there is any chance of these two mean values having been obtained from the 'same population' the analysis of variance test may be applied. That there is a significant difference between the two means is almost self-evident; nevertheless the test will be applied here to provide a worked example, and to justify the order of confidence with which the null hypothesis (that the two means have been drawn from the same population) may be rejected.

In the analysis of variance test it is necessary to examine the variance in the data. There are two components to this variance: the first is the variance within each sample; the second is the variance between the two samples. If both samples had been drawn from the same population then their mean values would, in all probability, still have been different from

each other. This is because of the variations present within the data values of the whole population. The important consideration is as to see whether or not the variation *between* the sample means is significantly greater than that *within* the samples themselves. To discover this it is necessary to separate these two out. A greater variation between the samples than within the samples would suggest that the two samples had been drawn from different populations. In the first instance it is necessary to calculate the sample means, and the number of degrees of freedom within the total data. The number of degrees of freedom is always one less than the total number of items, in this case basins, that are being considered. Next, the square is found of the difference between each value and the grand mean for all of the data values ($3 \cdot 227$ square miles $\times$ $100 \cdot$)

Table 8.4   To find the total sum of squares

	Slopes 1–10° $\left(\dfrac{T}{N} - A\right)^2$	Slopes 11–20° $\left(\dfrac{T}{N} - A\right)^2$
Basin 1	0·123	0·0018
2	0·988	0·0021
3	3·764	0·0262
4	62·806	0·0625
5	26·978	0·0104
6	82·247	0·0368
7	0·686	0·0433
$\Sigma \left(\dfrac{T}{N} - A\right)^2$	177·592	0·1831

Total sum of squares $= 177 \cdot 7751$

The total sum of squares ($177 \cdot 7751$) and the total number of degrees of freedom (13) can now be divided, or partitioned, into parts related to the *between* sample and *within* sample means respectively. To obtain the between sample effect, the within sample effect has to be eliminated by replacing each item by its own sample mean (i.e. $6 \cdot 221$ in the case of the 1–$10°$ slopes and $0 \cdot 232$ for the 11–$20°$ slopes). The between sample sum of squares is the square of the difference between the mean of each group and the grand average T/N (see Table 8.5).

The within sample sum of squares and degrees of freedom are obtained from an analysis of the variability within each sample. The within sample sum of squares is found as in Table 8.6. The results of the calculations in Table 8.6 can be compiled together in one analysis of variance table (Table 8.7).

This *F*-value may be compared with tabled values to assess its significance. To do this it is necessary to bear in mind that it was obtained

Table 8.5 To find the between samples sum of squares

Slopes 1–10° $\left(\dfrac{T}{N} - A\right)^2$	Slopes 11–20° $\left(\dfrac{T}{N} - A\right)^2$
8·964	8·970 for the seven basins in each group

Between samples sum of squares $= (8·964)^2 + (8·970)^2 = 125·538$
Between samples degrees of freedom $=$ number of samples $- 1$
$$= 2 - 1 = 1$$

Table 8.6 To find the within samples sum of squares

	Slopes 1–10° $A - \bar{A}$	Slopes 11–20° $A - \bar{A}$	Slopes 1–10° $(A - \bar{A})^2$	Slopes 11–20° $(A - \bar{A})^2$
Basin 1	−4·103	0·043	16·835	0·00185
2	−2·000	−0·046	4·000	0·00212
3	−4·934	−0·162	24·344	0·02624
4	4·931	0·250	24·315	0·06250
5	2·200	−0·1020	4·840	0·01040
6	6·075	−0·1920	36·906	0·03686
7	−2·166	0·2080	4·692	0·04326
	0·0	0·0	115·932	0·18323

Within samples sum of squares $= 116·1152$
Within samples degrees of freedom $=$ (number of basins $- 1$) multiplied by the number of samples $= (5 - 1)2 = 8$

Table 8.7 Table of analysis of variance

Source of variation	Sums of squares	Degrees of freedom	Variance estimate
Between samples	125·538	1	125·538
Within samples	116·115	8	$\dfrac{116·115}{8} = 14·514$
Total	241·653	9	

$$F = \frac{125·538}{14·514} = 8·65$$

with one and eight degrees of freedom. The tabled F-value (see Appendix Table D) is 5·32 at the 95% confidence level. This is conveniently written as F 1, 8, 0·05, 5·3177 (where the 0·05 represents the 95% confidence level). In this particular example, therefore, the null hypothesis may be rejected with 95% confidence since 8·65 is greater than the tabled F-value. In other words it is most unlikely that the mean value of slope area for slopes over 10° in the Weymouth lowland are drawn from the same population as those for the slopes of less than 10 degrees. The latter are on average significantly larger.

This analysis related to the size of the morphological units (slopes). The *density* of these within a basin can also be examined. An estimate of the density of different morphological units within a drainage basin can be found by dividing the total area of the basin by the number of morphological units which it contains (Table 8.8). In the Weymouth area,

Table 8.8   Mean size of all units within each of the seven third-order drainage basins measured in the Weymouth lowland

Basin number	1	2	3	4	5	6	7
Mean size of morphological unit (thousandth of a square mile)	10·3	11·0	5·0	10·3	10·5	12·1	9·0

basin 3 contains units which on average are only half the size of those in adjacent basins. Compared with the other basins this is a particularly compact basin with frequent changes in slope angle and slope aspect. It is much the most irregular of the basins in terms of its surface undulations. Conversely, basin 6 is appreciably smoother in its appearance in the field and this is reflected in the larger mean area of the morphological units which it contains. The mean size of morphological units is therefore a measure of the smoothness of the slopes within a drainage basin. This index is not complete however, for it is also necessary to take into account the strength of the discontinuity between morphological units in defining the degree of smoothness attained within a basin. It has already been shown that the highest changes in angle of slope between adjacent units occurs when the absolute angle of the units is greatest (Table 8.1) and analysis of variance has indicated that the steepest angles of slope are associated with the smallest of the morphological units. There will thus be a tendency for a basin which includes a large number of steep slopes to have more morphological units than a basin with the same area but composed predominantly of low angle slopes. The former will have a lower ratio of total basin area to the number of morphological units than will the latter.

8.4 Debris fans and drainage basins

In semi-arid areas the basins of a mountain front frequently provide the debris which forms alluvial fans at the basin mouth. A study of these features in California allowed Bull (1962) to consider the relationship between the size and slope of these fans with respect to the area of the drainage basin supplying detrital material. Through geological mapping Bull was able to divide the drainage basins into two types. One set of basins was being cut into predominantly mudstones and shale, while the other was underlain by sandstone. Although several distinct relationships emerged from this study, Bull found it necessary to analyse each set of basins separately. For example, the area of the alluvial fans (Af) was related to the area of the drainage basin behind it (A), but the precise nature of the relationship was different for the mudstone and shale basins as compared with the sandstone basins:

for mudstone and shale:

$$Af = 2{\cdot}4A^{0{\cdot}88} \text{ square miles}$$

for sandstone:

$$Af = 1{\cdot}3A^{0{\cdot}88} \text{ square miles}$$

where 0·88 is the slope of the line. The difference between the coefficients (1·3 and 2·4) shows that the fans derived from the mudstone basins are almost twice as large as those derived from sandstone. The rate of increase with basin area, however, is the same in each case. In general Bull also found that there is a decrease in the slope of the alluvial fan when there is an increase in the size of the basin behind it:

$$Sf = 0{\cdot}023A^{-0{\cdot}16} \text{ in mudstone and shale}$$
$$Sf = 0{\cdot}022A^{-0{\cdot}32} \text{ in sandstone}$$

Thus, in all but the smallest basins, for basins of comparable size the fans developed from mudstone and shale basins slope more steeply than those which are derived from sandstone basins.

In a similar type of study Mammerick (1964) was concerned with bedrock pediments (as opposed to alluvial fans) in the Mojave and Sonoran deserts of the south-western United States. In this area she discovered no significant correlation between the angle of slope of the pediments and the size of the drainage basin backing the pediment. Nor indeed was there a significant relationship between the slope of the pediment and its length, or the bedrock across which it was developed. Local tilting and the cyclic development of the pediments were thought to be much more important in influencing the slope of the pediment than any of these other variables. Further numerical analysis of many more examples will serve to show whether the laws governing the relationships between depositional slopes and erosional slopes and other features of their environment really are different.

8.5 Schumm's models of land-form development

Through the examination and measurement of slope and drainage basin characteristics in the field, at Perth Amboy in New Jersey, Schumm (1956) was able to suggest three models of land-form development. These models relate to areas similar in lithology, structure and climate, within which there are differences in the value of drainage density, length of overland flow, and relief ratio (available relief divided by length of drainage basin).

1   The first model relates to an area of high relative relief in which the relief ratio is 0·15, the drainage density is 20, and the length of overland flow is 0·025 miles. In such an area Schumm found that degradation by the stream channel is rapid. Straight valley side slopes develop rapidly, and parallel retreat of the slopes maintains a steep maximum angle. Because of this retreat, slopes of adjacent valleys quickly meet at their summits and thus eliminate any flat area which may have been present on the intervening divide. After this local relative relief tends to remain constant through an equal rate in the lowering of the stream beds and the intervening divides. A cessation of channel downcutting may result in lateral planation, and the subsequent undercutting of slopes helps to maintain the characteristic steep angle of slope. At the same time, however, divides will continue to be lowered, and there is therefore a general reduction in relative relief. As deposition takes place at the base of slopes their inclination is decreased. Continued deposition leads to the development of convex–concave hillside profiles.

2   In an area of moderate relative relief, where the relief ratio is 0·04, drainage density 10, and the length of overland flow 0·05 miles, degradation by the stream is less rapid than in the case described above. In the area of moderate relative relief the intersection of straight slopes, retreating parallel to themselves, will not eliminate interfluve crests as quickly as before. The stage of maturity, when maximum relief is present in the basin, will be briefer. As the value of the drainage density suggests, stream channels will be twice as far apart. Under conditions of moderate relief, the end stage is that of an area of rounded slopes with slope angles decreasing with time.

3   When the initial relative relief is low and the relief ratio is only 0·009, drainage density 5·0, and the length of overland flow 0·1 miles, then the distance between adjacent channels is, by definition, four times as great as in the case of the first model. Channel degradation tends to be slow and there is little stream incision. Straight valley side slopes never develop and broad convexities are characteristic of the area.

These three models are held by Schumm not to summarize a sequence of drainage basin development with respect to any one basin. They are

intended to describe three separate cases in which it can be seen how diverse slope-profile forms are rational parts of sequences in which both parallel retreat and slope decline play a part. The progress of development in each case depends on the relative relief at the start of denudation and on the stage which the reduction of relief reached.

Table 8.9 Characteristic and limiting angles of slope on several rock types in parts of the British Isles

Outcrop		Characteristic angles			Limiting angle
		'Flat' units	Sloping units	High angle units	
1 Sandstones:	Kellaways	0 –½	2½–4	6½–9	9
	M. Deltaic	1 –2	4½–6		
	Eller Beck	1 –2	6½–9	14–17	
	L. Deltaic	2½–4	6½–9	17½–21½ (and over 28)	
	Dogger	0 –½	6½–9	14–17 (and 22–27½)	
	M. Lias	1 –2	6½–9		17
Limestones:	Cornbrash	0 –½	2½–4	6½–9	11
	Grey limestone	2½–4		11½–13½	14
Shales:	U. Deltaic	1 –2			14
	U. Lias	2½–4	6½–9	17½–21½	
	M. Lias	2½–4			22
	L. Lias	0 –½	4½–6	9½–14	17
Gritstone:	Moor grit	2½–4		11½–13½	14
2 Sandstones:	Hangman grits	4 –5	13–15	28–29	
Shales:	Ilfracombe beds	3 –4	9–10	23–25	
	Silurian (glaciated area)		6–8 and 15	21–22 and 25	
Discontinuous soil cover, with projecting rock outcrops					41–49
Continuous soil, with bare gashes in the vegetation cover					36–40
Continuous soil and vegetation cover, with terracettes					33–36
Smooth soil and vegetation cover, no surface irregularities					below 33

	Bare bedrock slopes			Regolith or regolith and soil-covered slopes						
3 Gritstone	45	40	37	32	28	20				2
Sandstone		40		32	26	20			5	2
Shale	45			32	29–26	20	17–15	11		
Phyllite	45			32	29–26	20	17–15	11		

Sources: (1) Gregory and Brown (1966) ; (2) Young (1961) ; (3) Savigear (1956)

8.6 Slopes and geology

The relationship between bedrock and slopes may be brought out by numerical analysis through comparing slopes on different rock types. It is extremely difficult to allocate a numerical value either to rock hardness or to its resistance to weathering. Comparison between slopes on different rock types can take several forms. Slope units within a drainage basin may be subdivided into sets related to its geology. The characteristic and limiting slope angles can then be found for each geological type. Alternatively, or in addition, analysis of variance (section 8.3) may be used to test for a significant difference between the mean angles of slope on two different rock types.

Characteristic and limiting slope angles were determined by Gregory

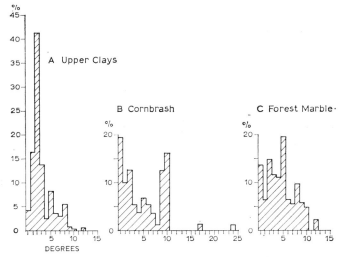

Fig. 8.1   Histograms for slope steepness on three different bedrocks in the Weymouth lowland.

and Brown (1966) for 13 different outcrops in Eskdale; they subdivided the slope units into three groups of flat, sloping and steepest slope units. Savigear (1956) distinguished not only between lithological types but also between those slopes which were composed of bedrock and those which were regolith and/or soil covered. Young (1961) made a similar distinction for determining limiting slope angles. The results of these investigations are presented in Table 8.9.

As was shown in section 7.2 characteristic and limiting slope angles are discerned through frequency distribution analysis. Figure 8.1 presents histograms for three different beds within the Weymouth lowland based on the ground area of individual morphological units. A significant point here is that the boundaries of the outcrops sometimes cut obliquely across a morphological unit. These units are thus erosional in origin. Their

actual steepness, however, is in part at least related to the beds on which they occur. The histograms show that low angle slopes predominate on the clays and the steepest slopes (17° and 24°) occur on the Cornbrash. The characteristic and limiting slope angles are defined in Table 8.10. Comparison with Table 8.9 shows that the values obtained for Cornbrash differ in the two areas analysed, despite the fact that similar methods of obtaining the data were used. This suggests that denudational history may play an important part in determining characteristic angles. It certainly plays an important part as far as limiting angles are concerned, because the observation of a slope of 24° in Cornbrash could not have been made but for the presence of an undercut river bank at this locality.

The importance of erosion is stressed by Young (1961), who suggests that slopes of 30–40° are the result of rapid basal erosion undercutting a slope. As soon as this erosion ceases these slopes are transformed into slopes of less than 30°, and often lie in the range 25–29°. Young concludes that the majority of characteristic slope angles of an area are related to local morphological history.

Table 8.10 Characteristic and limiting slope angles within the Weymouth lowland

Outcrop	Characteristic angles			Limiting angle
	'Flat' units	Sloping units	High angle units	
Upper clays	2	5–8		12
Cornbrash	0–2	5–7	9–10	24
Forest marble	0	2–5 and 8	12	12

An absolute classification of bedrock–slope relationships through characteristic and limiting slope angles may be possible for one area of similar geomorphological history, but not for areas having different histories or different climates. More than one classification will exist if history and climate are taken into account. In any one area, however, the effect of lithology on slope steepness may also be assessed through testing, by analysis of variance, for a significant difference between the mean angle of slope on each rock type. Melton (1957) carried out such a test for shales, schists, clastic, granitic, limestone and acid volcanic beds in Arizona, Colorado, New Mexico and Utah. He found a significant difference between the mean slope angle for each lithology at the 0·001 level.

The relationship between angle of slope and the material of which a scree is composed is generally thought to be related to the stable angle of repose of the constituent material. Some studies support this idea because characteristic slope angles constantly recur for scree forms in an area (usually 32–37°). Other studies, however, cast doubt on the universal truth of this concept. Detailed measurements of both material and angle

of slope by Melton (1965) in southern Arizona showed that at best only a low correlation exists between the size of scree material and its angle of slope. As Melton suggests, much more detailed fieldwork and numerical analysis is required before definite conclusions can be reached.

One study of both bedrock and debris slopes in relation to angle of slope which employs a consideration of slope stability, characteristic angles, and theories of its internal physical (engineering) properties is that of Lohnes and Handy (1968) on friable loess in Iowa. They found that the most frequently recurring slope angles were 38°, 51° and 70–85°. By measuring properties such as the angle of internal friction (ϕ), the angle of the exposed shear planes (θ), and the angle of slope before slope failure took place (i), they were able to establish that:

$$\theta = \frac{i}{2} + \frac{\phi}{2}$$

The angle of the shear plane became the new angle of slope when slope failure took place. Thus if 77° is taken as the median value of the high angle slopes, then when i = 77° (ϕ was found to be 25°):

$$\theta = \frac{77 + 25}{2} = 51°$$

Similarly:

$$\theta = \frac{51 + 25}{2} = 38°$$

This fits the observations on the most frequently recurring values (modes) mentioned above. The high angle slopes have a relative relief of less than 40 feet; the lower angle slopes of 51° have a vertical extent of 40–100 feet; and the 38° slopes stand up to 200 feet high. Lower slope angles are to be found, but none of these is due to shear failure. Lohnes and Handy also found that the natural angle of repose of this loess was 38°. This means that some of the 38° slopes were exposed shear planes while others were debris slopes.

There are many areas in which lithological conditions are much more complex than in this study of loess. Nevertheless, despite the difficulties which these areas present, the geomorphologist must become more concerned with the engineering properties of both bedrock and debris before angle of slope and slope form can be fully analysed. The data which engineering studies provide are capable of numerical analysis. An important property of bedrock and regolith is infiltration capacity. The lower the infiltration capacity the more water is available for flow as surface run-off. Melton (1957) discerned a high positive correlation between mean valley side slope (θ) and infiltration capacity (f).

$$\theta = 1\cdot312f + 20\cdot7 \quad (r = 0\cdot746)$$

Thus where infiltration capacity is high, run-off is small and the rate of slope erosion is slow. In such areas once valley side slopes have become steep they tend to remain steep.

8.7 Slopes and aspect

The relationship between aspect and slope steepness has often been examined. In general, discussion has been concerned with differences in the steepness of north- and south-facing slopes. In the higher latitudes of the northern hemisphere the south-facing slopes will receive a greater amount of sunshine and warmth, which especially under glacial and periglacial conditions may have a considerable effect on slope processes such as freeze-thaw activity and the development of meltwater channels down the hillside. Melton (1960) discusses some of the earlier work which has taken place on this subject. There are three possibilities:

1   North-facing slopes are steeper than south-facing slopes (Emery, 1947; Russell, 1931).

2   South-facing slopes are steeper than north-facing slopes (von Engeln, 1942; Büdel, 1953).

3   North- and south-facing slopes have approximately the same steepness (Strahler, 1950).

Melton's study in Wyoming showed (through the analysis of variance) that there was a significant difference between the steepness of north- and south-facing slopes. North-facing slopes were, on average, twice as steep as south-facing slopes ($4 \cdot 42°$ as compared to $2 \cdot 21°$). He concludes that in arid to subhumid temperate climates, excluding the extremes of dryness and wetness, low channel gradients favour the development of valley asymmetry in east–west trending valleys. This is because material moving down the south-facing slope 'pushes' the stream into the foot of the north-facing slopes, which hence become steeper.

The drainage basins of the Weymouth lowland also tend to follow an east–west trend, but they occur in a moist temperate climate. Although the north-facing slopes are, on average, steeper than the south-facing slopes ($\bar{\theta} = 4°$ and $5°$ respectively), a t-test showed that the null hypothesis—that the two samples come from the same population—could not be rejected. In other words, there is no significant difference between the means. It may be noted that, except in a few isolated instances, the streams in the Weymouth lowland are sluggish, and flow across aggraded alluvial valley floors. They are not effective in undercutting one bank rather than another and tend to follow the central axis of the valley floor.

A very rapid and easily applied non-parametric test, that is useful for analysing significant differences in slope steepness, is the *sign test* described by Siegel (1956, 69). This may be used for relatively small samples (i.e. less than 20). Larger samples can be analysed using the sign test described in Cole and King (1968, 121). The small sample method may be illustrated by reference to 20 measurements across one of the east–west trending valleys of the Weymouth lowland. A simple table was constructed (Table 8.11) in which a positive sign indicates that the north-facing slope is steeper than the south-facing slope, and *vice versa* for the

Table 8.11   The relative steepness of north- and south-facing slopes in a third-order basin of the Weymouth lowland

N-facing steeper than S-facing: $+ + - - + + + + + - - + - - -$
$- - - - +$
i.e. 9 ($+$) signs and 11 ($-$) signs

negative sign. The almost equal frequency of occurrence of $+$ and $-$ signs in Table 8.11 suggests that the north-facing slopes are not significantly steeper than the south-facing slopes. The lesser number, x, is 9, and the number of pairs for which a difference has been recorded, N, is 20: x and N may be compared with the tabled values (Siegel, 1956, 250) and it will be found that $p = 0.412$. Thus, if the null hypothesis states that there is no significant difference between the north- and south-facing slopes, it may be rejected with only 58.8% (i.e. $100 - 41.2\%$) confidence. That is to say, the null hypothesis could not be rejected with any certainty. Once again the analysis suggests that there is no significant difference between the steepness of north- and south-facing slopes in the Weymouth lowland.

Clark (1965) finds support for the idea that valley asymmetry reaches a maximum in valleys of medium dimensions, while smaller and larger valleys do not display asymmetry so strongly. He found that in the chalk country of southern England, valleys deeper than 150 feet showed negligible asymmetry. Gloriod and Tricart (1952) found little asymmetry in valleys less than 25 feet deep, but that with a depth of between 70 and 100 feet it reached a maximum. It is possible that inadequate climatic differential is produced across very shallow valleys, whereas in the largest valleys asymmetry is destroyed by concentrated fluvial erosion.

Melton's analysis (1960) went beyond just an analysis of north-facing and south-facing slopes in that each of these groups were subdivided into three categories:

1   No alluvial fan in the neighbourhood of the slope
2   Hillside measured lies opposite an alluvial fan
3   Hillside whose foot meets the top of an alluvial fan.

Analysis of variance was used by Melton to detect significant differences between the mean values of these groups. He concluded that valley side slopes are sensitive indicators not only of the intensity of gradational processes operating on their surfaces, but in addition display the effects of the proximity and activity of nearby channels, that is, of their erosional environment. Numerical studies, such as this by Melton, are bringing to light a close relationship between present-day slope form and valley floor processes. These studies suggest that the response of slope form to its erosional environment may be fairly rapid, and that the legacy of former conditions may not survive for very long in some areas.

Variation in both slope angle and aspect of individual morphological units within selected drainage basins was studied by Gregory and Brown (1966) for parts of Eskdale, in north-eastern England, where periglacial activity is likely to have taken place. The most interesting results came from subdividing the data into three altitudinal groups:

1   Below 700 feet, where south-facing and west-facing slopes were steepest
2   700–1000 feet, where both east-facing, west-facing and north-west-facing slopes were found to be the steepest
3   Over 1000 feet, where the west-facing group was steepest.

The fact that slopes steeper than average are orientated in several directions, but at different heights, is interpreted by Gregory and Brown as showing either that there are differences in the intensity of several processes at distinct heights, or that they are the result of different phases of landscape evolution. For example, above 1000 feet prevailing westerly winds during the later part of the Pleistocene would have allowed snow cover on the east-facing slopes, while solifluction processes were active on the west-facing (windward) slopes. On the west-facing, and especially the north-west-facing slopes, there are many benches whose steep fronts have higher slope values than those in the areas where slopes face in a different direction and where these benches are not to be found. The high proportion of south-facing slopes below 700 feet is attributed by Gregory and Brown to the effect of a northward shift of streams as a result of intense solifluction and deposition on, and at the foot of, the north-west-facing slopes. This example is similar to Melton's explanation of a shift in the position of the stream in the valley floor through an excessive supply of waste material from one valley side compared with the other. There is an implied difference, however, in that where Melton suggests a dependence of slopes on recent processes active at their base, Gregory and Brown interpret their findings in terms of processes most active during the late Pleistocene.

8.8 The multivariate nature of slopes

From the foregoing it will be clear that slopes are related to many environmental factors. The numerical analysis of slope steepness and form in relation to other environmental variables is a major field for future research. The availability of techniques of multivariate analysis, such as those described in part I, makes it possible to disentangle the complexities to be found through any study of slopes and their environment. One of the studies which has already proceeded along these lines is Carson's (1969). He presents an analysis of the relationship between the angle of straight-slope portions of a hillside (facets) and nine other variables at 46 sites. The significant correlation coefficients found by Carson are listed

in Table 8.12. Not only is the angle of slope most highly correlated (negatively) with the thickness of the waste mantle, but the latter also contributes most (76·2%) to the reduction of the variance of slope angle.

Table 8.12 Significant correlation coefficients (at the 99% level) between the angle of straight slopes and nine other variables in Exmoor and the Pennines

	A	B	C	D	E	F
Angle of slope	55	67	58	−87	43	69

The values listed = $r \times 100$
where A = vertical extent of straight slope; B = the proportion (by weight) of the regolith just under the vegetation mat containing material coarser than a $\frac{3}{8}''$ sieve; C = mean size of the intermediate axis of rock fragments; D = mean thickness of waste mantle; E = estimate of bankfull discharge of the stream at slope base; F = average boulder size on stream bed.
Source: After Carson (1969).

Little extra reduction in the variance of slope angle occurs when variables B, C, E and F (of Table 8.12) are included. The second most important variable is the gradient of the stream at the base of the slope (contributing another 5·5% to the reduction of the variance of slope angle) despite its low correlation coefficient of 0·28 with angle of slope.

8.9 Conclusion

The foregoing shows that in any study of the relationship between slope form and environment valuable results will only come from fieldwork that has been well conceived, organized with an efficient sampling design, and with knowledge in advance of the requirements of the numerical techniques to be employed in analysing the acquired data. This chapter has been concerned to show the type of work that has been done so far, and to indicate the methods of numerical analysis which have been used. The greatest openings in the future lie in the use of multivariate techniques. Nevertheless, such simple techniques as the comparison of histograms should not be snubbed for they can provide invaluable clues in the identification of slope form problems quite apart from their value in summarizing the statistics of the sample being analysed.

References

BLOOM, A. L. 1969: *The surface of the earth.* Englewood Cliffs, NJ: Prentice Hall.

BUDEL, J. 1953: Die 'periglazial'—morphologischen Wirkungen des Eiszeitklimas auf der ganzen Erde. *Erdkunde* **7**, 249–66.

BULL, W. B. 1962: Relations of alluvial fan size and slope to drainage basin size and lithology in western Fresno county, California. *Geol. Surv. Res. Papers, Short Paper* 19; *U.S. Geol. Surv. Prof. Paper* **450**-B, 51–3.

CARSON, M. A. 1969: Models of hillslope development under mass failure. *Geogr. Analysis* **1**, 76–100.

CARTER, C. S. and CHORLEY, R. J. 1961: Early slope development in an expanding stream system. *Geol. Mag.* **98**, 117–30.

CLARK, M. J. 1965: The form of chalk slopes. *University of Southampton, Department of Geography, Research Series* **2**, 3–4.

COLE, J. P. and KING, C. A. M. 1968: *Quantitative geography.* New York and London: John Wiley.

EMERY, K. O. 1947: Asymmetric valleys of San Diego county, California. *Bull. S. Calif. Acad. Sci.* **46**(2), 61–71.

ENGELN, O. D. VON 1942: *Geomorphology.* New York: Macmillan.

GLORIOD, A. and TRICART, J. 1952: Étude statistique des vallées asymétriques de la feuille St. Pol au 1/50,000. *Rev. de Géomorph. Dynamique* **3**, 88–98.

GREGORY, K. J. and BROWN, E. H. 1966: Data processing and the study of land form. *Zeit. für Geomorph.* NF **10**, 237–63.

HORTON, R. E. 1945: Erosional development of streams and their drainage basins. *Bull. Geol. Soc. America* **56**, 175–370.

KRUMBEIN, W. C. and GRAYBILL, F. A. 1965: *An introduction to statistical models in geology.* New York: McGraw-Hill.

LOHNES, R. A. and HANDY, R. L. 1968: Slope angles in friable loess. *J. Geol.* **76**, 247–58.

MAMMERICK, J. 1964: Quantitative observations on pediments in the Mojave and Sonoran deserts (south-western United States). *Am. J. Sci.* **262**(4), 417–35.

MELTON, M. A. 1957: An analysis of the relations among elements of climate, surface properties, and geomorphology. *Columbia Univ., Dept of Geol., Tech. Rept.* **11**.

1960: Intravalley variation in slope angles related to microclimate and erosional environment. *Bull. Geol. Soc. Am.* **71**, 134–44.

1965: Debris-covered hillslopes of the southern Arizona desert: consideration of their stability and sediment contribution. *J. Geol.* **73**, 715–29.

RUSSELL, R. J. 1931: Geomorphological evidence of a climatic boundary. *Science* **74**, 484–5.

SAVIGEAR, R. A. G. 1956: Technique and terminology in the investigation of slope forms. *Premier rapport de la commission pour l'étude des versants* (IGU, Rio de Janeiro), Amsterdam.

1967: The analysis and classification of slope profile forms. In L'evolution des versants, *Les Congrès et Colloques de l'Université de Liége* **40**, 271–90.

SCHUMM, S. A. 1956: Evolution of drainage systems and slopes in badlands at Perth Amboy, New Jersey. *Bull. Geol. Soc. Am.* **67**, 597–646.

SIEGEL, S. 1956: *Nonparametric statistics for the behavioral sciences.* New York: McGraw Hill.

STRAHLER, A. N. 1956: Equilibrium theory of erosional slopes approached by frequency distribution analysis. *Am. J. Sci.* **248,** 673–96; 800–14.

YATSU, E. 1966: *Rock control in geomorphology.* Tokyo: Sozosha.

YOUNG, A. 1961: Characteristic and limiting slope angles. *Zeit. für Geomorph.* **5,** 126–31.

9 Slope models

9.1 Analogue models
  [*trend-surface analysis, linear, quadratic and cubic surfaces and their equations, confidence surface, residual values, coefficient of determination, differentiation*]
9.2 Mathematical models
9.3 Experimental design models
  [*regression analysis*]
9.4 The use of models
  [*data banks, trend-surface analysis, mathematical models, experimental design models*]

The forms of hillsides are usually complex when they are measured on the detailed level made possible through the techniques of morphological mapping and slope profiling described in chapter 6. In many instances, therefore, it becomes necessary to generalize out the minor irregularities so that the dominant components of the hillside form may be appreciated. Levels of generalization of the slope form in profile were discussed in section 6.4. Generalization is required in comparisons between slopes, for if slopes are ever alike they will reveal this by the nature of their macro-form, not through the detailed irregularities which they contain. It is also necessary to generalize the forms of slopes if they are to be compared with an idealized form, or idealized model; that is to say with a form which it is thought typical of a particular environment or set of geomorphological circumstances. The objective generalization of the form of a slope in profile can only take place with reference to its numerical properties (section 6.4).

In its simplest form the idealized model is that of a single hillside, reaching from hill crest to valley floor. Most models relate only to single cycle slopes and do not include the complexities introduced during slope development by polycyclic river incision. More complex models arise, however, when an attempt is made to provide a generalized statement of the form of a whole landscape. Such models are used in describing the general form of land-systems. In a sense they are natural models because they are based on what has actually been observed within a whole landscape.

Both the idealized model and the natural model are usually represented in the form of block diagrams. Each can also be mapped, in which case the resulting map forms a *spatial model*. Through morphological mapping it is possible to define the ground pattern of the distribution of the main units of the idealized model; indeed it is also possible to define the

morphological units within each of these main units. Likewise the boundaries between land-systems can be mapped, and, though it is seldom done, within each of the land-systems the position can be defined of the land-form units of which it is composed.

The idealized model and the natural model are both types of *analogue models* (Chorley, 1967). Because they are derived from observations of land form they differ from those *mathematical models* which are derived from a theoretical consideration of how, under a set of defined processes, slope development will proceed. Some of the studies which qualify as mathematical models are based on field observation and measured slope form, but then theoretical considerations are involved in order to provide *deductive models* of slope form. Mathematical models are frequently concerned with the development of slopes through time. Analogue models are more concerned to show the form of the hillsides as they are now. Whereas analogue models are usually represented by block diagrams and mathematical models by a series of slope profiles, a third and important type of slope model, the *experimental design model*, is represented by regression equations. These models result from the application of statistical techniques to a data matrix of slope variables in order to find relationships which have meaning in terms of slope form and slope development. Experimental design models have great potential in the development of process–response models. Through numerical analysis it becomes possible to relate slope form to slope processes in a manner not allowed by the other types of models.

Numerical analysis is relevant in each of these types of models, in the sense that they all deal with quantitative information. Statistics in the strict sense are more relevant to experimental design models than they are to any of the others. The following sections deal with each of the models in turn, concentrating on those to which numerical analysis has the most to contribute.

9.1 Analogue models

The nature of an idealized model is summed up by the definition given by Savigear (1960) of a 'normal slope':

Any fully developed slope is divisible normally into three sections: (i) the *crest-slope* extending from the interfluve to where the slope steepens, (ii) the *backslope* consisting of the steepest morphological units of the profile, and (iii) the *footslope*, a gently inclined section extending from the backslope to the valley centre. In places the backslope may be cliffed.

The form of such a slope is indicated in Figure 9.1. It does not differ from the idealized models suggested by Wood (1942) or by King and Fair (1944), or King (1953). Wood places his concept of this form within a cycle of slope development (Fig. 9.2), while King and Fair relate the four units of their model to slope processess and regolith cover, as well as

to a verbal discussion of the changes which take place in each of these units through time. The four units of the hill slope (Fig. 9.1) are related to some of the environmental influences upon them by Schumm (1966). These are summarized in Table 9.1.

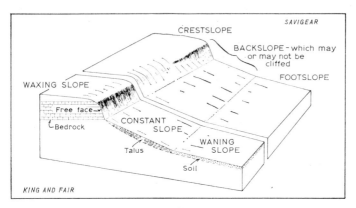

Fig. 9.1 Analogue model— 'normal' hillside.

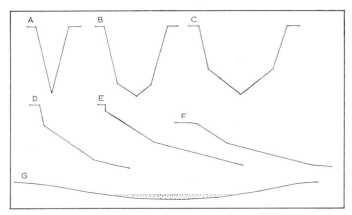

Fig. 9.2 Cycle of slope development as proposed by Wood (1942): **A:** Free face only, cut in flat, recently raised land surface; **B, C:** Constant slope formed; **D, E:** Constant slope developes; **F:** Wasting slope formed, some alluvial filling; **G:** Constant slope consumed, alluvial fill deepends, slope gradually flattens.

Not every hillside conforms to the idealized model (Fig. 9.1). Many contain only a part of this idealized crest to stream sequence, while other slopes carry this sequence of forms but it is repeated, either in part or as a whole, more than once down the hillside. Thus, the profile shown in Figure 9.3, for example, carries in sequence from the crest a crestslope, a backslope (not cliffed), and a footslope which passes across a convex

Table 9.1 Some of the factors influencing the development of hill slopes

	Erosion	Deposition
Convex crestslope:	Creep; rainwash (especially on permeable soils and on lower steeper slopes); mechanical and chemical weathering of massive rocks; incision of widely spaced streams in areas of low relative relief	Mantling of existing slopes, e.g. by loess or volcanic ash
Cliff:	Mass movement; basal erosion; incision of streams in areas of high relative relief and resistant strata	None
Backslope (constant slope):	Rainwash (on impermeable soils); incision of closely spaced streams in areas of moderate or high relative relief	Talus; sand; ejected volcanic material resting at the angle of repose
Concave footslope:	Rilling (erosion greatest half-way down the slope); creep	Accumulation of slope waste; colluvium at the base of the slope

boundary into another backslope which in turn gives way to a footslope. The slope in this example is annotated is accordance with the suggestions made in chapter 6. The numerical method of slope unit organization employed defines with considerable precision the nature of each of the

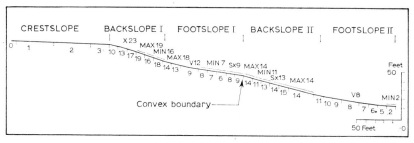

Fig. 9.3 A hillside classified with respect to the analogue model of Figure 11.1.

major units that can be related to the idealized model. Numerical analysis in this sense makes possible a description of slope form which can be used in characterizing the nature of each of the major units of the idealized model, and therefore also enables the comparison of one slope with another and of both slopes with the idealized model.

Dalrymple, Blong and Conacher (1968) found the four unit model too

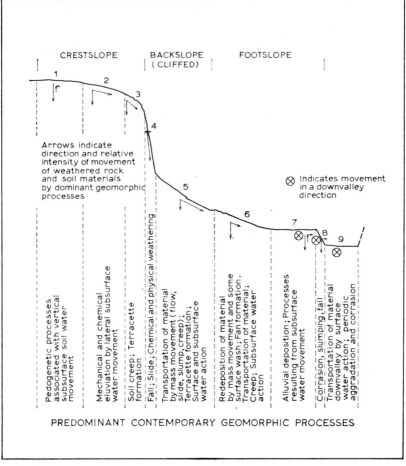

Fig. 9.4 Diagrammatic representation of a hypothetical nine-unit land surface model (*after Dalrymple, Blong and Conacher, 1968*) : **1** interfluve; **2** seepage slope; **3** convex creep slope; **4** fall face (>45°); **5** transportational midslope (usually 26–35°); **6** colluvial footslope; **7** alluvial toeslope; **8** channel wall; **9** channel bed.

generalized for the classification of the slope forms in a part of New Zealand. Instead they carried out a further subdivision of the slope into a total of nine units. The units they recognized were, in down-slope succession (Fig. 9.4):

1	interfluve	6	colluvial footslope
2	seepage slope	7	alluvial toeslope
3	convex creep slope	8	channel wall
4	fall face	9	channel bed.
5	transportational midslope		

Both by definition and by implication (Fig. 9.4) these terms involve slope processes as well as slope form. The application of the nine units to slope form description is illustrated by Figure 9.5, where three different types of landscape are described in profile form by reference to them. Not all the nine units occur in each case, and in some instances the same unit occurs more than once on the profile. Since each of the nine basic units is capable of numerical definition this system, as well as making the description of land form more precise, can provide the type of data which may be subjected to statistical analysis. As Dalrymple *et al.* (1968) imply,

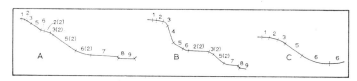

Fig. 9.5 The application of the nine-unit model of Figure 9.4 to three different landscapes (*after Dalrymple, Blong and Conacher, 1968*): **A**: South Auckland Greywacke land surface; **B**: Ignimbrite Valley with pumice aggradational surface **C**: South Auckland ash-mantled land surface.

however, it is not necessarily the case that in every area of the world only these nine units are to be found. In some areas it may be that additional units will require definition, while in others some of these will need to be replaced by other units.

Land-systems deal with whole regions and not just individual hillsides. Land-systems mapping has been developed in Australia by the Commonwealth Scientific and Industrial Research Organization. Land-systems define those areas within which there is a repeated pattern of land forms associated with specific soil and vegetation characteristics. They are frequently related to particular geological outcrops, and, if defined for very large areas, may also be related to climate. The block diagrams which accompany the description of the land-system always, by their very nature, concentrate on portraying the relief character of the land surface; they define the characteristic assemblage of land-form units rather than form a pictorial presentation of an actual portion of landscape. They are, therefore, visual analogue models. A typical block diagram of this type is shown in Figure 9.6. Different models are drawn for different regions, or land-systems, but within each land-system individual portions of country will bear a close resemblance to the general model. The land-form units of which the land-system is composed can be defined in numerical terms. For example, characteristic and limiting slope angles can be listed, together with the properties of the surface materials to be found on these slopes. Statistical analysis can be applied to this data, for example, in testing to see whether or not there is a significant difference between two land-systems.

The block-diagram form of the analogue model can be replaced by, or used in conjunction with, a spatial model. For example, morphological mapping can be used to define the extent of the major units of a hillside.

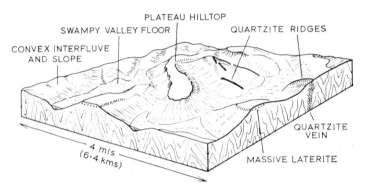

Fig. 9.6 The Masaka land-system, Uganda (*after Ollier, Lawrence, Webster and Beckett, 1967*).

Figure 9.7 is a morphological map, drawn in the field, of a slope which occurs in the lowland near Weymouth. Along the length of this slope the higher slopes are all gentle, predominantly of less than 3°. Likewise the lower slopes are gentle, being generally less than 4°, and concave in

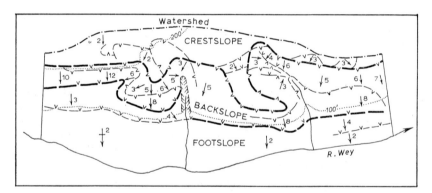

Fig. 9.7 Spatial model—a morphological map of part of the Weymouth lowland.

profile. Between these two sets of slopes the hillside displays a steeper portion which reaches a maximum of 12°. These three parts of the hillside correspond to the crestslope, backslope and footslope of Savigear. Changes in slope occur within each of the three main units. In other words, they are not as regular in plan as the model (Fig. 9.1) suggests. Most of the internal irregularities occur within the backslope, which has been incised

by spring and stream action, forming embayments in the line of the back-slope. This backslope is not associated with one particular rock type, nor can all of the changes of slope occurring within it be related to geological boundaries. It is, therefore, a product of denudation which appears to be retreating and consuming the hill crestslope. The crestslope also crosses geological boundaries and cuts across the dip of the Jurassic beds under-neath. It is thus a planation surface, attributed by Sparks (1952) to marine planation during the Pleistocene.

The major units of the analogue model can, therefore, be found through morphological mapping. At the same time this provides a state-ment of some of the details within each of these main units, and provides information which can be used for numerical analysis. Examples of the use of this type of data were provided in chapter 8. A different use for this information relates to the construction of a general model of slope form for a particular hillside through trend-surface analysis.

Trend-surface analysis is especially designed for spatial data. Its purpose is to detect both the amount and direction of any trend in the data values. For example, if an area contains varying values of drainage density, and it is suspected that these tend to decrease in value southwards, then trend-surface analysis may be used to find the general trend, both in direction and amount, of the drainage density values. The only in-formation required is the location of each observation, by means of co-ordinates on a grid system, and the drainage density value. Just as in linear regression a 'best-fit' line is computed for the points on a graph, so a 'best-fit' surface may be computed. This surface may be a plane, or a curved surface. The trend surface may be represented by equations of the following type, where U and V define the location of the point and Z is the variable, then, for a linear surface:

$$Z = a + b\text{U} + c\text{V}$$

the value of b is the slope of the surface in the U direction, and c is that in the V direction. This equation is very similar to that of the simple linear regression equation, the difference being that a cV element is added to take account of the third dimension. Just as a quadratic curve (Fig. 7.4B) may be used in regression analysis so a quadratic surface can be calculated in trend-surface analysis. In this case:

$$Z = a + b\text{U} + c\text{V} + d\text{U}^2 + e\text{UV} + f\text{V}^2$$

Higher-order surfaces can also be computed (see Chorley and Haggett, 1965), but the equations which define these become progressively more complex. The mathematics involved in calculating the equation of the trend surface lies beyond the scope of this book. In any case, when a large number of data points is being used it is necessary to perform the analysis with the aid of a computer.

Trend-surface analysis can also be used to generalize the form of a hillside to different, but known, degrees. Through the fitting of a linear

trend surface the complexities and details of the slope are reduced to a single best-fitting plane. It is likely that a quadratic trend surface will provide a better fit to the slope than a linear surface, for in the case of the former the mathematical surface has an increased flexibility with increased power. The quadratic surface can introduce one flexure, or curve, in the slope in its plan and another in its profile form. Similarly increasing the power of the surface to the cubic will give it even greater flexibility and allow a still closer fit to the slope. Ultimately, with no limit provided on the power of the surface being fitted it is likely that the trend

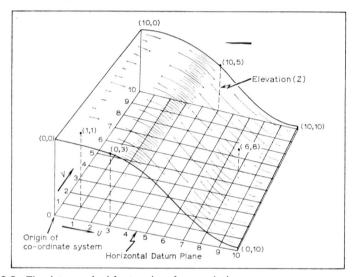

Fig. 9.8   The data required for trend-surface analysis.

surface and the actual hillside will have the same form. If this happens all that will be achieved is that the trend surface will have a mathematical equation too complex to understand, and the surface itself will be as complex as the original slope. In practice it is sufficient to fit trend surfaces up to the level of the cubic surface. Up to that level different trend surfaces result from different slopes and it is still possible to compare one slope with another, through the form of their trend surfaces.

The data required for trend-surface analysis consist of the coordinates that define the position of points on the hillside together with the height of each of these points above a common datum plane (Fig. 9.8). In practice the procedure, which is illustrated by Figure 9.9, is as follows:

1   A morphological map is produced in the field and checked by means of slope profiling
2   A grid is drawn across the surface of the morphological map to cover the portion of hillside for which a trend-surface map is required (Fig. 9.9A)

Table 9.2 A slope conversion table whereby angle of slope and horizontal distance (in feet) can be used to find the increase in altitude along a slope

Horizontal distance (feet)

Angle of slope (degrees)	2	4	6	8	10	12	14	16	18	20	22	24
1	0·0	0·1	0·1	0·1	0·2	0·2	0·2	0·3	0·3	0·3	0·4	0·4
2	0·1	0·1	0·2	0·3	0·3	0·4	0·5	0·6	0·6	0·7	0·8	0·8
3	0·1	0·2	0·3	0·4	0·5	0·6	0·7	0·8	0·9	1·0	1·2	1·3
4	0·1	0·3	0·4	0·6	0·7	0·8	1·0	1·1	1·3	1·4	1·5	1·7
5	0·2	0·3	0·5	0·7	0·9	1·0	1·2	1·4	1·6	1·7	1·9	2·1
6	0·2	0·4	0·6	0·8	1·1	1·3	1·5	1·7	1·9	2·1	2·3	2·5
7	0·2	0·5	0·7	1·0	1·2	1·5	1·7	2·0	2·2	2·5	2·7	2·9
8	0·3	0·6	0·8	1·1	1·4	1·7	2·0	2·2	2·5	2·8	3·1	3·4
9	0·3	0·6	1·0	1·3	1·6	1·9	2·2	2·5	2·9	3·2	3·5	3·8
10	0·4	0·7	1·1	1·4	1·8	2·1	2·5	2·8	3·2	3·5	3·9	4·2
11	0·4	0·8	1·2	1·6	1·9	2·3	2·7	3·1	3·5	3·9	4·3	4·7
12	0·4	0·•9	1·3	1·7	2·1	2·6	3·0	3·4	3·8	4·3	4·7	5·1
13	0·5	0·9	1·4	1·8	2·3	2·8	3·2	3·7	4·2	4·6	5·1	5·5
14	0·5	1·0	1·5	2·0	2·5	3·0	3·5	4·0	4·5	5·0	5·5	6·0
15	0·5	1·1	1·6	2·1	2·7	3·2	3·8	4·3	4·8	5·4	5·9	6·4
16	0·6	1·1	1·7	2·3	2·9	3·4	4·0	4·6	5·2	5·7	6·3	6·9
17	0·6	1·2	1·8	2·4	3·1	3·7	4·3	4·9	5·5	6·1	6·7	7·3
18	0·6	1·3	1·9	2·6	3·2	3·9	4·5	5·2	5·8	6·5	7·1	7·8
19	0·7	1·4	2·1	2·8	3·4	4·1	4·8	5·5	6·2	6·9	7·6	8·3
20	0·7	1·5	2·2	2·9	3·6	4·4	5·1	5·8	6·6	7·3	8·0	8·7
21	0·8	1·5	2·3	3·1	3·8	4·6	5·4	6·1	6·9	7·7	8·4	9·2
22	0·8	1·6	2·4	3·2	4·0	4·8	5·7	6·5	7·3	8·1	8·9	9·7
23	0·8	1·7	2·5	3·4	4·2	5·1	5·9	6·8	7·6	8·5	9·3	10·2
24	0·9	1·8	2·7	3·6	4·5	5·3	6·2	7·1	8·0	8·9	9·8	10·7
25	0·9	1·9	2·8	3·7	4·7	5·6	6·5	7·5	8·4	9·3	10·3	11·2
26	1·0	2·0	2·9	3·9	4·9	5·9	6·8	7·8	8·8	9·8	10·7	11·7
27	1·0	2·0	3·1	4·1	5·1	6·1	7·1	8·2	9·2	10·2	11·2	12·2
28	1·1	2·1	3·2	4·3	5·3	6·4	7·4	8·5	9·6	10·6	11·7	12·8
29	1·1	2·2	3·3	4·4	5·5	6·7	7·8	8·9	10·0	11·1	12·2	13·3
30	1·2	2·3	3·5	4·6	5·8	6·9	8·1	9·2	10·4	11·5	12·7	13·9
31	1·2	2·4	3·6	4·8	6·0	7·2	8·4	9·6	10·8	12·0	13·2	14·4
32	1·2	2·5	3·7	5·0	6·2	7·5	8·7	10·0	11·2	12·5	13·7	15·0
33	1·3	2·6	3·9	5·2	6·5	7·8	9·1	10·4	11·7	13·0	14·3	15·6
34	1·3	2·7	4·0	5·4	6·7	8·1	9·4	10·8	12·1	13·5	14·8	16·2
35	1·4	2·8	4·2	5·6	7·0	8·4	9·8	11·2	12·6	14·0	15·4	16·8
36	1·5	2·9	4·4	5·8	7·3	8·7	10·2	11·6	13·1	14·5	16·0	17·4
37	1·5	3·0	4·5	6·0	7·5	9·0	10·5	12·1	13·6	15·1	16·6	18·1
38	1·6	3·1	4·7	6·3	7·8	9·4	10·9	12·5	14·1	15·6	17·2	18·8
39	1·6	3·2	4·9	6·5	8·1	9·7	11·3	13·0	14·6	16·2	17·8	19·4
40	1·7	3·4	5·0	6·7	8·4	10·1	11·7	13·4	15·1	16·8	18·5	20·1

Matrix of increase in altitude (feet)

Horizontal distance (feet)

26	28	30	32	34	36	38	40	42	44	46	48	50	
0·5	0·5	0·5	0·6	0·6	0·6	0·7	0 7	0·7	0·8	0·8	0·8	0·9	1
0·9	1·0	1·0	1·1	1·2	1·3	1·3	1·4	1·5	1·5	1·6	1·7	1·7	2
1·4	1·5	1·6	1·7	1·8	1·9	2·0	2·1	2·2	2·3	2·4	2·5	2·6	3
1·8	2·0	2·1	2·2	2·4	2·5	2·7	2·8	2·9	3·1	3·2	3·4	3·5	4
2·3	2·4	2·6	2·8	3·0	3·1	3·3	3·5	3·7	3·8	4·0	4·2	4·4	5
2·7	2·9	3·2	3·4	3·6	3·8	4·0	4·2	4·4	4·6	4·8	5·0	5·3	6
3·2	3·4	3·7	3·9	4·2	4·4	4·7	4·9	5·2	5·4	5·6	5·9	6·1	7
3·7	3·9	4·2	4·5	4·8	5·1	5·3	5·6	5·9	6·2	6·5	6·7	7·0	8
4·1	4·4	4·8	5·1	5·4	5·7	6·0	6·3	6·7	7·0	7·3	7·6	7·9	9
4·6	4·9	5·3	5·6	6·0	6·3	6·7	7·1	7·4	7·8	8·1	8·5	8·8	10
5·1	5·4	5·8	6·2	6·6	7·0	7·4	7·8	8·2	8·6	8·9	9·3	9·7	11
5·5	6·0	6·4	6·8	7·2	7·7	8·1	8·5	8·9	9·4	9·8	10·2	10·6	12
6·0	6·5	6·9	7·4	7·8	8·3	8·8	9·2	9·7	10·2	10·6	11·1	11·5	13
6·5	7·0	7·5	8·0	8·5	9·0	9·5	10·0	10·5	11·0	11·5	12·0	12·5	14
7·0	7·5	8·0	8·6	9·1	9·6	10·2	10·7	11·3	11·8	12·3	12·9	13·4	15
7·5	8·0	8·6	9·2	9·7	10·3	10·9	11·5	12·0	12·6	13·2	13·8	14·3	16
7·9	8·6	9·2	9·8	10·4	11·0	11·6	12·2	12·8	13·5	14·1	14·7	15·3	17
8·4	9·1	9·7	10·4	11·0	11·7	12·3	13·0	13·6	14·3	14·9	15·6	16·2	18
9·0	9·6	10·3	11·0	11·7	12·4	13·1	13·8	14·5	15·2	15·8	16·5	17·2	19
9·5	10·2	10·9	11·6	12·4	13·1	13·8	14·6	15·3	16·0	16·7	17·5	18·2	20
10·0	10·7	11·5	12·3	13·1	13·8	14·6	15·4	16·1	16·9	17·7	18·4	19·2	21
10·5	11·3	12·1	12·9	13·7	14·5	15·4	16·2	17·0	17·8	18·6	19·4	20·2	22
11·0	11·9	12·7	13·6	14·4	15·3	16·1	17·0	17·8	18·7	19·5	20·4	21·2	23
11·6	12·5	13·4	14·2	15·1	16·0	16·9	17·8	18·7	19·6	20·5	21·4	22·3	24
12·1	13·1	14·0	14·9	15·9	16·8	17·7	18·7	19·6	20·5	21·5	22·4	23·3	25
12·7	13·7	14·6	15·6	16·6	17·6	18·5	19·5	20·5	21·5	22·4	23·4	24·4	26
13·2	14·3	15·3	16·3	17·3	18·3	19·4	20·4	21·4	22·4	23·4	24·5	25·5	27
13·8	14·9	16·0	17·0	18·1	19·1	20·2	21·3	**22·3**	23·4	24·5	**25·5**	26·6	28
14·4	15·5	16·6	17·7	18·8	20·0	21·1	22·2	23·3	24·4	25·5	26·6	27·7	29
15·0	16·2	17·3	18·5	19·6	20·8	21·9	23·1	24·2	25·4	26·6	27·7	28·9	30
15·6	16·8	18·0	19·2	20·4	21 6	22·8	24·0	25·2	26·4	27·6	28·8	30·0	31
16·2	17·5	18·7	20·0	21·2	22·5	23·7	25·0	26·2	27·5	28·7	30·0	31·2	32
16·9	18·2	19·5	20·8	22·1	23·4	24·7	26·0	27·3	28·6	29·9	31·2	32·5	33
17·5	18·9	20·2	21·6	22·9	24·3	25·6	27·0	28·3	29·7	31·0	32·4	33·7	34
18·2	19·6	21·0	22·4	23·8	25·2	26·6	28·0	29·4	30·8	32·2	33·6	35·0	35
18·9	20·3	21·8	23·2	24·7	26·2	27·6	29·1	30·5	32·0	33·4	34·9	36·3	36
19·6	21·1	22·6	24·1	25·6	27·1	28·6	30·1	31·6	33·2	34·7	36·2	37·7	37
20·3	21·9	23·4	25·0	26·6	28·1	29·7	31·3	32·8	34·4	35·9	37·5	39·1	38
21·1	22·7	24·3	25·9	27·5	29·2	20·8	32·4	34·0	35·6	37·3	38·9	40·5	39
21·8	23·5	25·2	26·9	28·5	30·2	31·9	33·6	35·2	36·9	38·6	40·3	42·0	40

Angle of slope (degrees)

Matrix of increase in altitude (feet)

3   The height is calculated for every grid intersection with the base line with the lowest point designated as zero (Fig. 9.9B)

4   The height is then calculated for every intersection within the grid above the lowest point on the base line. This may be done by measuring the horizontal distance from the base of the slope to the grid intersection in question, and converting this to a height value by multiplying the horizontal distance by the tangent of the angle of slope (Table 9.2 and Fig. 9.9B)

5   The coordinates of each point are listed together with the appropriate height value

6   Trend-surface analysis is carried out.

The actual form of the trend surface will depend on criteria which can be turned to advantage in comparing hillsides. For example, the number of points can be increased or decreased by changing the mesh of the grid. This allows the investigator to filter through either more or less detail (Nettleton, 1954). When a direct comparison is to be made between the equations of a trend surface for two slopes it is important that the ground distance between the points is the same in each case and that the same number of points (or nearly so) is being used. If however, the intention is to compare trend-surface maps of the two slopes, without reference to scale, then it is not necessary to maintain the same ground distance between points in each case. In fact, controlling the difference between the ground distances allows a comparison to be made between slopes which exist at two quite distinct scales (Fig. 9.11, see below).

Trend-surface analysis is only feasible through the use of a computer. The computer may be used to provide the trend-surface map as well as the coefficients of the trend-surface equations, but care has to be taken that the computer-printed map does not distort the shape of the trend surface when compared with the area for which it is being compiled. This can be allowed for by knowing in advance the dimensions of the printed map. In the examples described below the printed map's margins had a ratio to each other of 10 : 8·5: knowing this, the grids were constructed with this same ratio between the lengths of their margins. In erecting the grid across the morphological map it is important that the U-axis (Fig. 9.9) is aligned parallel to the direction of steepest slope. If these conditions are followed in all cases the resulting trend surfaces are comparable with each other and will reflect differences in the surface form of the slopes.

An example of the compilation and use of a trend-surface map is given in Figure 9.9. The morphological map here is of a part of an escarpment developed in alternating resistant sandstone and less resistant shale beds, near Sheffield. The scale of the map is indicated by the scale of the spacing between the grid squares drawn across the morphological map. The values given in Figure 9.9B are the heights, to the nearest foot above the lower right-hand corner of the grid network, of each grid intersection. Information was supplied to the computer concerning the grid location and height

of each grid intersection and a linear trend-surface map was produced (Fig. 9.9c). The equation for this surface is:

$$Z = 287 \cdot 71 - 19 \cdot 33U - 7 \cdot 7V$$

which means that when U = o and V = o—i.e. at the origin of the data grid—the trend surface stands 287·71 feet above the lowest point on the slope, at (10, 10). The negative sign attached to the coefficient of U indicates that the surface slopes from the top of the grid down towards the bottom. This will be a feature common to all trend-surface slope models constructed in this way, for one of the conditions is that the base

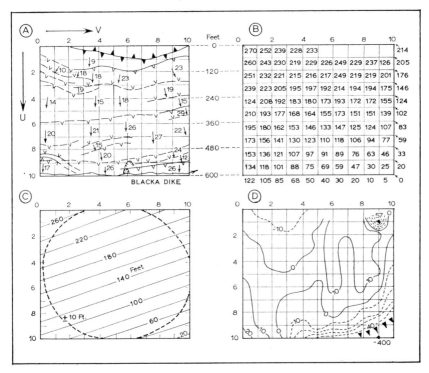

Fig. 9.9 Collecting data for a trend-surface analysis of slope form.

line of the grid should always follow the base of the slope, or portion of slope, being examined. The negative sign of the coefficient of V indicates that when U = o increasing values of V result in a decrease in the height of the slope, Z. What this means in terms of Figure 9.9c is that the 'contours' on the trend surface show a 'V' shape upstream. This would normally be expected, but there are instances where slopes do not result in this pattern. For example, if there is a marked spur running down the slope somewhere between columns 5 and 10 then the resulting trend-surface 'contours' can be inclined as though 'V-ing' downstream. This only serves

to show how even at the level of the linear trend surface differences between slopes are apparent.

An advantage of trend-surface analysis lies in the fact that it presents the actual coefficients of the equation relating height to position on the slope. In this case (Fig. 9.9), the coefficient of U defines the gradient of the slope down the U-axis, and the gradient along the V-axis is defined by the coefficient of V. The grid spacing of Figure 9.9A is 60 feet, so that one unit of U = 60 feet in horizontal distance. In other words, for every 60 feet horizontal distance there is a fall of the slope, in general, of 19·33 feet along a direction parallel to the U-axis. This is equivalent to a drop of about 32·2 feet for every 100 feet (horizontal distance). The fall along the V-axis can be found in a similar manner. Neither in fact represents the true slope of the trend surface, and therefore the general gradient of the hillside in question. This can be calculated by drawing a vector diagram, as shown in Figure 9.10, or by measuring this gradient direct from the trend-surface map. In either case the general gradient of the

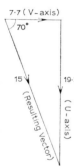

Fig. 9.10 Vector diagram to calculate true slope from the coefficients of a linear trend surface.

slope is almost 30 feet in 100 feet, at an orientation of 70 degrees to the V-axis.

The same coefficients would have been found if the data had been subjected to a multiple regression analysis of Z against U and V. These coefficients could then have been used in the form of a vector diagram (Fig. 9.10) and both the direction and amount of general slope could have been calculated. It is not possible, however, to do this with the same advantage for the higher-order trend surfaces. An addition provided by the computer program (written by M. J. McCullagh) used for Figure 9.9 is that it also provides a map of the *confidence surface* relating to individual trend surfaces. The confidence surface (at the 95% level) is shown on Figure 9.9C by a single 'contour' marked ±10 feet. Within this circle all of the calculated values for the height of the trend surface are likely to be within 10 feet of the actual ground values. Outside that circle the trend surface is less reliable, but is within ±20 feet. In this case the reliability is the same both for the high values in the top left-hand corner (e.g. 240 feet ±10 feet) of Figure 9.9C, and for the low values in the bottom right-hand

corner (e.g. 60 feet $\pm$ 10 feet). This implies that the trend surface actually is a better fit in the top left-hand corner as the likely error (in percentage terms) in predicting an unknown value there is less.

One advantage of the trend-surface output as compared with ordinary multiple regression, as referred to above, is that the end-product is a map which provides a visual impression of the general behaviour of the hillside in three dimensions. This visual impression is stronger when higher-order surfaces are examined and curvature is introduced into the plan of the trend-surface 'contours'.

In many studies, however, the most valuable results have come through an analysis of the *residual values* for each data point. These residuals are the differences between the actual height of the ground and the height it would have had if the point lay on the trend surface. These residual values may be expressed either as the difference in height in feet, or as a percentage difference between the observed and the calculated values. In the example (Fig. 9.9D) the percentage values have been used, and when plotted in the form of a map they bring to light the fact that the bottom right-hand corner of the hillside is much lower than the trend surface suggests it ought to be. Re-examination of the area shows that the stream downstream of the point $V = 4$, $U = 10$ passes over a knick point and into a more deeply incised section. This incision has led to the development of a short, but steep, lowest unit to the hillside. In addition, whereas the slopes higher up the hillside are formed within sandstone, at this bottom corner the stream has cut into shales that have allowed this relatively deep local incision.

Trend-surface analysis in itself, therefore, enables a description to be made of the general behaviour of a hillside. The analysis of residuals is a means by which it may be used as a search technique for finding those localities most different from the general trend in the area of the measured data. In other words it allows a selection of the 'odd-men-out' among the data values.

The use of quadratic trend surfaces for comparing the form of hillsides is illustrated in Figure 9.11. In examples A and B the data matrix was the same size in each case with respect to the horizontal distances involved. Example C is of a much smaller area. In each case, however, both hill crest and valley floor are included. The first two examples are drawn from the Jurassic lowlands west of Weymouth while example C comes from an area of Carboniferous shales and sandstone near Sheffield. A comparison of examples A and B (Fig. 9.11) shows that although they are both generally concave in profile they differ in plan. Example A tends to be in the form of a cove or embayment, while example B tends towards the form of a spur, especially in the lower part of the slope. In neither case, however, is the curvature very great. Nevertheless it is clear that in this one respect the two slopes are slightly different. Several questions still need to be asked before this matter can be taken further. How reliably do the trend surfaces reflect the form of the hillsides? The confidence

surfaces show that the reliability of a predicted value of ground height from the trend surface would be greater in example B than in example A. In neither case, however, would there be a very high chance of the error being greater than − 10 feet. When the sum of the squares of the residuals is examined and a coefficient of determination (R^2) calculated, as in regression analysis (section 3.6), both surfaces appear to fit the original data very well. Example A has an R^2 value of 96·3%, while example B has an R^2 value of 98·7%. These values are highly significant: in both

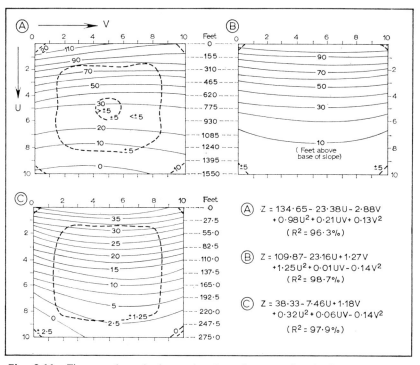

Fig. 9.11   The use of quadratic trend surfaces in comparing the form of hillsides, (see text)

instances, the quadratic trend surface appears to come very close to describing the general form of the two hillsides (the same may also be said for example C).

An important question relates to the meaning and interpretation of the coefficients of the equations which describe these surfaces. These equations are much more difficult to interpret than in the linear case. The first coefficient in each equation is the value of Z when U = 0 and V = 0. The negative coefficient of the U value indicates that the higher values of Z are associated with the lower values of U, i.e. the top of the data matrix. This again will apply in all those instances when the base of the

data matrix is made to coincide with the foot of the hillside. A negative coefficient for V suggests that the 'contours' of the trend surface turn downwards as they approach the U-axis (Fig. 9.11A); conversely a positive coefficient for V suggests that these 'contours' turn upwards as the U-axis is approached (Fig. 9.11B). In the former case the quadratic trend surface must tend towards a cove (embayment) form. Not only does this appear to occur when the V coefficient is negative, it seems that there is also a positive V^2 coefficient (example A) in this situation. The spur form of example B not only has a positive coefficient for V, but it has a negative coefficient for V^2. The same combination occurs with the spur form of example C. All three examples have a concave profile form, and this is reflected in the positive coefficient for U^2. Whether these relationships between slope form and the sign of the coefficients hold true in all cases has not yet been fully tested.

Example C (Fig. 9.11), although from a different area and developed on a different set of rocks, is very similar in form to example B. The use of trend-surface analysis has made possible their direct visual comparison despite the fact that the two slopes are quite different in size. The difference between the first coefficient in each case (109·87 in example B and 38·33 in example C) indicates at once that the crest of slope C is not as high as that of slope B. On the other hand, converting the coefficient for U by the scale shows that, in a direction parallel to the U-axis, slope C is nearly twice as steep as slope B. When the size of two original data matrices is different, in terms of ground distances, then the absolute value of their coefficients cannot be directly compared. Regardless of scale, however, the sequence of signs associated with the coefficients is the same in each case. Further research must be directed towards assessing the value of comparing slopes in this way. Potentially it seems that by trend-surface analysis the main characteristics of the gross form of a slope can be found and, possibly, used as a basis for slope classification.

The trend-surface technique described above uses power series polynomials which have the limitation that the coefficients for low orders of U and V change when higher-order polynomials are added to the series. The example given in Figure 9.9 whose linear surface is defined by:

$$Z = 287 \cdot 71 - 19 \cdot 33U - 7 \cdot 7V$$

has for its quadratic surface the equation:

$$Z = 252 \cdot 16 - 7 \cdot 95U - 1 \cdot 44V - 0 \cdot 69U^2 - 0 \cdot 89UV - 0 \cdot 18V^2$$

Thus, adding the quadratic terms (U^2, UV and V^2) has changed the value of the first three coefficients. This limitation may be overcome by the use of orthogonal polynomials. The coefficients of orthogonal polynomials may provide a better statement concerning surface form than those of the power series polynomials. Preston (1966) develops a further possibility which is based on the concept that the surface of a hillside, for example, can be considered as a continuous random variable whose value at any point in space is described by a probability distribution. In this

sense it is similar to time-series analysis. Much more work needs to be done with these techniques, however, before their potential value will be fully appreciated.

Contour maps can also be used for deriving equations which express the nature of a hillside, or portion of a hillside. For example, Troeh (1965) develops the concept of fitting mathematical equations to specific land forms by reference to an alluvial cone. This is a feature in which there is a tendency for regularity of form to be present, with a similarity in various directions radiating from the uppermost tip of the cone. The typical form of an alluvial cone has contours which are convex outwards, but a slope profile which is concave towards the base. Troeh describes the nature of the equations that can be used to define the form of alluvial cones. On a geometrically right circular cone, with similarity in all radial directions:

$$Z = P + ST$$

where Z is the elevation at any point on the surface of the cone; P is the elevation of the central point of the cone; S is the slope gradient of the sides of the cone; and T is the radial distance from P to the point whose elevation is Z. An alluvial cone differs from the right circular cone in that its slope profile has a concave curvature. To accommodate this a T^2 term needs to be added to the above equation:

$$Z = P + SR + LT^2$$

The meaning of L can be arrived at by calculating, through *differentiation* (section 7.1) the equation for the gradient of the slope at any point:

$$\frac{dZ}{dT} = S + 2LT$$

if dZ/dT, the gradient, is replaced by the letter G, then:

$$G = S + 2LT$$

In other words, the rate of change of slope gradient with radial distance is 2L. (NB. Compare the above equation with the general form $Y = a + bX$ where b is the gradient of the line.) From these equations the slope gradient (G) can be calculated for any point. The radius R indicates the curvature of the contours, and the curvature of the slope profile is controlled by the coefficient L. Troeh then describes an application of these equations to a specific example, a pediment form near Gila Butte, Arizona. He arrives at the equation:

$$Z = 1078 \text{ feet} - (102 \cdot 8 \text{ feet/mile})T + (5 \cdot 06 \text{ feet/mile}^2)T^2$$

This describes the pediment as a parabolic cone, which (by placing $T = 0$) has an origin at 1708 feet. It has an initial slope of 102·8 feet/mile, the negative sign implying that it is downhill, and a rate of decrease of slope of $2 \times 5 \cdot 06 = 10 \cdot 12$ feet/mile for each radial mile. Testing this equation against data for the slope shows that discrepancies seldom exceed 10 feet (for calculations of Z) and on average are only 3–4 feet, compared with a total relative relief of 400 feet.

9.2 Mathematical models

Theoretical models of slope development have been proposed by Bakker and Le Heux (1946), Scheidegger (1961), and Culling (1965) amongst others. These fall outside the scope of this book for it is our contention that slope studies and development theories should be related to field measurement and not solely to theoretical abstraction. This is not to deny the value of the theoretical approach, for, if nothing else, it is a means whereby it is possible to pause and think, and thus to find stimulus.

The more valuable mathematical models are the deductive models of the type discussed by Young (1963). These deductive models are based on field observation of slope process and the measurement of slope form,

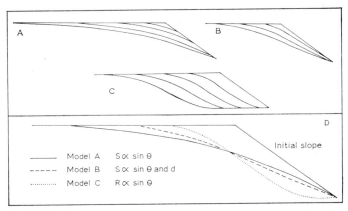

Fig. 9.12 Mathematical models—some slope development case studies (after Young, 1963, see text). (θ is the angle of slope and d the distance from the crest.)

and it is from these that deductive reasoning is pursued. Young considers the development of slopes under two quite distinct conditions, namely:

1 The retreat of a rectilinear slope when there is no erosion at its base
2 The development of a slope when vertical river erosion occurs at its foot, either continuously or at staggered intervals.

Consideration is also given by Young to the transport of waste material across these slopes. For example, in the first of 16 different models Young considers the development of a slope on which denudation is caused by the down-slope transport of soil rather than through its direct removal. Young assumes here that the rate of removal of material is proportional to the sine of the angle of slope (Fig. 9.12A). At first there is a rounding of the initial break of slope which extends progressively to the base of the slope. Following this, the angle of the lowest part of the slope decreases, but no concavity is developed. Progressive retreat of the upper limit of

the convexity takes place into the level ground above the slope. Thus the evolution of this slope results from slope decline, without the development of a concavity, and through the curvature of the convexity, which ultimately is long and smoothly curved, progressively decreasing as time goes on. In another model Young considers the development of a hillside in which the rate of transport of material across the slope is a function not only of the sine of the angle of slope but also of the distance from the crest (Fig. 9.12B). Yet another type of slope results when material is removed direct from the slope and the rate of removal is related to the sine of the angle of slope (Fig. 9.12C). For comparative purposes the form of these three slopes at an equivalent stage of their development can be placed on one diagram (Fig. 9.12D). This shows that the two slopes which developed by the down-slope transport of material have declined in steepness, and differ from each other in that model A shows a greater length to the crest convexity, while the other (B) has a lower concavity. Both of these slopes are different from the one (C) which has developed by the direct removal of material for this is only slightly gentler in angle than the initial slope, and it has a strongly developed basal concavity.

The study of models similar to these leads Young to the following conclusions:

1   Slope decline tends to be caused by processes that involve the down-slope transport of soil
2   Processes involving the direct removal of material from the slope tend to cause parallel retreat
3   The case of direct removal is equivalent to that in which the rate of weathering is the limiting factor in slope retreat. Consequently where weathering controls slope evolution, parallel retreat will occur; where transport is the limiting factor, slope decline will result.
4   A long and smoothly curved convexity is formed by processes of down-slope transport, particularly when the rate of transport varies only with slope angle
5   When river erosion takes place at the foot of a slope it is usually steepened until it reaches an angle at which rapid mass-movement occurs. An exception to this is possible if the rate of transport increases with distance from the crest, in which case the angle of the foot of the slope is related to the relative rates of river erosion and down-slope transport.
6   Where there have been two major periods of river incision (epicycles of erosion) evidence of the first epicycle will be distinguishable in the form of the hillside only when the duration of the second has been less than half that of the first
7   Where relative relief exceeds 300 feet, processes of direct removal are relatively more efficient in producing slope retreat than processes of down-slope transport.

In producing his deductive models Young proceeds from field observation
to his mathematical models. In this way a sense of reality is maintained
in examining the models produced.

9.3 Experimental design models

Experimental design models result from the type of numerical analysis
discussed in chapter 8. These models are based on the assumption that
within a given range of data there exist certain meaningful component
parts (Chorley, 1967). Regression equations express the relationship
between the variables of an experimental design model. Through such
analysis it becomes possible, as has already been shown, to establish the
general trend within the data and to separate out local variation from the
general trend. In terms of angle of slope, the relationships described in
chapter 8, and defined there in terms of regression equations (either
simple or multiple) represent experimental design models. Other experimental design models are to be found in other sections of this book, as
for example in the analysis of drainage basins (part I) which concluded
with an experimental design model for the third-order drainage basins
of south-west Uganda.

9.4 The use of models

One of the great advantages of having an analogue model is that it provides a quick way of summarizing the general form of a hillside, or a
landscape in the case of land-systems, from direct field observation. An
ancillary use of analogue models is their provision of a ready means of
comparing newly observed slopes with models developed for the slopes
already observed. Likewise it is possible to compare newly observed slopes
with the analogue model of the 'normal' hillside shown in Figure 9.1.
Such a comparison makes for a rapid description of the hillside being
observed simply by pointing out its similarity to, or its differences from,
the analogue model. In a sense the analogue model of Figure 9.1 is a
synthetic model because it does not represent the form of an actual hillside. More complex models are provided, for example, by the 'nine-unit
model' approach (Dalrymple et al., 1968). This provides an objective
classification of the surface with respect to recognizable and defined
components. It consists of a rigid definition of each of the individual
units, but it is also flexible in that any suitable combination of units is
possible, depending entirely on the nature of the hillside.

Both the nine-unit model and the land-systems approach provide
models which can be used for storing information about slopes. For example, for any land-system there are a defined and limited number of
land units. Each of these units can be classified in terms of its limiting

slope angles, its position with respect to other units, and the proportion of space it occupies within the land-system. These values can be recorded not only for the whole of one land-system but also for different parts of it. This information can be stored in a computerized slope data bank, and may subsequently be recalled for a variety of uses. Further information can be added to the bank as it becomes available. This may include information about the regolith cover on each of the land units, or its drainage properties. The data store can also be used for comparing land-systems with each other, not only for one country or continent but also for areas which are much further apart. Again, it is possible to search through the data bank for a land-system with particular characteristics by supplying a list of the required properties to the computer and leaving it to sort through the data store to find the land-system which has characteristics most similar to those being sought.

Slope form classification and objective generalization become possible through the use of trend-surface analysis. Comparisons between different slopes, at different scales, can also be made by the use of this technique. In addition, through an examination of the residuals of the individual data points, trend-surface analysis may also be used as a search technique to distinguish those portions of a slope which differ most widely from the general trend of the slope as a whole. This technique of trend-surface analysis is dependent on a spatial model already having been established in the form of either a morphological map, or alternatively a very detailed contour map. Through morphological mapping, however, the investigator has the most valuable technique available for seeing what is actually present in a landscape. It is a technique by which he may become familiar with the ground, and at the same time he is supplied with a spatial model and a source from which many measured variables may be drawn.

The establishment of a mathematical model is a salutary exercise in learning how to think about the general implications of what has been observed. At its best it provides a deductive model which becomes more than simply a hypothesis, for it is related to field measurement and therefore has already been tested in at least one area. Alternatively a mathematical model may be established and used as a starting point, being based on theoretical reasoning, and leading to a hypothesis for subsequent testing in the field.

Experimental design models are established through the analysis of numerical data. If the reliability of the data is known, statistical methods of regression analysis make possible a determination of the relationship between various aspects of slope form, and in addition enable an analysis to be made of the relationship between slope characteristics and processes. This is a field of numerical analysis which is continually being developed, not so much in terms of the statistical techniques used but in terms of refining methods of obtaining accurate information about slope processes. Once this information has been obtained, however, it is through numerical

analysis that the geomorphologist will be led to a clear understanding of slope form and slope development.

References

BAKKER, J. P. and LE HEUX, J. W. N. 1946: Projective-geometric treatment of O. Lehmann's theory of the transformation of steep mountain slopes. *Proc. Koninklijke Nederlandsche Akademic van Wetenschappen* **49**(5), 533–47.

CHORLEY, R. J. 1967: Models in geomorphology. In Chorley, R. J. and Haggett, editors, *Physical and information models in geography*, London: Methuen, 59–96.

CHORLEY, R. J. and HAGGETT, P. 1965: Trend-surface mapping in geographical research. *Trans. Inst. Brit. Geogr.* **37**, 47–67.

CULLING, W. E. H. 1965: Theory of erosion on soil-covered slopes. *J. Geol.* **73**, 230–54.

DALRYMPLE, J. B., BLONG, R. J. and CONACHER, A. J. 1968: A hypothetical nine unit land surface model. *Zeit. für Geomorph.* NF **12**, 60–76.

KING, L. C. 1953: Canons of landscape evolution. *Bull. Geol. Soc. Am.* **64**, 721.

KING, L. C. and FAIR, T. J. D. 1944: Hillslopes and dongas. *Trans. Geol. Soc. S. Africa* **47**,

NETTLETON, L. L. 1954: Regionals, residuals, and structures. *Geophysics* **19**, 1–22.

OLLIER, C. D., LAWRANCE, C. J., WEBSTER, R. and BECKETT, P. H. T. 1967: The preparation of a land classification map at 1 : 1,000,000 of Uganda. *Actes du IIᵉ Symposium Internat. de Photo-Interpretation, Paris, 1966* **IV**-1, 115–22.

PRESTON, F. W. 1966: Two-dimensional power spectra for classification of landforms. In Merriam, D. F., editor, Computer application in the earth sciences: colloquium on classification procedures, *Univ. of Kansas, State Geol. Survey, Computer Contribution* **7**, 64–9.

SAVIGEAR, R. A. G. 1960: Slopes and hills in West Africa. *Zeit. für Geomorph.* Suppl. **1**: Morphologie des versants, 156–71.

SCHEIDEGGER, A. E. 1961: *Theoretical geomorphology.* Berlin: Spring-Verlag.

SCHUMM, S. A. 1966: The development and evolution of hillslopes. *J. Geol. Education* **14**(3), 98–104.

SPARKS, B. W. 1952: Stages in the physical evolution of the Weymouth Lowland. *Trans. Inst. Brit. Geogr.* **18**, 17–29.

TROEH, F. R. 1965: Landform equations fitted to contour maps. *Am. J. Sci.* **263**, 616–27.

WOOD, A. 1942: The development of hillside slopes. *Proc. Geol. Assoc.* **53**, 128–40.

YOUNG, A. 1963: Deductive models of slope evolution. *Nachrichten der Akad. der Wissenschaften in Göttingen, II Mathematisch-Physikalische Klasse* **5**, 45–66.

Part III Coastal forms

10 Wave action

10.1 Wave refraction
[*vector methods, orientation data (mean, standard deviation and dispersion), area under the normal curve for circular normal distribution, F-values, Spearman rank correlation test, t-test*]
10.2 The beach form in profile
[*chi-square test, null hypothesis, degrees of freedom*]
10.3 Beach gradient
[*product-moment correlation, partial correlation, log normalization*]
10.4 Beach ridges
[*time series, regression analysis, trend equations, autocorrelation*]

The purpose of part III is to illustrate the application of numerical techniques to the *form* of coastlines. Earlier parts have dealt with the form of drainage basins and hillsides respectively. In these cases a part of the problem of numerical analysis has been that of measuring and defining the form to be analysed. In the case of coastal studies the measurement of form is frequently not the basic problem. Standard surveying techniques are usually adequate for its measurement. The main interest lies in relating the measured form to environmental influences, especially coastal processes.

There are three important dynamic processes which affect beach form: these are wave action, wind and tide. Of these the most important is wave action. This chapter illustrates the application of numerical techniques by considering the influence of wave action on the form of the beach in profile, beach gradient, the formation of beach ridges and the nature of beach material. To begin with, however, consideration is given to the effect of wave refraction with respect to the orientation of a coastline and its offshore characteristics.

10.1 Wave refraction

The aspect of a beach is important in that it determines the angle of approach of waves towards the shore. On any one coastline, however, aspect must be considered alongside the general outline of the coast. A crenulate and a smooth coast will respond in different ways to the same waves. The actual influence of the waves will also be related to the nature of the waves themselves. For example, there is a difference between sea and swell. The latter consists of the longer, lower waves which have travelled from a more distant source. These waves differ from storm waves in

that they are not associated with the wind at their point of impact with the coast. Storm waves, on the other hand, are being actively generated by strong winds and are thus directly related to the winds blowing at the time the waves reach the coast. The actual effect of these different wave types depends on their dimensions, especially their steepness or height to length ratio, on the time interval between waves (the period) and on their direction of approach. An additional influence on the effect of waves on the shore is the relief character of the offshore zone, for this directly influences the depth of the water, which, in turn, influences the refraction of the waves. The wave steepness determines to a considerable extent the response of a particular beach to the approaching waves. Wave length decreases as the water becomes shallower.

The relationship between beach form in plan and the nature of the offshore area can be illustrated by an analysis of beaches on the west coast of Ireland between Dingle Bay in the south and Donegal in the north (Fig. 10.1). This is an indented coast with long, deep bays in the south and stretches of crenulate coast farther north. The data to be analysed consist of:

1   the orientation of the beaches
2   the orientation of the 10 fathom (approx 20 m) depth contour
3   the gradient between the beach and this depth contour
4   the exposure of the beach, as measured by the angle, in degrees, between the orientation of the beach and the direction from which the longest swells and most effective waves may be expected to come. This direction of dominant wave approach is 225°, that is to say, the south-west, for this case study.

Three of these variables (1, 2 and 4) are directional in character. They are, therefore, rather different from the data so far analysed in this volume. Directional data are frequently analysed by means of *vector methods*, from which, for example, a mean value for the data may best be determined.

A *vector* is defined as a line which has both direction and length. Vector strength is indicated by the length of the line. Ordinary methods of finding means and standard deviations are not applicable to this orientational data. The use of vector methods, however, leads not only to a measure of the mean value for the data but also provides a measure of dispersion, and establishes whether a pattern is a random one. They may also be used to test if two patterns have significantly different mean vectors.

For the case study 23 beach orientations were measured from large-scale maps (Table 10.1). These values were converted into their sine and cosine values, each of which were summed for all 23 orientations. The results were:

$$\Sigma \cos \theta_1 = 4 \cdot 1661$$
$$\Sigma \sin \theta_1 = -12 \cdot 075$$

Each of these values can be used to relate the 23 original readings to a single cartesian coordinate system (Fig. 10.1, inset B). The object is to

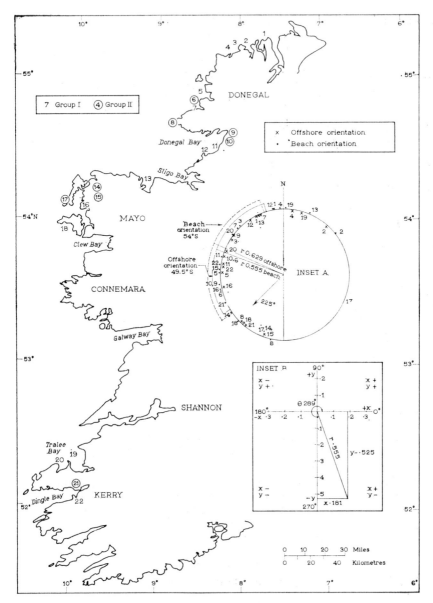

Fig. 10.1 Map of western Ireland to show the sites of beach measurement. Inset **A** shows the pattern of beach orientation and offshore 20 m depth orientation. The mean and standard deviation of the data are indicated. Inset **B** shows the relationship between rectangular Cartesian coordinates, x and y, and polar coordinates, which give the vector, r.

Table 10.1   West Irish beach data

No.	Beach	Orientation at coast	Orientation at 20 m depth contour	Difference in orientation between shore and 20 m depth contour	Difference in orientation between shore and 225°
1	Ballyhirnan Bay	350	333	17	125
2	Dunfanaghy	050	040	10	175
3	Dooros Point	315	302	13	90
4	Inishbofin Is.	355	005	10	130
5	Dooey	270	270	0	45
6	Loughros More*	256	285	29	31
7	Loughros Beg	313	335	22	88
8	Malin Beg*	190	220	30	35
9	Bell's Isle*	260	306	46	35
10	Rossnowlough*	260	285	25	35
11	Tullan	283	275	8	58
12	Inishnagar	349	325	24	124
13	Kilala Bay	020	334	46	155
14	Doolagh Point*	234	197	37	9
15	Gweesalia*	272	197	75	47
16	Blacksod	255	257	2	30
17	Elly Bay*	112	197	85	113
18	Trawmore	223	220	3	2
19	Stradbally	000	015	15	135
20	Lough Gill	305	289	16	80
21	Inch*	244	218	26	19
22	Rossbehy	276	275	1	51

Group II are shown by *.

find, on such a system, the position of the vector or line which provides a mean value for all of the measurements. This is done by dividing both $\Sigma \cos \theta_1$ and $\Sigma \sin \theta_1$ by the number of observations (n_1). In this case $4 \cdot 1661 \div 23 = 0 \cdot 18113$, and this is a measure of the length of the x_1 coordinate of the mean vector, on the cartesian system being established. Similarly $-12 \cdot 075 \div 23 = -0 \cdot 52501$, which is the y_1 coordinate. By using Pythagorus's theorem the length (r) of the vector from the origin to the point ($0 \cdot 18113$, $-0 \cdot 52501$) can be calculated:

$$r_1 = \sqrt{x^2 + y^2}$$

in this case
$$r_1 = \sqrt{0 \cdot 18113^2 + (-0 \cdot 52501)^2}$$
$$= 0 \cdot 555$$

The higher the value of r the greater will be the clustering of the orientations (the vector strength), the lower the value of r, the greater their dispersion. These calculations have provided a statement of the length of the mean vector ($0 \cdot 555$), but it is also necessary to know its orientation if it is to be precisely fixed. This orientation can be found by using the con-

ventions of vector analysis illustrated in Figure 10.1 (inset B). From this
it is seen that the axes from which orientations are calculated do not
coincide with normal compass bearings. For example, 0° lies to the right
(instead of the top) and the 90° bearing (usually east) is at the top of the
diagram. This convention is adopted so that when the trigonometrical
derivatives of angular measures (i.e. sine and cosine) are calculated
positive and negative signs may be associated with particular quadrants
of the diagram. The upper right-hand quarter of the diagram is known
as the first quadrant (i.e. that between 0° and 90°), the second quadrant
is that between 90° and 180°, and so on. The positive and negative signs
for x and y associated with each quadrant are marked on Figure 10.1
(inset B). The signs associated with the trigonometrical calculations are:

Quadrant		Sine	Cosine
First	(0°– 90°)	+	+
Second	(90°–180°)	+	−
Third	(180°–270°)	−	−
Fourth	(270°–360°)	−	+

The following relationships also hold (where A is any acute angle):

$$\sin(90 + A) = +\cos A \qquad \cos(90 + A) = -\sin A$$
$$\sin(180 - A) = +\sin A \qquad \cos(180 - A) = -\cos A$$
$$\sin(180 + A) = -\sin A \qquad \cos(180 + A) = -\cos A$$
$$\sin(270 + A) = -\cos A \qquad \cos(270 + A) = +\sin A$$

In the case study the mean vector lies in the fourth quadrant. Its orientation from 0° is 270° + A, where A is the acute angle between the vector
and the y-axis. From the above:

$$\cos(270 + A) = +\sin A$$

and
$$\sin A = \frac{\text{opposite}}{\text{hypotenuse}} = \frac{x_1}{r_1} = \frac{0.18113}{0.555} = 0.3263$$

The angle (A), whose sine is 0.3263, is 19°. The orientation of the mean
vector = 270 + A = 270 + 19 = 289° from zero. To convert this
value into a compass bearing it is necessary to remember that on the
vector diagram north is represented by the axis on the right and west is
the axis at the foot. The value 289° is therefore equivalent to N 289° E,
and is marked as the beach vector ($r_1 = 0.555$) in inset A.

The dispersion of the orientation values may be calculated from the
equation:

$$s_1 = \sqrt{2(1 - r_1)}$$

where the angular deviation (dispersion) s is measured in radians. In
this case study:

$$s_1 = \sqrt{2(1 - 0.555)} = 0.9434 \text{ radians or 54 degrees}$$

A similar analysis may be performed on data relating to the orientation

of the 20 m depth contour. In this case 20 observations (n_2) were made and the following values were obtained.

$$\Sigma \cos \theta_2 = \quad 4 \cdot 8774 \quad \text{and} \quad x_2 = \quad 0 \cdot 24387$$
$$\Sigma \sin \theta_2 = -11 \cdot 5946 \quad \text{and} \quad y_2 = -0 \cdot 57973$$
$$r_2 = 0 \cdot 629$$
$$\text{mean vector angle} = \text{N } 292° \text{ E}$$
$$s_2 = 49 \cdot 5°$$

The values derived for the beaches and the offshore contour may now be compared to see if there is a significant difference between them. First, however, it is necessary to see if the orientations themselves are random. If they are, then there is little point in proceeding with the analysis. The distribution will be non-random if r exceeds a critical value, which can be checked against values recorded in z_p tables (Batschelet, 1965). For this purpose r is converted into a value R by the relationship:

$$R = rn \quad \text{(where n is the number of observations)}$$

In the case study:

$$\text{for the beaches, } R_1 = 0 \cdot 555 \times 23 = 12 \cdot 765$$
$$\text{for the offshore contour, } R_2 = 0 \cdot 629 \times 20 = 12 \cdot 58$$

The value z_p which is checked in significance tables is calculated from:

$$z_p = \frac{R^2}{n}$$

$$\text{for the beaches, } z_{p1} = \frac{12 \cdot 765 \times 12 \cdot 765}{23} = 7 \cdot 1$$

$$\text{for the offshore contour, } z_{p2} = \frac{12 \cdot 58 \times 12 \cdot 58}{20} = 7 \cdot 91$$

These z_p tables are a modification of normal z tables, which give the area under the normal curve in terms of the standard deviation. The value of z is sometimes given (Krumbein and Graybill, 1965, 415) as the total area under the curve to the left of the z value specified. For example, where $z = -3 \cdot 00$ the area is $0 \cdot 0013$, or for $-2 \cdot 0$ it is $0 \cdot 0228$, while for $+2 \cdot 0$ it is $0 \cdot 9772$. Other tables measure the area on only one side of the curve (Alder and Roessler, 1962). The area indicates the expected range of values in the normal distribution at the specified point. Thus if z is 2— that is 2 standard deviations from the mean—the expected percentage of observations between $z = 2$ and $z = -2$ is $(0 \cdot 9772 - 0 \cdot 0228) \times 100 = 95 \cdot 44$. Between $z = -1$ and $z = +1$, the expected percentage of observations that are one standard deviation from the mean is given by $(0 \cdot 8413 - 0 \cdot 1587) \times 100 = 68 \cdot 26\%$. The z_p tables are adjusted to deal with the circular normal distribution, which defines a normal spread of values around a point on the circumference of a circle or a sphere. In the case of the beaches, for 23 observations, a value of z_p which is greater than $4 \cdot 47$ may be considered, with 99% confidence, as indicating a non-random set of orientations. For the offshore contour, having 20 observa-

tions, the 99% confidence value is 4·45, which again is exceeded and indicates that the orientations are non-random.

To test whether or not the two sets of data are derived from the same population it is necessary to refer back to earlier values, re-written as follows:

$$\text{let } V_1 = \Sigma \cos \theta_1 = \quad 4\text{·}1661 \text{ for the beaches}$$
$$\text{let } V_2 = \Sigma \cos \theta_2 = \quad 4\text{·}8774 \text{ for the offshore contour}$$

and:

$$\text{let } W_1 = \Sigma \sin \theta_1 = -12\text{·}075 \text{ for the beaches}$$
$$\text{let } W_2 = \Sigma \sin \theta_2 = -11\text{·}5946 \text{ for the offshore contour}$$

The test is based on the calculation of a combined vector (R) for the two sets of data, where $R^2 = V^2 + W^2$, and $V = V_1 + V_2$, and $W = W_1 + W_2$. Thus:

$$R^2 = (4\text{·}1661 + 4\text{·}8774)^2 + [-12\text{·}0753 + (-11\text{·}5946)]^2$$
$$R^2 = \sqrt{642\text{·}0491} = 25\text{·}338$$

The whole test has to be based on a null hypothesis, which in this case states that there is no significant difference between the two sets of orientations. To establish a level of confidence with respect to this null hypothesis it is necessary to calculate an F-value from:

$$F = \frac{(N - 2)(R_1 - R_2 - R)}{N - R_1 - R_2}$$

where $N = n_1 + n_2$

and $$F = \frac{41(12\text{·}774 + 12\text{·}5787 - 25\text{·}338)}{43 - 25\text{·}3527} = 0\text{·}03253$$

The tabled value of F with which this result is compared is that for F 1, 41, 0·05 and equals 4·08 (i.e. the F-value for 1 and 41 degrees of freedom at the 0·05, or 95%, confidence level equals 4·08). The calculated F-value falls very far short of the required value of 4·08, and the null hypothesis therefore cannot be rejected. In other words, the result shows that no significant difference can be demonstrated between the two sets of measurements. This suggests that the orientation of the offshore submerged 20 m contour is related to the orientation of the beach.

In terms of coastal morphology this result has certain implications. The beaches analysed are sandy beaches. The swell, because of the indentations of this coastline, has already been considerably refracted by the time it reaches the 20 m depth contour, and its orientation is already similar to that of the beach it is approaching. The close relationship between beach orientation and that of the 20 m depth contour therefore supports the view that sandy beaches are built up to face the direction from which the refracted swells arrive. Some additional refraction takes place after the 20 m depth contour has been passed by the swell. The amount of this refraction depends partly on the direction from which the waves are approaching relative to the offshore relief and the shoreline pattern. Further

analysis is therefore necessary to assess the effect of these influences. This may be done by calculating the difference in orientation between the beach and the direction of most effective wave approach (which for this coast may be taken to be N 225° E), and the difference in orientation between the beach and the 20 m depth contour (Table 10.1). In practice these orientations are not easy to measure because of the curvature of

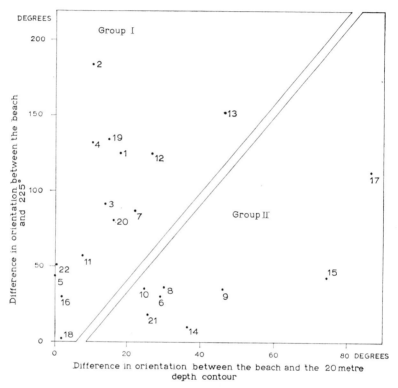

Fig. 10.2 Graph to show the relationship between beach orientation relative to south-west and the 20 m depth orientation for two groups of beaches in western Ireland.

both the beach and the depth contour. For this reason it is better, for correlation purposes, to rely on a rank test than on a parametric test.

In Figure 10.2 the beaches listed in Table 10.1 have been plotted, by their reference numbers, on axes relating to the difference in orientation between the beach and N 225° E, and to the difference between the orientations of the beach and the 20 m depth contour. This plot suggests that the data falls into two groups, as shown in Figure 10.2. These two groups may be analysed separately, and the members of the smaller group examined to see in what way they differ from those of the larger group.

The larger group (group 1) consists of 14 pairs of measurements and the smaller group has eight members. The two sets of difference values were ranked from smallest to largest for each of these two groups. The Spearman rank correlation coefficient (section 3.3) has been calculated for the relationship between the difference in orientation between the beach and N 225° E and the difference in orientation between the beach and the depth contour. The rank correlation coefficient (ρ) was calculated separately for each of the two groups. For the larger group:

$$\rho = 0 \cdot 770 \quad (\text{where } n = 14)$$

For the smaller group:

$$\rho = 0 \cdot 603 \quad (\text{where } n = 8)$$

The value for the larger group is significant at the 99% confidence level, and it only just fails to be significant at the 95% level in the case of the second group. This means that as the difference in orientation between the beach and the direction of maximum wave effectiveness (N 225° E) increases, so the beach turns at a higher angle to the offshore contours. This in turn indicates that as the difference in orientations increases, so wave refraction becomes more effective and the wave crests bend through a larger angle between the 20 m depth contour and the shore. This result conforms to existing knowledge of the properties of wave refraction, for as the waves bend to fit the coast they in turn mould the coast to fit their pattern.

Another look at either Figure 10.2 or Table 10.1, shows that in the case of the smaller group (group II) more extreme values are reached for the difference between the orientation of the beach and the 20 m depth contour. For group I the mean angle of difference is 12·6°, for group II it is 39·2°. The difference between the two means may be examined by a t-test, which is designed to see if the two means could have been drawn from the same population (see section 3.1). The t-test in this case shows that there is a significant difference between the two means at the 99·9% confidence level. This implies that the beaches belonging to group II are to be found in areas where the refraction of the approaching waves is greater. A re-examination of the character of these beaches shows that they are situated where a great deal of shelter is provided by headlands. Three of the beaches, numbers 14, 15 and 17 (Fig. 10.1), are in one very sheltered embayment where the waves are greatly refracted before they reach the shore. Others are also sheltered, such as Inch spit beach (number 21) which lies inland of Rossbehy spit beach, and those in Donegal Bay. Two, numbers 6 and 8, are in very small bays, where there is not much space for the beach to become aligned to fit the swells on account of rocky headlands.

The results of this analysis show, therefore, that the form of the coast, in terms of the orientation of reasonably exposed and wide, sandy beaches, is related to the offshore relief and to the direction of wave approach. The beaches face in a similar direction to the offshore relief, as expressed by

the 20 m submerged contour, but the details of this relationship are shown to depend in part on the pattern of wave refraction. This provides the link between the offshore area and the coastal form.

10.2 The beach form in profile

The form of a beach in profile is dependent on several variables, one of which is the nature of the waves that affect the beach. In particular, steep storm waves tend to be destructive, causing erosion on the foreshore. Low flat swells on the other hand are usually constructive, building up the foreshore. These two wave types may occur seasonally, in which case the beaches are cut during the stormy season and fill again during the months when winds are less intense and long swells predominant. The effect of different wave types, and of their associated wind patterns, can be analysed numerically. A case study is presented below using observations made on the east coast of England, at Marsden Bay, County Durham. This small bay is exposed to waves coming across the North Sea. The fairly coarse sand in this bay can only move up and down the beach profile, as rocky headlands prevent very much lateral movement. In order to assess changes in the beach profile it was surveyed at weekly intervals, using levelling methods, at both its northern and its southern ends (Fig. 10.3). A record was also kept of the main wave types and the wind directions between each of the surveys. For purposes of numerical analysis the beach may be considered in two parts, namely, the upper and lower beach, with the boundary between them fixed at the mean tide level.

The measured data can be analysed to see if there is a significant difference between the amounts of cut and fill which take place under different wave conditions. Secondly, the analysis can be extended to see if there is any difference between the behaviour of the upper and lower parts of the beach during periods of cut and fill. To carry out this analysis a statistic known as (chi square) may be calculated. The χ^2 test is one of the most useful of the statistical tests in geomorphology. It is suitable for any data which enables objects to be placed into categories in the form of frequencies. The number of categories may be 2 or more.

The χ^2 test is used to see if there is a significant difference between the *observed* number of objects in each category and the number which would have been *expected* to fall into that category. It is based on the null hypothesis (H_0) that there is no significant difference between the observed and expected values. Its fundamental criterion for success is that the number of expected values in a particular category can be calculated. When one sample is being analysed, and assuming an equal spread through all categories is anticipated, then the expected number of cases in a category E is given by:

$$E = \frac{n}{k}$$

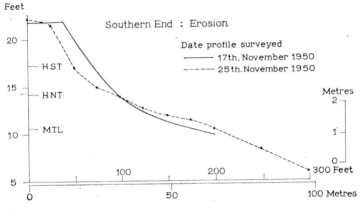

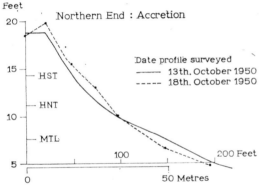

Fig. 10.3 Profiles of the beach at Marsden Bay, Co. Durham, to illustrate erosion and accretion on the foreshore.

where n is the total number of cases, and k is the number of categories. The statistic χ^2 is defined by:

$$\chi^2 = \sum_{i=1}^{k} \frac{(O_i - E_i)^2}{E_i}$$

where O_i is the observed number of cases in the ith category; E_i is the expected number of cases in the ith category and

$$\sum_{i=1}^{k}$$

means the sum over all categories of the value of:

$$\frac{(O_i - E_i)^2}{E_i}$$

If O_i and E_i are similar in value, then χ^2 will be small. The larger the difference between the observed and the expected values the larger

χ^2 will be, and the greater the likelihood that H_0 must be rejected. A requirement of this test, which should be noted, is that all of the observations should be independent of each other. The significance of χ^2 needs to be considered in relation to the number of degrees of freedom (df) (section 2.5); $df = k - 1$, when one sample is being analysed.

The application of the χ^2 test may be illustrated by a case study of onshore and offshore winds. In this case two independent groups of data are involved. The test is used to see if there is a significant difference between the frequency of cut and fill which takes place with onshore and offshore winds respectively. Table 10.2 lists the observed and expected

Table 10.2 Data for the chi-square test for beach cut and fill on the upper beach at Marsden Bay, County Durham

	Onshore winds		Offshore winds		Row total of O_i
	Observed	Expected	Observed	Expected	
Fill	4	8·76	13	8·24	17
Cut	13	8·24	3	7·76	16
Column total of O_i	17		16		33

frequency values for fill and cut both with onshore and offshore winds on the upper part of the beach. The null hypothesis for this test is that no difference is expected to result from variations in wind direction. The expected values are found for each cell of the table by multiplying its row total by its column total, and dividing by the total number of cases. Thus for the top left-hand expected frequency cell of Table 10.2,

$$E_{ij} = \frac{17 \times 17}{33} = 8\cdot76$$

and for the right-hand cell on the top row:

$$E_{ij} = \frac{17 \times 16}{33} = 8\cdot24$$

When, as in this case, two independent samples are being examined, the equation for χ^2 has to be modified, from that given above, and is:

$$\chi^2 = \sum_{i=1}^{r} \sum_{j=1}^{k} \frac{(O_{ij} - E_{ij})^2}{E_{ij}}$$

where O_{ij} is the observed number of cases in the ith row of the jth column; E_{ij} is the number of cases expected, under H_0, in the ith row of the jth column; and

$$\sum_{i=1}^{r} \quad \sum_{j=1}^{k}$$

means that the right-hand part of the equation should be summed over all cells of the table. When two or more samples are being analysed, the value of $df = (r - 1)(k - 1)$, where r is the number of rows and k the number of columns. In the case of Table 10.2 there are 2 rows and 2 columns, and $df = 1$.

To assist calculation it is worth setting the data out as in Table 10.3, in which χ^2 for this example is calculated and found to be 11·01. This value has to be compared with the values listed in a χ^2 significance table (see Appendix, table B). At the 99·9% confidence level the tabled value

Table 10.3   Calculation of χ^2 for the data in Table 10.2

O	E	O − E	$(O - E)^2$	$\dfrac{(O - E)^2}{E}$
4	8·76	−4·76	22·66	2·59
13	8·24	4·76	22·66	2·75
13	8·24	4·76	22·66	2·75
3	7·76	−4·76	22·66	2·92

$$\sum_{i=1}^{r}\sum_{j=1}^{k} 11\cdot01 = \chi^2$$

for χ^2 with one degree of freedom is 10·827. The calculated value exceeds this so the null hypothesis may be rejected at this level of confidence. This suggests, therefore, that on the upper beach the wind conditions determine whether cut or fill takes place. A similar analysis was carried out for the lower beach and H_0 can be rejected in this case at the same confidence level.

The data can be reconsidered with the null hypothesis that when onshore winds prevail there is no significant difference between the amounts of cut and fill on the upper and lower parts of the beach respectively. The data relating to this analysis are set out in Table 10.4, from which the χ^2 value was calculated to be 11·0. Likewise the calculation can be carried out for offshore as opposed to onshore winds, and χ^2 was found to be 13·39. In both cases, therefore, the null hypothesis that there is no significant difference between the behaviour of the upper and lower beaches can be rejected with 99·9% confidence.

Using the χ^2 test, it has thus been shown that onshore and offshore winds, through wave type, exert a significant effect on the number of occasions on which cut and fill occur on this beach. In addition, it has been shown that the upper and lower parts of the beach behave differently under one set of wind conditions. This particular example conforms to the general rule that onshore winds are accompanied by destructive waves, while offshore winds and their associated constructive swells move sand on to the upper beach from lower levels. This relationship is one of the

Table 10.4 Data for the chi-square test for beach cut and fill with onshore winds, Marsden Bay

	Upper beach		Lower beach		Totals for O
	O	E	O	E	
Fill	4	8·5	11	6·5	15
Cut	13	8·5	2	6·5	15
Column total for O	17		13		30

most fundamental aspects of the influence of waves and their accompanying winds on the form of the beach in profile.

10.3 Beach gradient

In general the gradient of a beach may be dependent on at least three variables. These are the size of the material of which the beach is composed, wave length, and wave steepness. The analysis of beach gradient in relation to each of these three variables is a problem of multivariate analysis. It is possible to study the separate effect of each of these variables on beach gradient, for example, in a wave tank in the laboratory. On the other hand, in section 4.5 it was shown how the influence of a variable could be 'controlled' by the use of partial correlation analysis. By this method the influence of wave length, for example, can be controlled while the correlation between wave steepness and the gradient of the beach is examined.

A case study is provided once again by measurements made in Marsden Bay (see section 10.2). Wave length and steepness were established at the same time as the beach gradient was measured. Graphs have been drawn of the relationship between beach slope and both wave steepness and the wave period (Fig. 10.4). The scatter of points around the two regression lines probably results from the fact that the beach gradients were unlikely to be at all times in equilibrium with the waves measured. It is particularly difficult to determine the dominant wave period in the field, because the wave pattern normally consists of a whole spectrum of wave periods combined: this also leads to scatter in the data. Wave period was therefore measured instead of wave length as this is an easier task in the field and, in any case, there is a direct relationship between wave period and wave length in deep water. It will be noted that a log normalization (section 2.4) has been used on the wave data.

In all, 40 observations were made. Initially it is relevant to note the

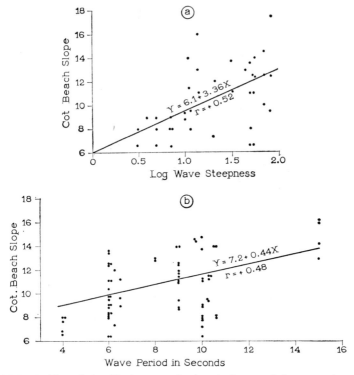

Fig. 10.4 A: The relationship between the beach slope and the wave steepness;
B: The relationship between the beach slope and the wave period at
Marsden Bay, Co. Durham.

product-moment correlation coefficient between each of the variables.
These are shown in Table 10.5.

Table 10.5 Correlation coefficients for wave steepness, wave period and beach slope

	A	B	C
A	1·00	−0·11	**0·52**
B		1·00	**0·48**
C			1·00

where A is the log of wave steepness; B is the log of
wave period; C is the cotangent of the beach slope
(i.e. its gradient).

In the case of 40 observations, if the correlation coefficient r exceeds 0·40,
then the value of r is significant at the 99% level. There is thus a high

positive correlation between beach slope and wave steepness, as well as between beach slope and wave period (i.e. wave length).

The degree of correlation between each pair of variables is considerably improved when partial correlation coefficients are found. For example, if beach slope is compared with wave period while the effect of wave steepness is controlled, then the partial correlation coefficient $r_{C,\,B.A} = 0.63$ (where A, B and C refer to the variables defined above and the controlled variable is given last). Partial correlation analysis controlling each of the other two variables in turn brings out the following relationships:

$$r_{C,\,A.B} = 0.66$$
$$r_{A,\,B.C} = -0.48$$

The last of these partial correlation coefficients is negative and indicates that there is a tendency for wave period and wave steepness to act in opposite directions to each other. The longer waves, which tend to flatten the slope, are associated with less steep waves, which tend to steepen the slope. (See King, 1970, for an expanded discussion of this effect.)

10.4 Beach ridges

Some beach features move fairly regularly with time, others are static or move only very slowly, while some change irregularly or randomly. In order to analyse some of these *time-series* situations beach profiles on two beaches will be examined. Both beaches are wide and sandy, and have ridge and runnel profiles. There are, however, significant differences in the movement of the ridges on the two sets of profiles. The first beach is at Blackpool, where the observations analysed were made at fortnightly intervals over a period of eight months. During this time there were three ridges present on the beach all the time, while towards the end of the period other minor ridges developed.

The first ridge crest was situated between 700 and 800 feet from the top of the beach. During the 17 periods of two weeks the ridge moved 100 feet generally towards the shore. A study of the trend of its movement was made by regression analysis. This gave the equation:

$$Y = 733 - 5.44X$$

The coefficient 5.44 indicates that the movement averaged an approach to the shore of 5.44 feet every two weeks. The coefficient of 733 defines the mean position of the ridge, that is to say its position in the middle of the period analysed. The method used to calculate this *trend equation* is included in the discussion of the second series of observations, as it is easier to appreciate this method of trend analysis in time series when the rate of movement is greater, giving a higher trend coefficient. In the analysis under discussion the coefficient of trend (-5.44) is negative, indicating a shoreward movement. However, if monthly observations are

taken, covering a longer period of 13 four-week spells, the equation for the trend equation becomes:

$$Y = 744 + 5 \cdot 98X$$

The second equation shows that the movement was not regularly land-wards. Over the longer period the measured distance which the ridge moved seawards was 60 feet from a point 700 feet from the top of the beach. The movement was oscillatory and not regular. During the period of measurement there were 27 turning points in a total of 58 observations.

The second ridge lay seaward of the first, and the equation giving its trend for the fortnightly periods is:

$$Y = 1109 - 7 \cdot 07X$$

Again, however, during the longer period of observations the ridge moved seaward from 1000 feet to 1030 feet from the beach crest.

The third ridge was very consistent in position. Its trend equation is:

$$Y = 2800 + 0 \cdot 46X$$

The coefficient is very small, showing a movement of less than 6 inches per two weeks in a seaward direction. This ridge was the highest of the three, having a crest height of 4 feet above the landward trough.

The ridges may be considered to be essentially static, showing only rather slow and irregular movements, with reversals of their low trend coefficients over fairly long periods of time. The three ridges, which were the most permanent on the beach, were situated at the mean tide heights—the first near the mean high spring tide, the second near mean high neap tide, and the largest and lowest near the mean low neap tide. There was another near mean low spring tide, but its record of surveys was too intermittent to analyse as it was only exposed at fortnightly intervals at low spring tide.

These ridges appear to bear a definite relationship to the tide levels. It is at these levels that wave action will be most prolonged on the beach for any one part of the wave, such as the swash and backwash. The reason why the ridges do not move consistently on this beach is that they lie parallel to the coast. They are built up by the waves where they can act for the longest period on this beach, which has a very high tidal range reaching nearly 30 feet at spring tide.

A very different pattern of ridge movement has been observed on the second beach on the Lincolnshire coast at Gibraltar Point. The pattern of ridge movement here, over a period of more than ten years, is shown in Figure 10.5. The ridges on this beach moved shorewards for a period of about three years. This covers the time from their initial formation to their stabilization, which occurs by the growth of further ridges to sea-ward of an elevation sufficient to protect the landward ridge from further wave action. During the period from 1957 to 1968 three ridges have be-come stabilized and two more have become welded onto the last of the three to be stabilized. Another ridge is still moving actively landwards and another has recently developed on the lower foreshore.

The ridges develop about 800 feet seaward of the seaward limit of the stabilized beach. This point has moved offshore from 100 feet to 860 feet seaward of the dunes during the period analysed. The latest ridges to form have started to develop at a point 1680 feet seaward of the dunes. The ridges on Figure 10.5 are labelled 1 to 7 in order of their formation

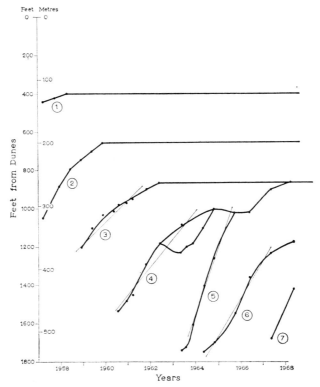

Fig. 10.5 The movement inshore of the beach ridges on profile 4, south Skegness, Lincolnshire. Trend lines are shown.

and proximity to the shore. Trend equations have been calculated for ridges 2–6 and are:

$$
\begin{aligned}
\text{Ridge 2} \quad & Y = 798 - 45\!\cdot\!4\mathrm{X} \\
\text{Ridge 3} \quad & Y = 995 - 37\!\cdot\!2\mathrm{X} \\
\text{Ridge 4} \quad & Y = 1210 - 48\!\cdot\!6\mathrm{X} \\
\text{Ridge 5} \quad & Y = 1222 - 81\!\cdot\!3\mathrm{X} \quad \text{(whole period)} \\
\text{Ridge 5} \quad & Y = 1330 - 113\!\cdot\!4\mathrm{X} \quad \text{(early period)} \\
\text{Ridge 6} \quad & Y = 1500 - 66\!\cdot\!0\mathrm{X}
\end{aligned}
$$

As was noted above, these equations are in the form $Y = a + b\,\mathrm{X}$, where a is the mid-point of the trend and b is the trend coefficient, indicating the rate of movement. a is given by $\Sigma\,Y/n$ and b is given by $\Sigma\,\mathrm{XY}/\Sigma\,\mathrm{X}^2$.

The trend is found by least-squares, so the b value is similar to the regression coefficient (section 3.5). The method of calculating the trend for ridge 6 is given in Table 10.6 as an example. The trend line is entered on Figure 10.5. The X values refer to values from the plotted curve taken at four-monthly intervals as the times of survey were not exactly equal. Most of the curves start fairly steeply and then flatten off as the ridge becomes increasingly sheltered by new ones growing up to seaward. The trend thus underestimates the early ridge movement and overestimates its later movement. The fastest movement normally occurs when the

Table 10.6 Calculations for the trend equation for ridge 6 of Figure 10.5

Y	X	XY	X²	
				n is 9
1730	−4	−6920	16	
1690	−3	−5070	9	$a = 13500/9 = 1500$
1650	−2	−3300	4	
1600	−1	−1600	1	$b = \dfrac{-3960}{60} = -66.0$
1510	0	0	0	
1440	+1	+1400	1	
1350	+2	+2700	4	$Y = 1500 - 66.0X$
1290	+3	+3870	9	
1240	+4	+4960	16	
$\Sigma Y = 13500$	$\Sigma X = 0$	$\Sigma XY = -3960$	$\Sigma X^2 = 60$	

ridge is on the lower foreshore. In this position it can be reached by the waves without much loss of energy. The older ridges moved landwards at about 150 feet/year, but the greatest rate was reached by ridge 5, which moved landwards at over 340 feet/year during 1964.

The reason why these ridges move landwards consistently until they become stabilized, or welded onto other ridges, is that they lie at a small angle to the shore, diverging from it southwards. The direction of movement of material along the shore is southwards and the ridges also move bodily southwards. This has the effect of an apparent landward movement of the ridges on any one profile, such as the one analysed. The movement of the ridges, although it is a function of time and bears a close relationship to time, is in fact due to the southerly transfer of material along the coast. This longshore movement must, therefore, be regular with time.

In this section the numerical analysis of ridge movement has shown that on one beach the ridges are essentially static and only move a small distance and irregularly in direction, while on the other they move with considerable regularity in a landward direction. The reason for this difference is the alignment of the ridges relative to the shore. On the first beach there is little longshore movement as the ridges are parallel to the

shore, but on the second the ridges lie at an angle to the shore and move southwards along it.

One of the difficulties of analysing time series is that of *autocorrelation*. This term refers to the situation where each variate is in turn influenced by the preceding one, so that the series cannot be considered as random values. A simple test has been devised by Hart (1942) to assess whether a time series is affected by autocorrelation. This test is applied to two of the trends of beach ridge movement that have been analysed in this section. The method is based on the calculation of the ratio between the mean squared successive differences and the variance of the set of values. This ratio, k, is found by means of the equation $k = m^2/s^2$, where s^2 is the variance, and m^2 is found by:

$$\sum_{i=1}^{i=N-1} (x_{i+1} - x_i)^2/n - 1$$

If k is small then there is positive autocorrelation between the successive values; but if k exceeds the upper confidence limit then there is negative autocorrelation. The former implies that successive values change in the same general direction, while a negative autocorrelation means that each successive value induces a great reversal of the trend. If the value of k falls between the upper and lower significance limits at the given level of confidence, then the values are not autocorrelated and can be considered to vary randomly. The limits for $n = 10$ are as follows—95% significance level: lower limit 1·18, upper limit 3·26; 99% significance level: lower limit 0·84; upper limit 3·61. The test can also be used to assess whether there is autocorrelation in the residuals from regression. An example of this application is discussed and illustrated in section 13.1, where it is applied to spatial rather than time data.

The data for the analysis are set out in Table 10.7. They refer to the two beaches analysed in this section and consist of the positions of ridge crests over a period of time. The upper set of values were obtained for the highest ridge on Blackpool beach, which is steady in position. The first column gives the positions of the ridge in tens of feet from the top of the beach. The second column gives the successive differences, which are squared in the third column. In the fourth and fifth columns the variance is calculated by summing the squared differences between each value and the mean, and dividing by n. The value of m^2 is 302/12, which is 25·17. The variance s^2 is 213·21/13, which is 16·40.

The value of k is therefore 25·17/16·40 which is 1·535. This value falls between the upper and lower significance levels and therefore confirms that there is no systematic movement of the ridge on the beach at Blackpool. This is due to the fact that the ridges in this vicinity lie parallel to the shore and their position is largely determined by the position of the tide levels, although there is a small random shift in ridge position with varying wave conditions and tide levels.

The second ridge tested is ridge 6 on profile 4 south of Skegness on the Lincolnshire coast. Again the first column gives the position of the ridge in tens of feet from the dune foot at successive time intervals. The trend equation for this ridge movement is $Y = 1500 - 65 \cdot 3X$ and there is a

Table 10.7 Autocorrelation of time-series observations

Upper ridge on Blackpool beach, n = 13

Ridge position x	$(x_{i+1} - x_i)$	$(x_{i+1} - x_i)^2$	$(x_i - \bar{x})$	$(x_i - \bar{x})^2$
70	0	0	4·54	20·61
70	−4	16	4·54	20·61
66	12	144	8·54	72·93
78	−2	4	−3·46	11·97
76	−4	16	−1·46	2·13
72	8	64	2·54	6·45
80	0	0	−5·46	29·81
80	−4	16	−5·46	29·81
76	0	0	−1·46	2·13
76	−4	16	−1·46	2·13
72	5	25	2·54	6·45
77	−1	1	−2·46	6·05
76			−1·46	2·13
Σ 969	−19 +25	Σ 302	+22·70 −22·68	Σ 213·21

$\bar{x} = 969/13 = 74 \cdot 54$

Ridge 6 on profile 4 south Skegness, n = 9

173	−4	16	23	529
169	−4	16	19	361
165	−5	25	15	225
160	−9	81	10	100
151	−7	49	1	1
144	−9	81	−6	36
135	−6	36	−15	225
129	−5	25	−21	441
124			−26	676
Σ 1350	Σ −49	Σ 329	+68 −68	Σ 2594

$\bar{x} = 1350/9 = 150$

Note: Values in the first column are given in tens of feet

fairly consistent movement of the ridge crest. The value for m^2 in this instance is $329/8$, which is $41 \cdot 125$, and the variance is $2594/9$, which is $288 \cdot 222$. The large variance is due to the rapid and consistent ridge movement. k therefore equals $41 \cdot 125 / 288 \cdot 222$, which is $0 \cdot 143$. This value is

far below the lower significance limit, showing that there is strong positive autocorrelation in the data. This is to be expected because of the steady landward shift of the ridges on this beach. This has been shown to be due to their orientation at a slight angle to the coast, so that as they move south they appear to move steadily inland on any one profile. This test, therefore, indicates very clearly one of the most important differences in the patterns of ridge movement on these two ridged beaches.

References

ALDER, H. L. and ROESSLER, E. B. 1962: *Introduction to probability and statistics*. San Francisco and London: Freeman.

ARTHUR, R. S., MUNK, W. H. and ISAACS, J. D. 1952: The direct construction of wave rays. *Trans. Am. Geophys. Union* **33**, 855–65.

BATSCHELET, E. 1965: *Statistical methods for the analysis of problems in animal orientation and certain biological rhythms*. A.I.B.S. Mono.

DAVIES, J. L. 1958: Wave refraction and the evolution of shoreline curves. *Geogr. Studies* **5**, 1–14.

1964: A morphogenic approach to world shorelines. *Zeit. für Geomorph.* **8**, 127*–142*.

DOLAN, R. and FERM, J. C. 1967: Temporal precision in beach profiling. *Prof. Geogr.* **19**, 12–14.

HART, B. I. 1942: Significance levels for the ratio of the mean square successive differences to the variance. *Ann. of Math. Stats.* **13**, 445.

KING, C. A. M. 1970: Feedback relationships in geomorphology. *Geogr. Ann.* **52**A, 147–59.

KRUMBEIN, W. C. and GRAYBILL, F. A. 1965: *An introduction to statistical models in geology*. New York: McGraw-Hill.

RUSSELL, R. J. 1967: Aspects of coastal morphology. *Geogr. Ann.* **49**A, 290–309.

SIEGEL, S. 1956: *Nonparametric statistics for the behavioral sciences*. New York: McGraw-Hill.

ZEIGLER, J. M. and TUTTLE, S. D. 1961: Beach changes based on daily measurements of four Cape Cod beaches. *J. Geol.* **69**, 583–99.

11 Accretion and erosion

11.1 Profile variability
11.2 Time-series correlation
  [*sign correlation and chi-square test*]
11.3 Discriminant analysis
  [t-*test, discriminant function, Mahalanobis' D^2, variance, degrees of freedom, F-values, Hotelling's T^2*]
11.4 Trend patterns
  [*trend equations*]
11.5 Analysis of covariance
  [*correlation tests, regression analysis, F-test, significance levels, degrees of freedom*]
11.6 Coastal erosion
  [*product-moment correlation, t-test, Poisson distribution*]

The major changes in coastal form are brought about through accretion and erosion. Changes in coastal form as a result of these processes can only be numerically assessed if records are kept of beach or cliff over a period of time. In this chapter case studies are presented illustrating the manner in which such data may be analysed. Sections 11.1–11.5 deal with accretion, and the last section describes the analysis of information relating to erosion. The influence of accretion on beach form is illustrated by three of the profile lines surveyed at Gibraltar Point, Lincolnshire (UK) (profiles 1, 2 and 4, Fig. 11.1). Observations covering a period of 17 years will be used. The profiles are analysed in five ways:

1 the variability and long-term changes of the profiles are considered
2 the degree to which the three profiles change together is analysed by time-series correlation
3 discriminant analysis is used to test for a significant difference in ridge growth on two profiles
4 the trends of changes on the three profiles are calculated
5 the analysis of covariance is used to study variations between the three profiles in terms of ridge height and the accumulation of sand.

11.1 Profile variability

The variability of the profiles over the 17 year period is compared by the analysis of sweep zone profiles. The sweep zone defines the variability of the beach over a period of years (Fig. 11.2). It is drawn by joining up all the highest points on the profiles to form an upper sweep zone profile

and all the lowest points to form a lower sweep zone profile. The sweep zone is the envelope within which all the profiles lie. The upper sweep zone profile is delimited on these beaches by the ridge crests and the lower by the runnel troughs. The width of the sweep zone, therefore, depends largely on the height of the ridges and their movement. The detailed observations on the beach near profile 2 (Fig. 11.2) showed that the major changes of sand level on the beach were related to ridge and runnel development and their movement (King, 1968). Other beach changes were

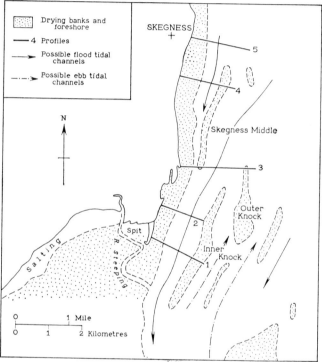

Fig. 11.1   Offshore features and the position of profiles at Gibraltar Point, Lincolnshire.

slow and small. Thus occasional surveys show well the development and movement of the ridges.

The processes that build these ridges act through time. The variations on neighbouring profiles indicate how the processes are operating in this area. The major variable affecting the processes and causing the differences between the profiles is concerned with the supply of material to the shore. Where the supply is plentiful the beach builds up rapidly, ridge growth is active, and the whole level of the beach is raised. Under these conditions the sweep zone will be wide, because the ridges will be high. Where the supply is meagre the beach will be static or lose material by

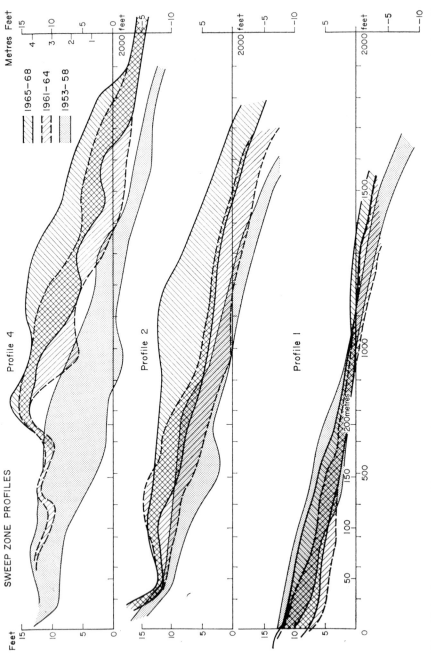

Fig. 11.2 Sweep zone for profiles 1, 2 and 4 south of Skegness, Lincolnshire for the periods 1953–58, 1961–64 and 1965–68.

longshore transfer. The ridges will become smaller in these conditions and the sweep zone will be narrow. Some of these effects are shown in Figure 11.2.

The sweep zones show that the trends of the 1950s have continued through the first half of the next decade and appear to be continuing steadily on some of the profiles. The three profiles, used in the case study below, have been selected from several measured and they are situated (Fig. 11.1) at Gibraltar Point spit (profile 1), about ½ mile farther north (profile 2) and a further three miles to the north (profile 4 of the surveyed profiles) near the south end of Skegness. The profiles are all situated in the zone of accretion at the southern end of the Lincolnshire coast. The ridge development is closely linked with the accretion, which is building the coast out eastwards.

The sweep zones of the three beaches give a useful indication of the relationship between accretion and ridge size. The table illustrates the differences on the three profiles during four separate periods:

Table 11.1   Maximum sweep zone width (in feet)

Period	1 1952–55	2 1957–59	3 1961–64	4 1965–68
Profile				
1	5	4	5·5	5·5
2	7	5	6·5	7·5
4	7	9	8	8·5

The values increase northwards and they are related to the amount of accretion on the beaches, as shown by the level of the sweep zone in Figure 11.2. On profile 1 the sweep zone of the second period is low and narrow, but it rises again throughout the beach in the third period, the rise being particularly marked on the lower beach. On profile 2 there is a steady rise of the whole sweep zone, through a total width of about 10 to 12 feet, on the middle and lower beach. The sweep zones for the succeeding periods lie above each other and are almost separate, so that the lower sweep zone of one period lies above the upper one of the previous period. The same changes are even more marked on profile 4, where the total width of the sweep zone for the whole period is 15 to 16 feet, over a horizontal distance of about 1000 feet. The sweep zone on the upper beach has become very narrow during the last two periods, owing to the stabilization of the ridges on this part of the beach. The maximum width of the sweep zone has moved seawards during the whole period, indicating that part of the beach where ridge development was most active. The stabilization was discussed in connection with ridge movement trends considered in section 10.4. The only changes now taking place in the

upper sweep zone are due to marsh development in the runnels and dune development on the stabilized ridges.

11.2 Time-series correlation

The processes that are operating on these profiles to produce the observed changes through time can be further analysed by comparing the way in which the profiles vary among themselves. In order to measure the approximate amount of accretion or erosion on the beaches two successive profiles were superimposed and the area between the two profiles was measured. The amount of erosion and accretion were both recorded. From these values the net change on the beach could be obtained by noting the difference between erosion and accretion. The net change was calculated for the upper beach above mean tide level, and separately for the beach between mean tide level and low neap tide level, this last being the lowest level to which all the surveys extended. The values were then added cumulatively. The values between each survey were used to test the correspondence between changes on the different profiles. Both the upper beach and the whole beach were analysed for the three pairs, profiles 1 and 2, 1 and 4, and 2 and 4.

The analysis uses a sign correlation test, in association with the chi-square test, to assess the degree of similarity between changes on different profiles. The data are set out as in Table 11.2 which refers to the comparison between profiles 1 and 2 for the upper part of the beach. The first column lists the dates at which the profiles were surveyed. In the next column, which refers to profile 1, the values are the amount of accretion which has taken place between two successive measurements of the same profile. Thus, -62 cubic feet per foot of material was lost between 20 April 1952 and the time of the previous profile (not listed). During the next interval of time, up to 17 June 1952, another 14 cubic feet per foot of material was lost, and so on. The same type of information is recorded in the next column for profile 2.

The fourth column, of negative and positive signs, is obtained by finding the difference between the p_i value on profile 1 and the q_{i+1} value on profile 2, and just recording its sign. Thus, the first negative sign is found by $-62 - (-2) = -60$, whose sign is negative, and so on. Column 5 is derived in the same way, except that from the q_i value for profile 2 is subtracted the p_{i+1} value in profile 1, and the sign recorded. Column 6 is obtained by recording the product of the two signs in columns 4 and 5. Thus for the first row, $-x - = +$. If a 'o' appears in any of the columns then the product is recorded as zero which, in any later analysis is ignored. In Table 11.2 there are 20 negatives and 10 positives in column 6, making a total of 31 observations relevant to the calculations. Of these 31 observations, and assuming there is no significant difference between the way in which the two profiles behave, it would have been

Table 11.2 Analysis of the amounts of accretion between successive profiles on each of two adjacent beach profiles (profiles 1 and 2 of Fig. 11.2)

Date of profile survey	Amount of accretion between two successive profiles cubic feet per foot unit		Sign of $p_i - q_{i+1}$	Sign of $q_i - p_{i+1}$	Product of previous two signs
	Profile 1 p	Profile 2 q			
20. 4.52	−62	−36	−	−	+
17. 6.52	−14	−2	−	+	−
17. 8.52	−4	−6	−	+	−
7. 2.53	−15	147	−	+	−
18. 4.53	51	24	+	−	−
20. 6.53	27	7	+	0	0
21.11.53	7	−31	−	+	−
24. 4.54	−49	45	−	+	−
12. 6.54	30	91	+	+	+
21. 6.55	−58	13	−	+	−
17. 3.57	−9	73	+	+	+
28. 4.57	25	−61	+	−	−
9.11.57	−26	−73	+	−	−
5. 5.58	39	−34	+	−	−
8.11.58	−11	129	−	+	−
11. 5.59	−10	17	−	+	−
1.11.59	−12	107	+	+	+
30. 5.60	46	−80	+	−	−
2. 7.60	−38	31	−	+	−
20.11.60	−76	40	−	+	−
19. 3.61	39	−12	−	−	+
24. 6.61	11	100	+	+	+
22.10.61	9	−69	+	+	+
5. 5.62	−113	−48	−	+	−
10.11.62	−79	84	−	−	+
27. 4.63	109	163	+	+	+
7. 7.63	7	−77	−	−	+
30.11.63	−9	79	−	+	−
26. 4.64	−5	81	−	+	−
10.10.64	−106	111	−	+	−
10.10.65	105	90	−	−	+
9. 6.66	135	270	−	+	−

expected that 15·5 of the observations would have been positive and 15·5 negative. That is to say, the expected frequency, E, is 15·5. This means that:

$$\chi^2 = \frac{\Sigma (O - E)^2}{E}$$
$$= \frac{(20 - 15·5)^2}{15·5} + \frac{(11 - 15·5)^2}{15·5}$$
$$= 2·7$$

(For explanation see section 10.2) The value of 2·7 is just significant at the 90% level, with one degree of freedom (section 10.2). Because there is a greater number of negative signs, this χ^2 value suggests a weak tendency for these two profiles to change in the reverse way to each other. A similar analysis on profiles 1 and 4, for the upper part of the beach, gave 16 minus and 11 plus signs, giving a chi-square value of 1·812, which is significant at only the 80% level. This again only suggests very tentatively that the two profiles might be responding in opposite ways to the processes which are operating. The results for profiles 2 and 4 (Fig. 11.2) gave eight minus and six plus signs, and this is definitely not significant.

When the whole beach is considered the results are rather different, and the values obtained are probably more important, for as the processes will operate over the whole profile, the results should be more representative of the total change in form over time. In correlating the trends for profiles 1 and 2 the results gave 16 minus and 15 plus signs, showing that the changes on profiles 1 and 2 appear to be independent of each other; sometimes one experiences accretion while erosion is taking place on the other. There is also no significant correlation between the changes on profiles 2 and 4 over the whole period, there being 11 minus and 14 plus signs; but in the early part of the period, 1953–59, there was a very weak tendency, significant at 80%, for the changes to correlate positively: there were eight plus and three minus signs in this period, and between 1957 and 1959 all the values were positive.

The highest correlation was found between profiles 1 and 4 over the period 1953 and 1963. During this nine year period there were 15 minus and two plus signs, giving a chi-square value of 9·94, which is significant at the 99% level. The preponderance of minus signs shows that the beaches were changing in the reverse way: one was accreting while the other was eroding. This correlation suggests that what was happening on profile 4 was affecting the changes on profile 1. The fact that all but one of the differences between the p_i value on one profile and the q_{i+1} value on the other were positive, shows that accretion was greater on profile 4, where nearly all the observations showed gains of material. The accretion on profile 4 is in part responsible for the losses taking place on profile 1. The sediment moves south on this coast, as demonstrated by the ridge movement analysed in section 10.4, so that if so much is accumulating at profile 4 there is not enough to maintain the level on profile 1.

Profile 2 is in an intermediate situation. It started by correlating positively, though only weakly, with profile 4. The lack of correlation between profiles 1 and 2 suggests that the effect of the accretion on profile 4 sometimes reaches as far as profile 2, but that at other times the accretion does not affect profile 2. There is a weak tendency for there to be more negative values (12 minus, seven plus) in the early period up to 1960, and more plus values in the later period (eight plus, four minus) in the correlation between profiles 1 and 2, although the results are not

statistically significant. This would agree with the positive results between profiles 2 and 4 for the early period up to 1960, and the non-significant results after this period.

11.3 Discriminant analysis

The application of discriminant analysis to third-order drainage basins was discussed in chapter 5. There it was shown how 130 basins could be classified into five groups, and that multiple discriminant analysis could be used to differentiate between the characteristics of each of these groups. Discriminant analysis can also be used, in the present context, to assess the degree to which two beach profiles differ in terms of the size of their beach ridges and the amount of accretion which takes place on them.

It will be recalled from chapter 5 that discriminant analysis is a multivariate technique that has some affinity with the t-test, in that its function

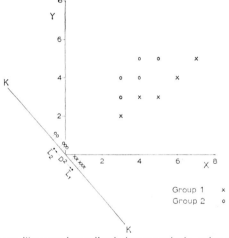

Fig. 11.3 Diagram to illustrate how discriminant analysis, using two variables, can differentiate between two groups. KK is the discriminant axis, and the distance between L_1 and L_2—the centroids of the groups—defines the discriminant function, D^2.

is to discriminate between two groups of data that are each characterized by several different variables. In the t-test, two groups of data can be shown to belong to two different populations, so long as the difference between their means divided by an estimate of the population standard deviation exceeds the appropriate value in the t-tables. In some examples, however, there is too much spread in the data, so that the degree of overlap is such that the two groups cannot be shown to belong to two separate populations on the evidence provided by the variable being tested. Figure 11.3 shows a hypothetical situation where in both the X and Y variables used to describe the two groups there is too much overlap for either variable to be suitable to differentiate between the two groups. With

discriminant analysis, however, it is possible to consider both variables together. The product of X and Y can be used for this purpose and the value of the product may be thought of as projected on to a 'discriminant axis'. This is an axis so placed that the separation of the two groups is brought to a maximum in terms of both the variables. The line K–K in Figure 11.3 indicates the axis on which the values for X and Y combined do not overlap at all, although on both the X- and Y-axes separately there is a large degree of overlap.

Two points may be obtained on the K–K axis that represent the mean centres (L̄) of the two groups of points differentiated by different symbols and belonging to the separate populations. The distance between the L̄ values for each of the two groups is called the *discriminant function* (Fig. 11.3) which is usually denoted by D^2. It is sometimes called *Mahalanobis' D squared*.

One of the problems of discriminant analysis is to decide how many variables are necessary or desirable to characterize the two groups of data. Usually some of the variables have a greater power to discriminate between the groups than others and it is these variables that should be included. The exact contribution of each variable cannot, however, be determined until the analysis is complete. It is, therefore, usually best to use as many variables as can be made readily available. When more than two variables are used, however, a computer is required for the calculations.

The computation of the discriminant function, D^2, may be illustrated by reference to the case study under consideration. The two beach profiles for which the data are analysed are profiles 1 and 4 of Figure 11.2. In this case X is the size of the beach ridges along these profiles, and Y the amount of accretion which is taking place. The data and calculations are set out in Table 11.3. There are six observations for profile 1 and seven for profile 4: thus the total number of variates N is 13. In the first place the totals and mean values of X_1, X_2, Y_1 and Y_2 are found. The differences between the means of the two X values and between the means of the two Y values, are represented by d_1 and d_2 respectively. A value known as the 'corrected sums of squares' is found by deducting a correction factor (C) from the uncorrected sum of squares. The value of C is found, for X_1, from $(\Sigma X_1)^2/n_1 = 8 \cdot 0^2/6 = 10 \cdot 66$, and so on for X_2, Y_1 and Y_2. The 'corrected sums of products' is found in a similar way. The sum of the products $X_1 Y_1$ is adjusted by another correction factor (c), where:

$$c = \frac{(\Sigma X_1 \times \Sigma Y_1)}{n_1} = \frac{8 \cdot 0 \times 14 \cdot 3}{6} = 19 \cdot 07$$

for the first profile, and similarly for the second profile. The value of the corrected sum of products need not be positive.

The 'pooled within group value' is then found by summing the corrected sums of squares and the corrected sum of products to give (xx), (yy) and (xy). The variance and covariance values can now be found by dividing

Table 11.3 Discriminant analysis of Profiles 2 and 4 of Figure 11.2

Total number of variates, $N = 13$
Let X = size of beach ridges, and Y = amount of accretion
Number in group 1, $n_1 = 6$; number in group 2, $n_2 = 7$

Group 1 (profile 1)					Group 2 (profile 4)				
X_1	$X_1{}^2$	Y_1	$Y_1{}^2$	X_1Y_1	X_2	$X_2{}^2$	Y_2	$Y_2{}^2$	X_2Y_2
1·6	2·56	3·0	9·00	4·80	1·0	1·00	3·0	9·00	3·00
1·2	1·44	2·3	5·29	2·76	3·7	13·69	7·3	53·30	27·00
1·8	3·24	2·2	4·84	3·96	3·7	13·69	9·0	81·00	33·30
1·6	2·56	2·2	4·84	3·52	4·6	21·16	10·7	114·50	49·31
0·8	0·64	2·4	5·76	1·92	4·0	16·00	10·8	116·60	43·20
1·0	1·00	2·2	4·84	2·20	6·8	46·24	14·2	201·60	96·50
—	—	—	—	—	4·6	21·16	16·2	262·50	74·50
8·0	11·44	14·3	34·57	19·16	28·4	132·94	71·2	838·50	326·81

Totals ΣX_1 8·0 ΣY_1 14·3 ΣX_2 28·4 ΣY_2 71·2
Means $\bar{X}_1$ 1·33 $\bar{Y}_1$ 2·38 $\bar{X}_2$ 4·06 $\bar{Y}_2$ 10·17

Differences between group means
$d_1 = \bar{X}_1 - \bar{X}_2 = -2·73$
$d_2 = \bar{Y}_1 - \bar{Y}_2 = -7·79$

Sums of squares

	x_1 A 11·44	y_1 A 34·57	x_2 A 132·94	y_2 A 838·50
	C 10·66	C 34·08	C 115·22	C 724·21
Corrected sums of squares:	0·78	0·49	17·72	114·29

Sums of products

	x_1y_1 A 19·16	x_2y_2 A 326·81
	c 19·07	c 288·90
Corrected sum of products:	0·09	37·91

Pooled within-group values
$(xx) = 0·78 + 17·72 = 18·50$
$(yy) = 0·49 + 114·29 = 114·78$
$(xy) = 0·09 + 37·91 = 38·00$

Variance and covariance values
$s_x{}^2 = ((xx)/N - 2) = 18·50/11 = 1·68$
$s_y{}^2 = ((yy)/N - 2) = 114·29/11 = 10·39$
$s_{xy} = ((xy)/N - 2) = 38·00/11 = 3·45$

the (xx), (yy) and (xy) values by $(N - 2)$. The resulting values are s^2_x, s^2_y and s_{xy}. In the example given these values are respectively $1 \cdot 68$, $10 \cdot 39$ and $3 \cdot 45$. The variance of y is considerably greater than that of the combined x values in this case. The values of the variance and the differences between the means are used to calculate D^2. This is the method that is also used in the simple t-test, and D^2 is found from:

1   Calculation of a_1 and a_2:

$$(xx)a_1 + (xy)a_2 = (N - 2)d_1$$
$$(xy)a_1 + (yy)a_2 = (N - 2)d_2$$
$$18 \cdot 50a_1 + 38 \cdot 00a_2 = 11 \times -2 \cdot 73 = -30 \cdot 03$$
$$38 \cdot 00a_1 + 114 \cdot 78a_2 = 11 \times -7 \cdot 79 = -85 \cdot 69$$
$$679 \cdot 43a_1 = -190 \cdot 62, \quad a_1 = -0 \cdot 28$$
$$679 \cdot 43a_2 = -44 \cdot 125, \quad a_2 = -0 \cdot 654$$
$$D^2 = a_1d_1 + a_2d_2 = (-0 \cdot 28 \times -2 \cdot 73) + (-0 \cdot 654 \times -7 \cdot 79)$$
$$= 0 \cdot 7644 + 5 \cdot 0947$$
$$= 5 \cdot 8591$$

2   The calculation of $\bar{L}$ proceeds as follows:

$$\bar{L}_1 = a_1\bar{x}_1 + a_2\bar{y}_1 = (-0 \cdot 28 \times 1 \cdot 33) + (-0 \cdot 654 \times 2 \cdot 38) = -1 \cdot 9$$
$$\bar{L}_2 = a_1\bar{x}_2 + a_2\bar{y}_2 = (-0 \cdot 28 \times 4 \cdot 06) + (0 \cdot 654 \times 10 \cdot 17) = -7 \cdot 8$$

3   From these values D^2 can, again, be calculated:

$$D^2 = \bar{L}_1 - \bar{L}_2 = (-1 \cdot 9) - (-7 \cdot 8) = 5 \cdot 9$$

Another check on D^2 is provided by:

$$D^2 = 1/\Delta(s^2_y d^2_1 - 2s_{xy}d_1d_2 - s^2_x d^2_2),$$
$$\text{where } \Delta = s^2_x + s^2_y - (s_{xy})^2 = 5 \cdot 552$$
$$D^2 = 1/5 \cdot 552(10 \cdot 39 \times 7 \cdot 45) - (2 \times 3 \cdot 45 \times -2 \cdot 73 \times -7 \cdot 79)$$
$$- (1 \cdot 68 \times 60 \cdot 86)$$
$$= 1/5 \cdot 552(77 \cdot 41 - 146 \cdot 74 - 102 \cdot 25)$$
$$= 5 \cdot 9$$

The value of D^2 checks with that found by the first method. As will be noted from the above, D^2 can also be found by calculating $\bar{L}_1$ and $\bar{L}_2$ and finding the difference between them. The value of L can be calculated for every value of x and y by the relation $L = a_1x + a_2y$. The mean value of L for each group of data can be found by summing the values of L and dividing by n. There is, however, a simpler method of finding $\bar{L}$ for the two variable case, since the value of $\bar{L}_1$ is given by $a_1\bar{x}_1 + a_2\bar{y}_1$ and the value of $\bar{L}_2$ is given by $a_1\bar{x}_2 + a_2\bar{y}_2$.

The next stage in the analysis is the calculation of *Hotelling's T squared*. The value of T^2 is found by:

$$T^2 = \frac{n_1 n_2}{n_1 + n_2} \times D^2 = 42/13 \times 5 \cdot 8591 = 18 \cdot 929$$

In order to check that there is a significantly greater difference between the groups than within the groups, considering all the variables, the T^2

value is converted into an *F*-value that can then be checked against *F*-tables. The *F*-value is given by:

$$F = \frac{N - m - 1}{(N - 2)m}T^2$$

where N is the total number of variates and m is the number of variables, which is two in this example. The degrees of freedom that are used for entering the *F*-tables are $v_1 = m$ and $v_2 = N - m - 1$. If the calculated value of *F* exceeds the tabled value then the discriminant analysis has shown that the two groups may be differentiated in terms of the two or more variables used in the analysis. In this case study:

$$F = \frac{N - m - 1}{(N - 2)m}T^2 = 10/22 \times 18\cdot929 = 189\cdot29/22 = 8\cdot604$$

Degrees of freedom for $v_1 = m = 2$, and for $v_2 = N - m - 1 = 10$. *F* 2, 10, 0·01 is 7·56; therefore *F* is significant at the 99% level. There is thus more variability between the two groups than within each of them when the variables X (size of beach ridges) and Y (amount of accretion) are combined in the analysis. The results confirm the difference of ridge growth on these two profiles.

11.4 Trend patterns

The amount of accretion which has taken place on a beach can be assessed from a set of continuous profile records in yet another way. The amount of material which has accumulated during the time period between two surveys along the same profile line can be calculated, as in Table 11.1, and cumulative totals can be found for each survey date. The cumulative totals can be plotted against the date and graphs drawn. This has been done for profiles 1, 2 and 4 at Gibraltar Point in Figure 11.4. The graphs of the actual data values are irregular, showing minor departures from the general trend. The latter has, however, been calculated (as in linear regression, section 3.5) for each of these three profiles. There is a considerable difference in the rate of accumulation of material along each of the profile lines, as indicated by the gradient of the trend lines. The trend for profile 4 is both the steadiest and the most rapid. In this case the trend equation is:

$$Y = 1321 + 205\cdot2X$$

where 1321 gives the value of the accretion in the middle of 1960, which is the central point of the analysis. The trend coefficient of 205·2 gives the rate of accretion in cubic feet per foot width of the beach between mean low neap tide and high tide level over each annual period used. During the whole period the beach, along profile 4, gained 2750 cubic feet of material, according to the trend line, and 2600 cubic feet according to the actual observations, which for the first time for several years showed

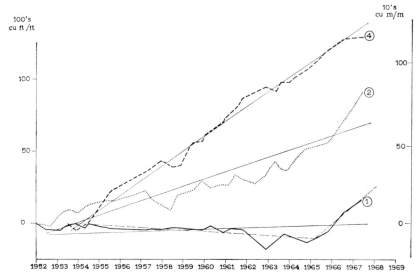

Fig. 11.4   Graphs to show the trend of accretion on profiles 1, 2 and 4 south of Skegness, Lincolnshire.

a slowing down of the rate of accretion during 1967. The difference of 150 cubic feet is the maximum divergence between the calculated and observed trend. The rate of accretion (i.e. the slope of the line) can be shown by this method to be very regular and rapid on this profile, at 205·2 cubic feet per foot per year. The rapid rate of accretion helps to explain the great width of the sweep zone (Fig. 11.2) and its gradual elevation.

As might be expected profile 2 shows an intermediate rate of accretion. The equation for its trend is:

$$Y = 673 + 100 \cdot 4X$$

The rate of accretion is therefore 100·4 cubic feet per foot per year, which is almost exactly half the value for profile 4. In 1960 the value of accretion had reached 673 cubic feet and the rate of increase was very nearly 100 cubic feet per year. The divergence between the regular calculated trend and the observed trend is rather greater, being 180 in the early period and increasing to −300 cubic feet during 1962 and +400 cubic feet in 1967, when the observed accretion was taking place more rapidly than suggested by the calculated trend. These changes indicate a slower trend during the early period, with a more rapid acceleration of the accretion since 1962 when the observed curve becomes steeper than the calculated one.

Two trends have been calculated for profile 1. The trend covering the whole period has an equation $Y = -55 + 13 \cdot 8X$. The trend coefficient of 13·8 is very low although it is just positive. The mean value in 1960

was -55 cubic feet, indicating a loss of material up to this time. There is, however, a marked divergence, amounting to 300 cubic feet between the observed and calculated trend. Trend lines can be made to fit the data more closely if the period is divided into two. The first part of the period has a trend $Y = -101 - 20 \cdot 1X$. This trend shows that the beach was losing sand during the period up to 1965 at the rate of 20 cubic feet per year, the loss amounting to 101 cubic feet in 1959 and to about 200 cubic feet by 1965. The trend appears to have altered during the last two years, when it changed to $Y = 113 + 250X$, showing a very rapid gain of sand which has raised the lower beach particularly. The mud on the lower part of the beach has been buried by new sand. The zone of accretion appears to have been moving south along the coast, as shown by the deceleration in movement on profile 4 and the rapid rate of accumulation on profiles 1 and 2. The reversal of the trend is particularly marked on profile 1.

The ridge size trends can also be calculated and compared with the trends for accretion. Their equations are as follows:

$$
\begin{array}{lll}
\text{profile 1} & Y = 1 \cdot 42 - 0 \cdot 0027X & \text{(whole period)} \\
\text{profile 1} & Y = 1 \cdot 33 - 0 \cdot 14X & (1953\text{--}1963) \\
\text{profile 1} & Y = 1 \cdot 44 + 0 \cdot 082X & (1961\text{--}1967) \\
\text{profile 2} & Y = 1 \cdot 86 + 0 \cdot 2X & \text{(whole period)} \\
\text{profile 4} & Y = 4 \cdot 40 + 0 \cdot 257X & \text{(whole period)}
\end{array}
$$

The slope of the trend lines is similar to those of accretion in that profile 4 has the steepest trend $(0 \cdot 257)$, followed by profile 2 $(0 \cdot 2)$. The trend on profile 1 is very slightly downwards over the whole period $(-0 \cdot 0027)$, consisting of an earlier steeper downward trend $(-0 \cdot 14)$, followed by an upward trend $(+0 \cdot 082)$.

11.5 Analysis of covariance

The relationship between the ridge height and accretion may be examined more closely by *analysis of covariance*. This method allows the correlation of two interval scale variables—the ridge height and cumulative accretion —to be calculated for all three profiles by controlling for differences between the profiles. There are thus two interval scale variables and one nominal scale variable in the analysis. The covariance test procedure allows the interaction between the nominal scale variable, in this case the three different profiles, to be calculated. If the interaction is not significant then the three nominal values can be pooled to give a stronger value for the correlation between the two interval scale variables. The method is rather similar to partial correlation in that the third variable is controlled, but it provides more information than partial correlation, as a test for interaction is included in the procedure. The details of the test procedure may be illustrated using the data for the three profiles.

The first part of the test is made to see if there is any interaction between

the profiles. This test indicates whether all the nominal scale variables can be pooled to give a justifiable correlation of the two interval scale variables for all the combined data. The interaction tested is based on the variation between the slopes (*b*) of the regression lines derived for each of the profiles. If the interaction is not significantly different from zero then all the data can be pooled. The average within-class correlation can then be obtained, and tested to see whether it also is significantly different from zero. The average within-class correlation coefficient so adjusts the means of each set of data that the differences between them are controlled. As a result they may be thought of as brought together to give an adjusted correlation between X and Y, the interval scale variables. Where the relationship is suitable the correlation becomes stronger as in Figure 11.5A; but it may become weaker under some circumstances, as in Figure 11.5B. The results of the analysis of covariance will be most interesting if the pattern of points is similar to one or other of the patterns shown in the

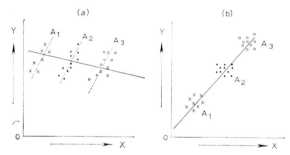

Fig. 11.5 Analysis of covariance. Total correlation (*r*) is increased in (**A**) if A_1, A_2 and A_3 are controlled by superimposition of means. Total correlation is decreased in (**B**) if A_1, A_2 and A_3 are controlled by superimposition of means.

diagrams. The procedure breaks up the covariance between the total set of points, into explained and unexplained parts.

The data, for the stretch of beach under consideration, consist of values for three different profiles. The profiles are the nominal scale variables. The interval scale data concerning the profiles are firstly the maximum ridge height, which is the dependent variable Y, and secondly the amount of cumulative accretion or loss on the beach, which is the independent variable X. The plot of the data (Fig. 11.6) shows some similarities with both the patterns given in Figure 11.5. Values for profile 1 cluster together showing little correlation, while the other two trend in the same general direction, but one line lies above the other. It is, therefore, worthwhile discovering whether there is a stronger relationship between the profiles considered separately than between the controlled data for all the profiles. The mean points and the regression trend through them are shown in Figure 11.6. The calculations from which they are derived are given in Table 11.4. The two columns headed 1 and 2 give the nominal scale variables

Table 11.4 Analysis of covariance for ridge height and accretion on the coast of Lincolnshire

Class Column (1)	N_j (2)	ΣY (3)	$(\Sigma Y)^2/N_j$ $(4) = (3)^2/(2)$	ΣY^2 (5)	Σy^2 $(6) = (5) - (4)$	ΣX (7)	$(\Sigma X)^2/N_j$ $(8) = (7)^2/(2)$	ΣX^2 (9)	Σx^2 $(10) = (9) - (8)$	ΣXY (11)
Profile 1	15	21·3	30·246	35·11	4·864	36·8	90·283	120·46	30·177	52·91
Profile 2	15	27·9	51·894	67·31	15·416	145·9	1419·12	1737·77	318·649	323·75
Profile 4	15	66·2	292·163	323·06	30·897	243·4	3949·571	5025·48	1075·909	1211·78
Sums	45	115·4	374·303	425·47	51·177	426·1	5458·974	6883·71	1424·736	1588·44
Totals	45		295·934	425·47	129·536		4034·694	6883·71	2849·02	1588·44
Between-class explained by A					78·369				1424·28	
Within-class unexplained by A					51·177				1424·736	

Correlations

Class Column (1)	$(\Sigma X)(\Sigma Y)/N_j$ $(12) = (3) \times (7)/(2)$	Σxy $(13) = (11) - (12)$	$b = $ slopes $\Sigma xy/\Sigma x^2$ $(14) = (13)/(10)$	$(\Sigma xy)^2/\Sigma x^2$ Explained by X $(15) = (13) \times (14)$	$\Sigma y^2 - (xy)^2/(\Sigma x^2)$ Unexplained by X $(16) = (6) - (15)$	$r^2 = (\Sigma xy)^2/$ $\Sigma x^2 \, \Sigma y^2$ $(17) = (15)/(6)$	$r = \sqrt{(17)}$ $(18) = \sqrt{(17)}$
Profile 1	52·256	0·654	0·0217	0·0142	4·850	0·00292	0·017
Profile 2	271·374	52·376	0·1644	8·6110	6·805	0·55857	0·747
Profile 4	1074·205	137·575	0·1279	17·5917	13·305	0·5694	0·7545
Sums	1397·835	190·605		26·2169	24·960	0·513	0·717
Totals	1092·710	495·730	0·174	86·257	43·279		
Between-class explained by A	305·125				17·743		
Within-class unexplained by A	190·605	190·605	$b_w = 0·134$	25·541	25·636	0·499	0·707

and the number of values in each. The columns 3 to 6 inclusive give the sums of squares for the Y variable and 7 to 10 inclusive give values for the X variable. The covariance is calculated in columns 11 to 13 inclusive, and column 14 gives the *b* coefficient, which is the slope of the regression line. Columns 15 and 16 give respectively the proportions explained and not explained by X. The final columns, 17 and 18, give the square of the correlation coefficient r^2 and the correlation coefficient *r*.

The first three rows in each column give the values for each A (the individual profiles) separately. The data from which these were calculated are not given. The steps necessary to derive the values are given under the column headings, where these are derived from previously calculated

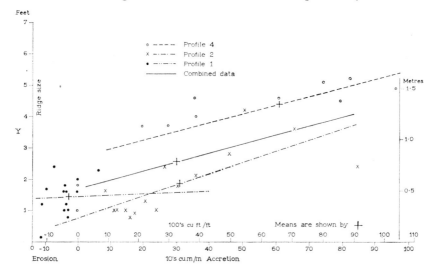

Fig. 11.6 Graph to show the relationship between ridge height and accretion. The combined data have been analysed by analysis of covariance.

values. The third column gives the sums of Y, the fourth gives the correction factor, which is the sum of Y squared and divided by N, and the fifth gives the corrected sum of squares for Y. The corrected sum of squares is found by deducting the correction factor from the uncorrected sum of squares. The X variable is treated in the same way. The corrected value for the covariance is found similarly in column 13, by deducting the correction factor calculated in column 12, from the uncorrected sum of products of X and Y found in column 11. The covariance divided by the corrected sum of squares for X, the independent variable, gives the *b* coefficient. The amount of variability explained by X is found in column 15 by dividing the squared covariance by the sum of squares of X. The amount of variability unexplained by X is the difference between the sum of squares of Y and the amount explained by X. The square of the correlation coefficient (r^2) is found by dividing the squared covariance by the

NAG—I

product of the sums of squares of both X and Y. The square root of this value gives r, the correlation coefficient.

The fourth row of the table gives the summed values for the three profiles together. The values are found simply by summing the three rows above. The values in the fifth row, labelled totals, are a little less straightforward to calculate. Column 4 is found by squaring the sum value in column 3 and dividing by the sum value in column 2, thus $(115\cdot4)^2/45$ $= 295\cdot934$. Column 6 is found by deducting $295\cdot934$ from $425\cdot47$ in column 5 to give $129\cdot536$. The same method is used for the X values in columns 8, 9 and 10. The total value in column 12 is found by $(115\cdot4 \times 426\cdot1)/45 = 1092\cdot710$. Column 13 is the difference between column 11 and column 12, which is $1588\cdot44 - 1092\cdot710 = 495\cdot730$. The value for b in column 14 is found by dividing the total value in column 13 by that in column 10. This value is $495\cdot730/2849\cdot02 = 0\cdot174$. The value in column 15 is the product of the two previous columns, $495\cdot730 \times 0\cdot174$ $= 86\cdot257$. This value is deducted from $129\cdot536$, in column 6, to give $43\cdot279$ in column 16.

The between-class, explained-by-A, value for Y in column 6 is the difference between $129\cdot536$ and $51\cdot177$, the total and sums values respectively. The value is $78\cdot369$. The difference between $129\cdot536$ and $78\cdot369$, which is $51\cdot177$, is the unexplained-by-A, within-class, value for Y. The values for X and XY are obtained in the same way in columns 10 and 13. The pooled data slope, b, is given in column 14 and is found by dividing $190\cdot605$ by $1424\cdot736$—the totals row of columns 13 and 10—to give $0\cdot134$. This regression line is entered on the graph in Figure 11.6 through the grand mean. The value of b ($0\cdot134$) $\times \Sigma$ xy ($190\cdot605$) gives the value explained by X ($25\cdot541$), which is entered in column 15. The unexplained part of X is found by deducting this amount ($25\cdot541$) from the total sum of squares of Y, $51\cdot177$, giving $25\cdot636$. X, therefore, explains very nearly half of the variation in Y, and this agrees with the value of $0\cdot499$ in column 17, r^2, found by dividing $25\cdot541$ by $51\cdot177$. The square root of this, given in column 18, is r, the correlation coefficient, which for the pooled data is $0\cdot707$. This value is rather smaller than those for profiles 2 and 4, but is much larger than that for profile 1.

The percentage variability that is explained by both X and A is found by subtracting $25\cdot636$ (the variation unexplained by X) from $129\cdot536$, which is the total sum of squares of Y, leaving $103\cdot900$. The ratio of $103\cdot900$ to $129\cdot536$ gives the percentage explained by both X and A. The value is $103\cdot900/129\cdot536 \times 100 = 80\cdot21\%$. The percentage explained by using both X and A is, therefore, considerably increased. The test for interaction can now be made (see Table 11.5). The F-value is $0\cdot507$, which is less than 1 and can, therefore, be considered insignificant. The value of $25\cdot636$ can now be used in the further tests for the error term, i.e. the amount of variation in Y not explained by X or A, assuming there is no interaction. The lack of interaction indicates that the slopes of the regression lines are sufficiently parallel for different A values to allow the

Table 11.5 Test for interaction

Source of variation	Sum of squares		df	Variance estimate	F
Unexplained by X and A, assuming no interaction	25·636	$N - (k + 1) =$ $45 - 4 = 41$			
Explained by interaction	0·676	$k - 1 = 3 - 1 = 2$		0·338	0·507
Error	24·960	$N - 2k = 45 - 6 = 39$		0·6656	

analysis to proceed. The figure shows that at least two of the individual profile regression lines are very close to the slope of the pooled regression line. This relationship explains the lack of interaction in this instance.

The next F-test is concerned with testing the significance of the within-class correlation (see Table 11.6). The value unexplained by A, 51·177, can be divided into that part explained by X and into that part left unexplained both by X and A, the error term. These values are 25·541 and 25·636. There are $(N - k)$ degrees of freedom for the part unexplained by A and $[N - (k + 1)]$ degrees of freedom for the error term. This

Table 11.6 Test for significance of average within-class correlation

Source of variation	Sum of squares		df	Estimate of variance	F
Unexplained by A	51·177	$N - k = 42$			
Unexplained by A but explained by X	25·541	1		25·541	40·866
Error	25·636	$N - (k + 1) = 41$		0·625	

leaves only one degree of freedom for the unexplained by A but explained by X portion. The F-value is 40·866, which is significant at the 99·9% level. This value is highly significant and shows that the correlation of X and Y, controlling for A, is strong. The correlation coefficient, $r_{XY.A}$, is the average within-class correlation, 0·707. The third F-test uses X as the control variable (see Table 11.7). The sum of squares used is the value unexplained by X, which is 43·293. This is associated with $(N - 2)$ degrees of freedom (43). The proportion explained by A of this amount is the difference between 43·293 and the error term, i.e. $43·293 - 25·636 = 17·653$. This has $k - 1 = 2$ degrees of freedom. There are 41 degrees of freedom left for the error term and the F-value is 14·1224, which is significant at the 99·9% confidence level, as F 2, 41, 0·001 is 8·25. The result shows that there is a significant difference among the three profiles analysed when the rate of accretion, the X value, is controlled. The correlation coefficient of variables Y and A, controlled for X, has also been calculated, and the value found is 0·467, which is a significant correlation.

The analysis has also provided a pooled estimate of the regression slope which can be used to predict the likely ridge height if the amount of accretion is known. It has also been shown that there is a significant relationship between the ridge height and the amount of accretion, and that the difference between the individual slopes of the regression lines is not significant, because the interaction value is very small. The individual profiles are significantly different when the X variable is controlled as shown by the third F-test.

Theoretically it would be expected that as a beach accumulates more material, its ridges would increase in height. The increase of material on a concave parabolic beach profile should have the effect of reducing the slope. A ridge is built by the waves in their attempt to establish their equilibrium gradient—which is steeper as the sand size increases—on a beach which is naturally too flat. If the flatness is increasing, as is likely to

Table 11.7 Test for significance of differences among adjusted means

Source of variation	Sum of squares	df	Estimate of variance	F
Unexplained by X	43·293	N − 2 = 43		
Unexplained by X but explained by A	17·653	k − 1 = 2	8·8265(v$_b$)	14·1224
Error	25·636	N − (k + 1) = 41	0·625(v$_e$)	

$$r_{iYA \cdot X} = \frac{V_b - V_e}{V_b + (n - 1)V_e} = \frac{8\cdot8265 - 0\cdot625}{8\cdot8265 + 14(0\cdot625)} = \frac{8\cdot2015}{17\cdot5765} = 0\cdot467$$

occur with accretion, the ridges would be expected to increase in height— the more so as the sand becomes coarser—if this hypothesis of ridge development is correct. Numerical analysis was carried out to test these hypotheses using the surveyed data from the south Lincolnshire coast. Each of the three profiles has a different mean ridge height and amount of accretion. Profile 4 has the coarsest sand. Profiles 2 and 4 show a fairly strong positive correlation between ridge height and accretion, with r values of 0·747 and 0·754 respectively. Profile 1 has not received nearly so much sand. Its ridges are lower and there is no correlation between ridge height and accretion, as shown in the scatter of points in Figure 11.6.

When the means of the three sets of data are combined by controlling for the different profiles, the value of the percentage variation in the data explained increases from about 50% to about 80%, using both X (the accretion) and A (the three different profiles) to explain the variation in Y (the ridge height). This is because the inherent variation between ridge size on profiles 2 and 4 is controlled. The lack of interaction, shown in the first F-test, between the profiles indicates that they all respond similarly to an increase in the accumulation of sand, in that their b coefficients are

significantly similar. The second F-test showed a very significant within-class correlation when A is controlled and X is left to do the explaining of the variation in Y. This result supports the hypothesis that ridge size is dependent upon the amount of material accumulating on the beach, although there are also other factors involved because a considerable percentage remains unexplained. This is partly accounted for by the rather haphazard operation of the tidal processes. The tides scour the runnels extra deep at times, thus making the ridges higher than normal. At other times deposition in the runnels, aided by salt marsh vegetation growth, causes a reduction in ridge height. These tidal processes in the runnels are unconnected directly with the rate of sand accretion and with the processes of ridge building. Both the latter are mainly dependent on wave action which brings the material to the beach and builds up the equilibrium slope.

The third F-test indicated a significant variation between the different profiles when they are controlled for the amount of accretion. The difference between profiles 2 and 4 is shown by the higher mean ridge height on profile 4 relative to 2, although their regression slopes are very similar (see Fig. 11.6). They react in the same way to sand accretion, but with significantly different ridge heights. This result supports the hypothesis that ridge height is also dependent on sand size. The coarser sand ($1\cdot35\ \phi$) on profile 4 needs a steeper slope for equilibrium to be established, and therefore, the ridges are higher for a given amount of accumulation than on profile two where the sand size is $2\cdot27\ \phi$. (NB if X is the diameter in mm, then $\phi = -\log_2 X$ (Inman, 1952).) The major criterion associated with variable A, the different profiles, is thus the sand size. The relationship between ridge height and accretion does not apply so clearly to profile 1, as it has not received enough extra sand for large ridges to form. Its sand size, $2\cdot35\ \phi$, is nearly the same as that on profile 2. This accounts for the similarity of mean ridge height, although the greater accretion on profile 2 has allowed larger ridges to grow. The results indicate that only on beaches where accretion is prolonged do large ridges form and grow, and that the ridges are larger as the sand becomes coarser (Fig. 11.6).

11.6 Coastal erosion

The loss of land to the sea only occurs in areas where certain conditions are fulfilled. There must be longshore drift operating so that more beach material is carried away from the coast than is brought to it, either from alongshore or offshore. This will be facilitated where the coastline is smooth in outline and the dominant waves reach it at a considerable angle. The lack of material will prevent an effective thick beach from forming, and the beach alone can protect the coast from attack by the waves. In attempting to make good the loss of material the waves will attack the cliffs or coast in their effort to build up their equilibrium slope.

One area where consistent erosion has been going on, probably since sea level reached its present height about 5000 years ago and certainly for the last 2000 years, is the Holderness coast in east Yorkshire. Material, moving along the coast from the north, cannot pass around Flamborough Head, which acts as a very large groyne, so that there is nothing to replace the material that is moved away from the Holderness beaches. The cliffs are composed of soft unconsolidated glacial material that yields readily to wave action in the absence of an effective beach. Material is

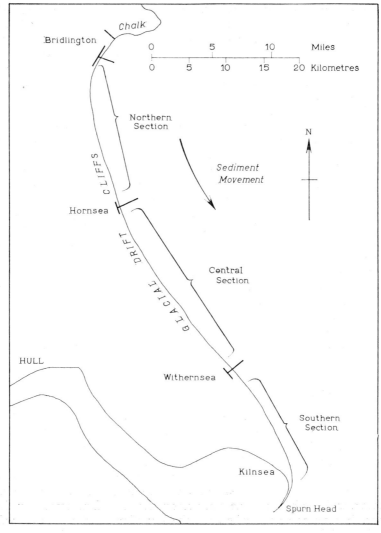

Fig. 11.7   Map of Holderness to show position of the three sections analysed.

also supplied to the foreshore by mass movement on the cliffs themselves. The rate of erosion on this coast has been studied in detail by H. Valentin (1954) over the period 1852–1952. The average rate of cliff retreat was found by him to be about 6 feet per year, although it varies along the coast.

The pattern of erosion is neither straightforward nor uniform. The analysis of erosion along this coast, which is presented below, has been carried out for three distinct sections, as indicated in Figure 11.7. The sea walls at Bridlington, Hornsea and Withernsea mark the northern point of each of these sections. Variations in erosion can be analysed with reference to these three sea walls by relating the amount of erosion to the distance south of the respective sea walls. The amount of erosion has been measured from Valentin's diagram at 1 km intervals from the relevant sea wall. The relationship between the amount of erosion and the distance south (of a sea wall) is shown in Figure 11.8A, for each of the three sections. The correlation coefficients are low for both the northern and southern sections, but for the central section $r = -0.944$ is very high and is significant at the 99% level. The central section is also interesting in that it is the only one where there is a *decrease* in the amount of erosion the further the cliff is located south of the sea wall. The northern section shows a slight tendency for erosion to increase southwards, although the result is not significant. This tendency could be due to the increasing shelter northwards as Flamborough Head is approached. The southern section shows no correlation at all with distance from sea walls, the highest values for erosion occurring at the southern end of this section. The effects of the sea-defence works are only felt consistently along the stretch of coast where exposure is most uniform. In this section erosion is most severe adjacent to the sea wall where the supply of material from the up-drift direction is least. Exposure seems to be the factor that counteracts the effect of the sea walls in the northern and southern sections.

The amount of erosion between 1852 and 1952 ranges from none in the north to 290 m in the south. The highest value is reached where the cliff is highest. Numerical analysis was therefore undertaken to assess the effect of cliff height on erosion. The relationships are illustrated in Figure 11.8B and c. The scatter of points (Fig. 11.8c) indicates an almost complete lack of correlation between cliff height, which ranges from 4 m to 43 m, and erosion. The correlation coefficient for the central section was found to be -0.109, which is not significant. The mean cliff height in this section was 18.57 m.

The mean amount of erosion in the three sections was compared by using the *t*-test (see Table 11.8). The *t*-value for a comparison of the values in the centre and south is $54.7/13.6 = 4.022$. The *t*-value for n = 37, at a confidence level of 0.001, is 3.60, so the difference is significant at the 99.9% level. The *t*-value for the south and north sections is 3.62 compared with the tabled value for *t*, n = 30, 0.01, 2.75 or 0.001, 3.646. The result is nearly significantly different at the 99.9% level. The *t*-value

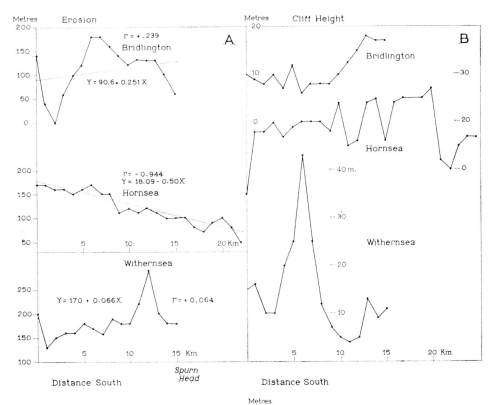

Metres — Erosion

$r = +.239$
Bridlington
$Y = 90.6 + 0.251 X$

$r = -0.944$
$Y = 18.09 - 0.50X$
Hornsea

Withernsea
$Y = 170 + 0.066X$ $r = +0.064$

Distance South   Spurn Head

Metres — Cliff Height

Bridlington

Hornsea

Withernsea

Distance South

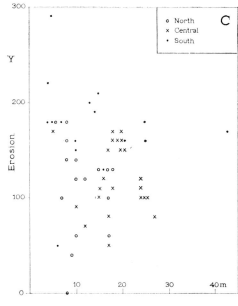

Metres — Erosion

o North
x Central
• South

Cliff Height   X

Fig. 11.8 Erosion along the coast of Holderness in relation to distance south of the sea defences in the three sections, and cliff height. The variation of cliff height along the coast is also shown.

for the north and central section is 0·644, which are not significantly different. The calculations show that the main difference lies between the high values for erosion in the south compared with the central and northern section of the coast. This is accounted for by the much greater exposure of the south, where deep water approaches close inshore. There is a resulting increase in the obliquity of wave approach and hence of longshore sand transport. The central section is saved to a certain extent because material eroded from the north is carried south to compensate for the increasing exposure southwards. The compensation is, however, outweighed in the south by a further and more rapid increase in exposure.

On the geological time scale rates of erosion such as those analysed are very great, and they are only attained locally and rarely exceeded on the world's coasts as a whole. Rates of erosion are normally much slower, as for example on the hard rock coasts of western Britain, although they are much more exposed. The erosion on the Holderness coast is on the whole fairly regular with time, although Valentin has shown that minor variations occur in association with oscillations of sea level.

Table 11.8 t-test data for erosion on the Holderness coast

Section	Mean	Number	Variance
North	111·9 m	16	235·0
Centre	120·9 m	23	123·4
South	175·6 m	16	228·33

A different type of erosion takes place on the Lincolnshire coast (Fig. 11.9). On this coast erosion has been operating for approximately the last 750 years, during which period the sea has driven between a quarter and half a mile inland, averaging about 3 feet per year. The erosion is, however, intermittent. The coast is low and its natural protection is a belt of sand dunes and a sandy beach overlying the clay foundation that was deposited during the post-glacial period behind a now-destroyed offshore barrier. Erosion is achieved under the exceptional conditions provided by the storm surges that periodically raise sea level to an abnormal height along this coast. The extra height of the sea allows the dunes to be breached and washed inland, while the clay foundation of the beach is stripped of its sand covering and permanently reduced in elevation by wave erosion. The dunes have been replaced along much of the coast by sea walls. These defences were severely damaged and breached during the storm surge of 1953, which is the latest occasion on which the coast has suffered erosion. The sea level rose during this surge to 8 feet above the predicted high tide level. The new sea defences have been built inside the old ones, which were destroyed, and the sea has thus gained on the land. The incidence of coast erosion in Lincolnshire follows

the Poisson distribution, which is useful for analysing rare events and provides a means of assessing the frequency of surges and the expectation of major surges along this coast.

The use of the *Poisson distribution* to predict the probability of occurrence of damaging storms may be illustrated by reference to the records of damage and flooding on the Lincolnshire coast. The historical data, for 1272–1952, used for the analysis were collected by A. E. B. Owen (F. A. Barnes and C. A. M. King, 1953). The last storm surge that caused

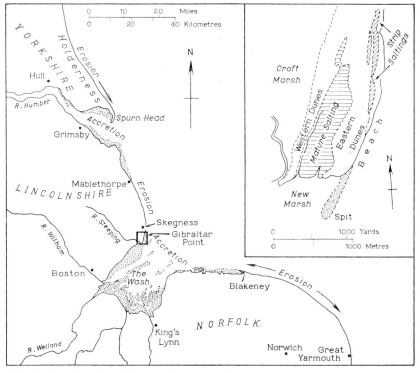

Fig. 11.9   Map of the Lincolnshire coast with an inset to show the morphological units at Gibraltar Point, Lincolnshire.

serious damage on the coast occurred in 1953. During the 69 decades from 1272 to 1962 there were 22 recorded reports of serious flooding and/or damage along the coast. These events will be assumed to have been the result of storm surges. The damage occurred occasionally and at irregular intervals. The timing of events that occur in this way can be described by the Poisson distribution, since it is suited to the description of isolated events in a continuum. This type of situation cannot be described as having a probability p of occurring and q of not occurring, so it cannot be adequately described by the binomial distribution. The

Poisson distribution is based on the natural growth as exemplified in compound interest, which is given by the exponential expansion e^z. The value of e, the *exponential function* is found by:

$$1 + \frac{1}{1!} + \frac{1}{2!} + \frac{1}{3!} + \frac{1}{4!} + \ldots = 2 \cdot 718282$$

(where 4! means factorial 4 and is found from $4 \times 3 \times 2 \times 1$, likewise factorial 3 is $3 \times 2 \times 1$.) The expansion of e^z can be written:

$$e^z = 1 + z + \frac{z^2}{2!} + \frac{z^3}{3!} + \frac{z^4}{4!} + \ldots$$

Table 11.9 Poisson distribution for storm damage in Lincolnshire

Number of damaging storms per decade	0	1	2	3	4
Probability p	0·727	0·232	0·37	0·0394	0·00314
Predicted frequency $p \times 69$ to nearest whole number	50	16	3	0	0
Observed frequency	50	16	3	0	0

To make this equation statistically more amenable both sides are multiplied by e^{-z}, because $e^z . e^{-z} = 1$. The equation now becomes:

$$1 = e^{-z}\left(1 + z + \frac{z^2}{2!} + \frac{z^3}{3!} + \frac{z^4}{4!} + \ldots\right)$$

Each term in the expansion gives the probability of 0, 1, 2, 3 etc. storms occurring in the time interval considered, where z is the mean number of storms during each time interval (a decade in this instance). The expansion is strongly skewed so that there is a much greater likelihood of no storm damage occurring than of many damaging storms occurring in one decade.

The number of recorded damaging surges during the 69 decades from 1272 to 1962 was 22. The mean number of surges per decade was 22/69, which gives a z value of 0·319. The value of e^{-z} is thus $e^{-0.319}$ which is equal to 0·727. The equation for the Poisson distribution can now be written:

$$0 \cdot 727\left(1 + 0 \cdot 319 + \frac{0 \cdot 319^2}{2} + \frac{0 \cdot 319^3}{3} + \frac{0 \cdot 319^4}{4} + \ldots\right)$$

The calculations are set out in Table 11·9. The observed and predicted frequency agree exactly, as shown in the table, so it can be concluded that the occurrence of damaging storms on this coast follows the Poisson distribution.

References

BARNES, F. A. and KING, C. A. M. 1953: The Lincolnshire coastline and the 1953 storm flood. *Geography* **38,** 141–60.

INGLE, J. C. 1966: *The movement of beach sand.* Amsterdam: Elsevier.

INMAN, D. L. 1952: Measures for describing the size distribution of sediments *J. Sed. Petrol.* **29,** 412–24.

INMAN, D. L. and BAGNOLD, R. A. 1963: Littoral processes. In Hill, M. N., editor, *The sea,* **3,** New York: Interscience.

KING, C. A. M. 1968: Beach measurements at Gibraltar Point, Lincolnshire. *East Mid. Geogr.* **4,** 295–300.

OWEN, A. E. B. 1952: Coastal erosion in East Lincolnshire. *The Lincs. Hist.* **9,** 330–41.

ROBINSON, A. H. W. 1964: The inshore waters, sediment supply and coastal changes of part of Lincolnshire. *East. Midl. Geogr.* **3,** 307–21.

VALENTIN, H. 1954: Den Landverlust in Holderness, Ostengland von 1852 bis 1952. *Die Erde* **3,** 295–315.

12 Beach material and beach form

12.1 General relationships

The nature of beach material exerts a strong influence on beach form. Beach material falls into three main types, shingle, sand and mud. Material that is intermediate between sand and shingle is very rare. There are, however, examples of this type of intermediate material on Looe Bar in Cornwall and at the western end of Chesil beach in Dorset. Most beaches consist of either sand or shingle or a mixture of both materials. On a worldwide scale sand is much more common than shingle, and mud will only occur where the beach is sheltered from strong wave action. A good example of the last is the lower foreshore at Gibraltar Point, Lincolnshire, where drying tidal banks, only half a mile offshore, protect the beach from wave action at low tide and allow mud to be deposited. Mud also accumulates within the sheltered runnels that lie landward of the ridges that form along parts of this coast.

The most obvious effect of material on beach form is the relationship between material size and beach slope. Shingle beaches are very much steeper than sand beaches. Coarse sand beaches are steeper than fine sand beaches. The relationship between material size and beach slope is due to the variation of porosity and permeability with material size. The fact that beach material is usually well sorted adds to the effectiveness of material size in determining permeability. A coarse material will let water pass through much quicker than a fine material, because of its greater permeability. The backwash of the wave, moving down the beach, will be reduced in volume relative to the swash, which moves up the beach. The backwash will, therefore, be relatively ineffective on a coarse beach. As a result the gradient will need to be steeper to maintain an equilibrium between the swash and backwash on a coarser beach.

This relationship has been studied by the Beach Erosion Board (1933) along the coast of New Jersey. A straight-line relationship was

found to exist between the tangent of the beach slope and the median diameter (in mm) of the sand. W. N. Bascom (1951) has demonstrated that the same relationship also applies to the coast of California. The gradients were measured at mid-tide level and ranged from 1 in 90 for a sand median diameter of 0·17 mm to 1 in 5 for a material of 0·85 mm median diameter. A similar analysis, on data from western Ireland, gave a highly significant product-moment correlation coefficient, r (section 3.1), between beach gradient and median diameter, of $r = 0.995$. Thus sand size explains 99% of the variation of the beach slope. The general similarity of the exposure of the beaches, in this case, ensures that wave lengths are comparable for each beach included in the analysis. The comparability of the wave length and wave steepness, which have already been shown to affect the beach gradient, ensures that the sand size is the major variable affecting the beach gradient on these beaches.

The type as well as the size of the beach material also affects the beach gradient. The effect of sand type is well illustrated in the steeper than normal beach gradients near Roundstone in Connemara, western Ireland. These beaches are formed of foraminiferal sand of much lower density than normal quartz sand. The size of the material is very similar to that on Inch spit in county Kerry, where the median diameter of the sand is 0·164 mm and the beach gradient is 1 in 89. Two of the beaches near Roundstone are built of sand with a median diameter of 0·152 and 0·150 mm. The slopes of these two beaches are 1 in 23·4 and 1 in 16·95, respectively. The beaches are much steeper than would be expected for the sand size. The light-weight material, which is composed almost entirely of calcareous foraminiferal remains, is more easily carried up the beach by the swash against gravity. The sand is also packed more loosely so that the backwash can penetrate more readily. A steeper than normal slope therefore results.

12.2 Material, tidal range and energy

The variables involved in beach processes are complex and numerous. In this section the effect of three variables on the response variable is studied. The form of the beach is taken as the dependent variable and the independent variables are two aspects of the material, the tidal range and an energy value. The form of the beach is indicated by its gradient, recorded by surveyed profiles over the whole width of the tidal section of the beach. The mean gradient over this stretch has been used, and has been modified to a logarithm of the cotangent of the beach slope to render the relationship between it and the other variables more nearly linear. This is the dependent variable Y. It is compared with four X values. The first variable X_1 is the mean sand size in ϕ units. This has been calculated by sieve analysis of a mid-tide sample in most instances. On a few beaches, where there is a very clear distinction between coarser

upper sand and finer sand lower down the beach, two samples and two slopes have been recorded. The mean value, M_z, has been used. The second variable, X_2, is also associated with the material and is the sorting of the sand, using the graphic measure of Folk and Ward (1957). The third variable, X_3, is the spring tidal range in feet. This is one of the process variables. The fourth variable, X_4, is concerned both with wave processes and with the area, in that it is a measure of the exposure of the beach in terms of the wave energy and wave length. An attempt was made to adjust the distance of the fetch, which in part determines the wave length, for the exposure of the beach by weighting the fetch length. The fetch distance was recorded in miles of open water. On some beaches, however, long swells could only reach the beach after considerable refraction, so that allowance was made by taking the exposure of the beach relative to the dominant swell direction into account. The energy factor, therefore, was calculated by multiplying the fetch by an exposure factor. The factors used were as follows:

Weighting factor	Orientation of line normal to beach
2	220–290°
1	290–045°
0·5	045–100°
0·25	100–160°
1	160–220°

The greatest weight was given to the beaches facing south-west, from which direction the largest and most powerful waves would come. The weighting factors seem reasonable in view of the distribution of the beaches concerned in the analysis. They are mainly around the British Isles and north France, but there are some from south France, Iceland and Baffin Island. The beaches have a very wide range of fetch, which is further extended by the weighting factors. To allow for this spread the logarithmic value of the energy factor was used in the analysis.

All the beaches analysed are sandy but they cover a wide range of types and conditions. They range from very coarse sand, mean size $-0\cdot44\phi$ (NB if X is the diameter in mm, then $\phi = -\log_2 X$ (Inman, 1952)) at Gateville near Cherbourg, to the fine sand, mean size $2\cdot92\ \phi$, of Inch Beach in west Kerry, Ireland. The sorting ranges between $0\cdot25$ for the best sorted sand on Redcar beach to $1\cdot32$ for the worst sorted sand on a beach near Skegness, where some shingle is mixed with the sand. Seven of the samples are very well sorted; 12 are well sorted; and all but one of the remainder are moderately sorted: only one is poorly sorted.

The tidal range also varied greatly, and the modifying effect on the beach slope due to this factor was investigated. The tidal range varied from negligible in the Mediterranean on the south French coast, through a low range on the south coast of Iceland and in Foxe Basin to the very

high ranges on parts of the west coast of England. Ranges over 26 feet occur at Blackpool and in south Wales. Intermediate values between 10 and 15 feet are found on the east coast of England.

The range of exposure is also considerable. Some beaches had short fetches across the English Channel or Irish Sea. Other beaches, such as those in western Ireland, south-west England and west Cherbourg Peninsula, are all exposed to long Atlantic swells. Not all these beaches face the direction of maximum fetch to the south-west, so allowance has been made by using the energy correction factor. The least exposed beach is that on west Baffin Island, where only small waves can approach over shallow water, which is frozen for most of the year.

The 27 beaches sampled have been analysed by simple product-moment correlation (section 3.1), partial correlation (section 4.5) and multiple correlation (section 4.4), as well as by trend-surface analysis (section 9.1), in order to assess the various interrelations between the beach slope and the four variables measured. The correlation coefficients are shown in Table 12.1. There are 100 possible correlations in the analysis. 17 of these are significant at the 99% level, ten at the 95% level, and 73 are not significant for the number of beaches tested. The results of the analysis show the relative importance of the factors in influencing the beach slope.

The simple correlation matrix is dominated by one significant correlation. This is the relationship between beach slope and sand size. This strong positive correlation shows that sand size explains 71% of the variability in the beach slope, as was suggested in the introduction to this chapter. The only other significant correlation at the 99% level is between sand size and energy, which explains 24% of the variability of the sand size. This relationship, which is positive, is of interest. It means that the finer sand (larger ϕ value) is associated with the more exposed environment. This is the reverse of the relationship that has been shown to apply on individual beaches on which mixed material has been sorted out in such a way that the largest material is associated with the greatest energy. On the beaches under discussion there is no connecting link between the individual beaches, most of which were relatively homogeneous along the length of the beach. The relationship may be due to the fact that most of the more sheltered beaches with the shorter fetches are on the east coast of England, in Baffin Island, north-east France and the Irish Sea, all of which are in areas where there is much glacial material, which contains a mixture of sizes, including much coarser material. Less glacial material has been available on the more exposed beaches on the west coast of Ireland, south-west England and west Cherbourg peninsula. The greater wave energy on these beaches has sorted out the material more effectively. This is indicated by the 95% significant negative correlation between sorting and energy. The well sorted beaches are those with higher energy.

A noticeable point in connection with the simple correlations is the low

correlation between tidal range and other variables. Tide has a negligible correlation with sand size, sand sorting and energy. This lack of correlation appears reasonable in view of the completely different controls that affect these variables. The tidal range is dependent on the response of

Table 12.1   Correlation coefficients for beach variables

Simple correlation coefficients

Slope	Size	Sorting	Tide	Energy	
1	**0·840**	−0·148	0·301	0·313	Slope
	1	−0·311	0·178	**0·489**	Size
		1	0·023	−0·397	Sorting
			1	0·055	Tide
				1	Energy

Correlations significant at 99% are bold

Partial correlations (variable in brackets is the controlled one)

Significant at 99%		Significant at 95%	
Slope and size (sorting)	+0·845	Size and energy (slope)	+0·438
Slope and size (tide)	+0·838	Size and energy (sorting)	+0·419
Slope and size (energy)	+0·829	Sorting and energy (tide)	+0·399
Size and energy (tide)	+0·487		

Multiple correlations (the first variable is correlated with the last two, all correlations are positive)

Significant at 99%		Significant at 95%	
Size—slope and energy	0·873	Energy—slope and sorting	0·473
Size—slope and sorting	0·861	Slope—tide and energy	0·423
Slope—size and tide	0·854	Sorting—size and energy	0·419
Slope—size and sorting	0·848	Energy—sorting and tide	0·402
Slope—size and energy	0·847	Sorting—tide and energy	0·400
Size—slope and tide	0·844	Sorting—slope and energy	0·398
Energy—size and sorting	0·553		
Energy—slope and size	0·520		
Size—tide and energy	0·512		
Size—sorting and energy	0·505		
Energy—size and tide	0·490		

the oceans to the tide-producing forces, while the sand type and energy are related to past sedimentation and configuration of the coast in relation to exposure and wave attack. The weak correlation between tide and beach slope suggests that a greater tidal range is faintly related to a flatter beach slope. This relationship would be reasonable in view of the

fact that a large tidal range spreads out the wave energy and would give waves less opportunity to build a steeper profile at the equilibrium gradient.

The weak correlation between energy and slope suggests that a flatter slope tends to be associated with greater exposure to long fetches. The longer waves would be expected to be associated with flatter slopes, so this relationship is reasonable. There is a weak negative correlation between sand size and sorting, the coarser sand being less well sorted. This relationship is often found within the sand range on beaches. The simple correlations show the great importance of size on slope, this being the only high correlation.

Three of the four partial correlations that are significant at the 99% level refer to the relationship between slope and size. They show the effect of controlling sorting, tide and energy, but only the control of sorting improves the correlation marginally. The other four significant correlations all refer to energy. The highest is energy correlated with size, controlled for tidal range, but this is slightly less than the simple correlation and the other correlations are also only changed marginally by controlling a third variable.

The paired multiple correlations have more significant combinations. The highest of these explains 76·2% of the variability in sand size in terms of beach slope and energy combined. There is a slight improvement in the correlation for slope when tide, sorting and energy are combined separately with sand size. The correlation of energy with sand size and sorting now increases to 0·553, and with sand size and beach slope to 0·520. These correlations are significant at the 99% level.

The multiple correlation coefficient of the beach slope taking all four independent variables into account together has also been computed, giving $r = 0.864$. The percentage variability explained is not very much greater than the explanation for sand size alone. The patterns of the simple and multiple relationships, some with the regression lines shown, are illustrated in Figures 12.1 and 12.2.

Trend-surface analysis is another useful method of analysis for this type of problem (section 9.1). Here it is used to illustrate the relationship between two variables and a third dependent variable. In this case the two variables are plotted along the U- and V-axes, and the surface is calculated for variations in the dependent variable Z. The trend-surface maps provide a valuable graphic means of illustrating the relationships beyond the simple linear correlation. The independent U and V variables in this analysis were the sand size in ϕ units and the logarithm of the energy, respectively. The dependent Z variable was the log cotangent of the beach slope. U and V, in this case, are the two independent variables that have been shown to exert most influence on the beach slope. When U, V and Z are plotted against each other, in the form of a trend-surface map the scatter of data values is such that there is a diagonal belt within which the accuracy of the trend surface is relatively high. The two

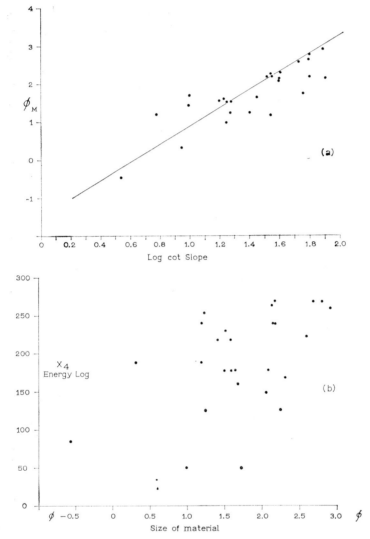

Fig. 12.1 **A:** Graph to relate the logarithm of beach slope cotangent to the mean beach material size in ϕ units; **B:** Graph to relate wave energy, in relation to fetch and exposure, with beach material size in ϕ units for 27 beaches.

remaining corners of the map are of little value in the analysis since few observations are available here from which the trend surface can be calculated. As a result the confidence limits for the surface are very large in these corners.

In this case study the linear surface accounts for $71 \cdot 82\%$ of the variability of the beach slope. The trend surface slopes down steeply almost

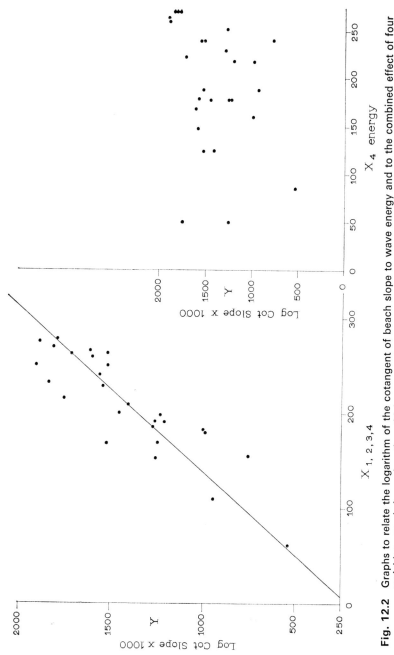

Fig. 12.2 Graphs to relate the logarithm of the cotangent of beach slope to wave energy and to the combined effect of four variables—sand size and sorting, tidal range and wave energy.

due left to right, indicating that the sand size is the most important variable. The linear surface equation is:

$$Z = 407.71 + 4.20U - 0.71V$$

The quadratic surface accounts for 75.76% of the variability in the beach slope, and this improves to 78.73% for the cubic surface. The equations for these two surfaces are:

Quadratic surface:
$$Z = 811.83 + 4.24U - 6.16V - 0.00066UV + 0.00088U^2 + 0.0156V^2$$

Cubic surface:
$$Z = -815.93 + 6.15U + 22.72V + 0.0291U^2 - 0.1005UV$$
$$- 0.08622V^2 - 0.00005816U^3 + 0.00004816U^2V + 0.000233UV^2$$
$$+ 0.00007952V^3$$

The pattern of the cubic surface, which is shown in Figure 12.3A, is of interest. Over the more reliable part of the map the pattern of isopleths shows well how both the material size and the wave energy control the beach foreshore slope. The importance of the sand size is brought out by the general north–south pattern of the isopleths over the central portion of the map. The slope reaches a minimum value of 1 in 79.4 or 43′ in the bottom right-hand corner. At this point the sand size is smallest and the wave energy and fetch is greatest. The high value of the wave energy is largely due to the great length of the waves, and the fetch factor will increase the energy in exposed situations, where long waves occur. These long waves would be expected to be associated with gentler gradients. This relationship on the better fitting cubic surface is the reverse of that on the linear surface. At the top left corner of the trend map the beaches reach their steepest gradient of more than 1 in 3.16 or 17° 33′. The sand size is largest at this point and the energy is lowest. The low energy implies sheltered conditions with low waves, which because of the fetch limitation will also be short. Short waves are associated with steeper equilibrium gradients for the same sand size. This relationship is not apparent on the linear or quadratic surface, owing in part to the sparseness of points in this part of the diagram. The trend towards flatter beaches with increasing energy is, however, apparent at the bottom right-hand corner of the quadratic surface, and the greater number of points in this part of the diagram gives the results greater confidence in this area. The confidence surface for the cubic surface could not be calculated as the matrix 'blew up' on inversion.

The relatively high degree of fit of the cubic surface through the points means that on the whole the residuals (section 9.1) are small. The largest residual of −41.79% of the actual value (i.e. with too steep an observed value) is on Cruit beach in western Ireland. At the time of observation this beach was being actively eroded and this may well account for the high steepness of the mid-tide gradient compared with the computed value. Another beach which was steeper than the computed value by

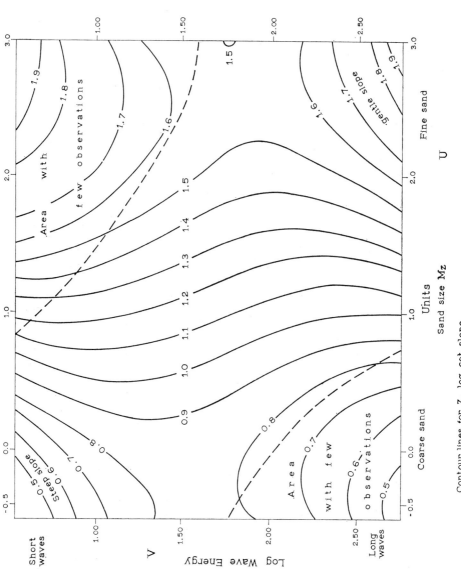

Contour lines for Z, log cot slope

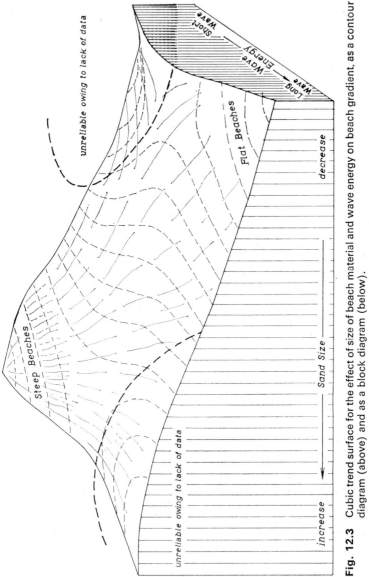

Fig. 12.3 Cubic trend surface for the effect of size of beach material and wave energy on beach gradient, as a contour diagram (above) and as a block diagram (below).

39·8% was that at Les Karentes in the Mediterranean. This is a tideless beach. The waves can, therefore, establish a more permanent swash slope. The gradient measured over this much shorter distance would be expected to be steeper than a mid-tide gradient on a tidal beach. The only other steeper beach, with a value of 23·1% steeper, was recorded on the cusp crest on Glen Bay beach, western Donegal. The gradient along the crest of the cusp is steeper than that along the trough, so this result is to be expected.

There are only three residuals giving a flatter slope in excess of 10%. The largest residual is 24·83% in the low tide gradient of Druridge Bay in Northumberland. This part of the beach was only under the swash of the waves at low tide, so an equilibrium swash slope gradient could not readily be established. The beach at Rhossili was flatter by 13·53% than the computed value. This can be accounted for by the fact that there is a belt of shingle at the top of this beach which was not included in the profile, so the slope value was rather lower than if the beach had been pure sand. The other beach, flatter than computed by 10·65%, was at Ingolfshöfdi in south Iceland. The great exposure of this beach to the long Atlantic swells could account for this effect.

This method of analysis, as well as providing polynomial equations of best fit relationship for two variables simultaneously in respect to a third dependent variable, also provides a useful visual representation of the three-dimensional relations of the three variables (Fig. 12.3B). It supplements and extends the same conclusions made from the multiple and partial correlations, in that it allows curved surfaces to be fitted to the data rather than just simple linear ones. All the stages in this analysis have shown the great importance of sand size in determining beach slope in a wide variety of beach types.

12.3 Material and time

Time influences beach material through the action of processes. One of the best situations for the assessment of the effect of time on material is when sea level is changing at a measurable rate so that the time factor is known. The effect on the material can be assessed by measuring the changing shape of beach pebbles as attrition by the waves rounds them to an increasing degree. The situation is usually not simple, however, as factors other than the rate of change of sea level determine the effectiveness of the waves. The complexities of the situation are increased by the effects of weathering on pebbles left above the sea on a rising coast. Some results of roundness studies on the coasts of Baffin Island will be used to illustrate the interaction between time and material on raised and modern beaches.

One series of observations was made on three beach ridges fairly close to the present sea level. The highest was 3 m above sea level, the middle

1·30 m and the lowest 1·25 m above present high tide level. Nine samples of 50 stones were taken, three on each ridge crest. The data were analysed by two-way analysis of variance to establish whether the variation between the ridges was significantly greater than the variation along each ridge. The stones were sampled randomly from points 10 m apart on each of the ridge crests. The results for the nine sample means are set out in Table 12.2. The figures give Cailleux's roundness index (Cailleux, 1947) which is found from the formula $2r/L \times 1000$, where r is the minimum radius of curvature in the principal plane and L is the length of the long axis in mm. The index reaches a maximum of 1000 for a completely circular pebble. The analysis of variance test (section 8.2) used all 50 values at each site so that it was possible to work out the interaction between the samples. The F-value for the interaction was 2·251, which is not significant at the 95% level. The interaction was, therefore, added to the residual term to test the F-value for the variation between the ridges and that along the ridges. The F-value for the variation along the ridges was 2·799, which is not significant at the 95% level. The variation

Table 12.2   Stone roundness on beach ridges

	Low ridge	Middle ridge	High ridge	Means
West	337	393	314	348
Central	381	450	272	368
East	366	447	361	391
Means	361	430	316	369

between the ridges was calculated to be $F = 19\cdot332$ and the table value for F 2, 450, 0·001 was 6·91. The F-value for the between-ridge variation is, therefore, highly significant. It is now necessary to account for this variation.

The upper ridge had the least round pebbles, the central ridge had the roundest and the lowest ridge had pebbles of intermediate roundness. The shape of the pebbles on the highest ridge could be due to two factors. First of all, the sea level was falling more rapidly when this ridge was built; hence the stones would not have been influenced by the waves for so long. In the second place, the stones have been above the waves for a longer period of time and some of them showed increasing angularity due to weathering. The material involved in all the ridges was granite-gneiss, with some schistose pebbles: both these materials appeared particularly liable to subsequent fairly rapid weathering, which increased their angularity. Size does not appear to be a significant factor as there is no significant difference in size between the groups of pebbles. The second ridge would have formed when sea level was falling more slowly. The marine limit in this area dates from about 8000 years B.P. and was at a

height of about 22 m. By the time the second ridge began to form the rate of uplift would be falling off rapidly. This would have allowed the waves longer to round the pebbles of the second ridge. Time for weathering subsequently has been limited. The lowest ridge is still being actively modified by the waves at present in times of storm, but this ridge protects the second from attack by the sea at its present level. The rather lower roundness of the stones on this ridge appears to be due to the short period the sea has been acting on it, as it is at almost the same height as the second ridge. It has probably been formed recently and has not yet had time to reach the degree of roundness of the stones of the second ridge.

Another test of variations of beach stone roundness was made, using samples from a wider range of sites and elevations. A simple one-way

Table 12.3 Roundness of beach stones

Eskimo Point terrace	Eskimo Point beach ridge	Cape Henry Kater ridges	Cape Henry Kater marine limit	Cape Henry Kater modern beach
8 m	10 m	2 m	22 m	0 m
172	293	337	375	591
168	350	381	408	
162	322	366	342	
206	328	393		
222	285	450		
233	336	447		
208	319	314		
229	317	272		
263	274	361		
means 207	314	369	375	591

Each of the values in the table refers to the mean of 50 pebbles

analysis of variance test was carried out to analyse the significance of the difference between the stone roundness. The data are set out in Table 12.3. The calculated F-value for these data was 27·73, and the tabled value for F 3, 26, 0·001 is 7·36. The result is, therefore, highly significant. It is also useful to work out which pairs are significantly different by means of the t-test (section 2.5). The modern beach at Cape Henry Kater is very different from all the others, but the difference between the Cape Henry Kater low beaches and the Eskimo Point ridges, which have mean values of 369 and 314 respectively, is less easy to assess. The t-test showed that these sets of pebbles were significantly different at the 98% level as t was 2·683 and t 16, 0·02 is 2·58. The low beaches at Cape Henry Kater do not differ from the marine limit ridges, but there is a very significant difference between the Eskimo Point ridges and terrace stones: t is 7·973 for these groups and t 16, 0·001 is 4·01.

These results cannot be explained solely by the time factor, because the similarity between the marine limit beaches and the low beaches at Cape Henry Kater shows that other factors must be involved. In this instance it is the exposure which explains the difference. The marine limit ridges face the open Davis Strait, with deep water offshore, while the low Cape Henry Kater beaches were in the more sheltered and shallower waters of Home Bay. The effect of time is shown, however, on the much greater roundness of the modern beach pebbles, which reached a high value of 591, when compared with the marine limit pebbles, with a roundness of 375. The modern beach pebbles are the roundest, and they and the marine limit pebbles, which are the next roundest, both occur on the beaches with the greatest, and with similar, exposures. The time factor can explain the difference in roundness in these two samples, as sea level is now only changing height very slowly compared with the change at the marine limit. The smaller roundness value of the Eskimo Point stones compared with the Cape Henry Kater beaches is partially due to the time factor. The Cape beaches are lower and have had longer to form. The lesser exposure of the Eskimo Point beaches also plays a part, as these beaches, which are 12 miles (19 km) farther up Home Bay, are sheltered from the open sea by a rock peninsula. The significant difference between the Eskimo Point ridge and terrace stones is due to at least two factors. Firstly unweathered stones were sampled on the ridge, while some of those on the terrace were weathered. The second reason is the form of the deposits. The stones on the terrace were on a flat surface where they must have been thrown by the waves, but they could not subsequently have been moved owing to the low gradient. The stones on the ridge, which were taken from both crest, distal and proximal sides, were in a situation in which effective wave action could have operated on them for a longer period of time. Their similar height shows that the time factor was similar in both samples. In general the results of this numerical analysis show that time and exposure are important factors in the rate of attrition by rounding of beach material.

In part III a number of numerical techniques that are of value in the analysis of coastal forms have been discussed. The case studies illustrate how these techniques are applied, though there are many situations in the study of coasts, not touched upon here, where they may also be useful. Part IV, the final part of this book, illustrates the application of these, and other numerical techniques, to glacial landscapes.

References

BASCOM, W. N. 1951: The relation between sand size and beach face slope. *Trans. Am. Geophys. Union* **32,** 866–74.

BEACH EROSION BOARD 1933: *Interim report.* Washington, DC.

1947

CAILLEUX, A. L'indice d'emoussé: definition et premiere application. *C. R. Somm Geol. France* **13,** 250–2.

CARR, A. P. 1969: Size grading along a pebble beach: Chesil Beach, England. *J. Sed. Petrol.* **39,** 297–311.

FOLK, R. L. 1966: A review of grain-size parameters. *Sedimentology* **6,** 73–93.

FOLK, R. L. and WARD, W. C. 1957: Brazos River Bar: a study in the significance of grain size parameters. *J. Sed. Petrol.* **27,** 3–27.

INMAN, D. L. 1952: Measures for describing the size distribution of sediments. *J. Sed. Petrol.* **22,** 125–45.

1963: Sediments: physical properties and mechanics of sedimentation. In Shepherd F. P., editor, *Submarine geology,* New York: Harper and Row, chapter 5. (2nd edn.)

KLOVAN, J. E. 1961: The use of factor analysis in determining depositional environments from grain size distribution. *J. Sed. Petrol.* **36,** 115–25.

KRUMBEIN, W. C. 1963: A geological process-response model for analysis of beach phenomena. *Bull. Beach Erosion Board* **17,** 1–15.

MCCAMMON, R. B. 1966: Principal component analysis and its application in large-scale correlation studies. *J. Geol.* **74,** 721–33.

MILLER, R. L. and ZIEGLER, J. M. 1958: A model relating dynamics and sediment pattern in equilibrium in the region of shoaling waves, breaker zone and foreshore. *J. Geol.* **66,** 417–41.

WHITTEN, E. M. T. 1964: Process-response models in geology. *Bull. Geol. Soc. Am.* **75,** 455–64.

Part IV Glacial forms

13 Forms of glacial erosion

13.1 Introduction
13.2 Glacial valley morphology
    [*curve-fitting, parabola, logarithmic forms, regression analysis, product–moment correlation coefficient, Spearman rank correlation coefficient, autocorrelation of residuals*]
13.3 Corrie form
    [*two-way analysis of variance, two-way analysis of variance with replications*]

13.1 Introduction

The different aspects of glacial and periglacial geomorphology are so numerous and varied that it would be quite impossible in a short account such as this, to consider them all. This part of the book is highly selective, and the emphasis has been placed on the analysis of a few characteristics of glacial and periglacial features in order to illustrate those numerical techniques which may most usefully be employed in this field of geomorphology. Some of the techniques used in earlier parts of this book are demonstrated once again to be of value. In addition some tests not used earlier, for example the fitting of lemniscate loops to drumlins and the Kruskal–Wallis and the Kolmogorov–Smirnov tests are introduced, with case studies, in this part.

The analytical methods considered in this and the next chapter are concerned with aspects of glacial landscape that relate to glacial forms and their development. The analyses are chiefly centred on the static characteristics of the features, rather than the processes which have been at work in their formation. The analytical methods used belong essentially to the empirical and areal approaches to land-form study, and these are the aspects which particularly lend themselves to statistical analysis.

When the morphology of glacial features is under consideration the whole form is usually studied and one example of the feature is compared with other individuals. It may in some instances be possible to measure all the features in one particular area, but it would rarely be feasible to study all the examples of any one feature, which will normally have a wide distribution in many parts of the world. The results obtained from numerical analysis of glacial features, as for drainage basins (part I), may only be applicable to one particular area. In some instances the numbers of individuals may be too great for all of them to be measured. This would apply to some of the larger drumlin fields, which may consist of 10,000 or more individual drumlins. Sampling is therefore essential in this type of study, and may be undertaken along the lines suggested in section

1.1. The study of erosional morphology of glacial valleys, which is the concern of this chapter, is rather different in that the valley is a continuous feature which can be measured anywhere along its course, giving an infinitely large number of possible measurements. In this situation sampling is again essential.

13.2 Glacial valley morphology

In chapter 7 mathematical curves were fitted to slope profiles. Curves can be fitted in a similar way to the cross-profile of glacial valleys. These valleys are usually described as U-shaped, but it is useful for purposes of comparison and also for the study of their formation to describe the form more precisely in terms of a specific curve. The simplest curve that can be considered to approximate to the form of a well developed glacial valley is a *parabola*. The equation for a parabola is $y = a\,x^b$. The y coordinate is taken as the elevation above the valley floor and the x coordinate as the distance from the centre line of the valley, measured in the same units. The constant, a, and the exponent, b, are adjusted to give the best-fit curve for the valley side slope. In the true parabola the exponent is 2 and this value is found to fit many glacial valleys closely if a is suitably adjusted. The higher the value of a the steeper the valley side. The goodness of fit of the parabola to the actual profile provides a useful indication of the extent of glacial modification of the valley. A well developed glacial valley fits the parabola very closely. Often the floor of the valley becomes infilled with later sediments. Thus, if the walls fit the parabola closely then it is possible to extrapolate the curve down to the valley bottom and hence obtain a measure of the depth of infill by a more objective method than merely continuing the sides of the valley down in a freehand curve (see also section 7.1).

The curve of a true parabola, $y = a\,x^2$, may be fitted by a simple process of trial and error, or alternatively the curve $y = a\,x^b$ may be fitted by the method of least-squares, as has been done by Svensson (1959). The method of fitting the curve $y = a\,x^b$ uses the logarithmic form of this relationship, which can be expressed as:

$$\log y = \log a + b \log x$$

Y can be used for log y, X for log x, and A for log a: the relationship can then be written as:

$$Y = A + b\,X$$

This is a linear equation and can be solved by the method of least-squares to obtain the best fit. The values of A and b then define the curve (*cf.* section 3.6). The logarithms of the height and distance measurements are used in the calculations to solve the equation. The solving of the equation, which gives unique values of A and b, indicates how closely the exponent of the curve is to 2. The correlation coefficient (section 3.6)

can also be calculated and this gives a measure of the goodness of fit of the curve to the observed data.

Before three examples of the application of this equation are considered, the approximate method will be applied to a number of valleys. The examples are drawn from the Lake District, the Pennines and north Wales. The profiles have been obtained from the 1 : 25,000 Ordnance Survey maps. The valley of Grisedale between Bleaberry Crag and St Sunday Crag near Helvellyn is a good example of a glacial valley. The cross-profile, which is shown in Figure 13.1, was drawn with an exaggeration of

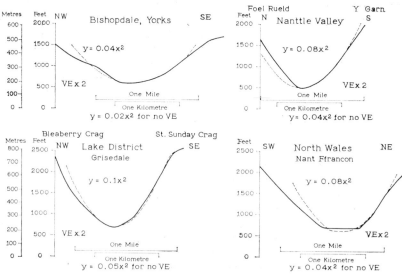

Fig. 13.1   Profiles and fitted parabolas in glaciated valleys in north Wales, the Lake District and Bishopdale, Yorkshire.

2 on the vertical scale. A parabola of the form $y = 0.1x^2$ fitted the valley side very closely. If the vertical exaggeration is allowed for, then the equation of the parabola which fits the form of the actual valley is $y = 0.05x^2$. The valley is 1800 feet (550 m) deep at the point studied, rising from 700 (214 m) to 2500 feet (760 m). The mean gradient of the slopes is 33°. The south-eastern side of the valley has the form of a smooth continuous curve throughout its length. The opposite side flattens off above 1200 feet (366 m) so here the upper part of a calculated best-fit curve could not be expected to fit so well. The other valley in the Lake District whose cross-profile is compared with the parabolic form is at Seathwaite in Borrowdale. The profile runs from Base Brown at 2120 feet (647 m) to Hind Crag. The equation of the parabola that best fits this profile is $y = 0.06x^2$. This curve fits closely from the valley bottom at 500 feet (152 m) to 1500 feet (458 m) on the east side, particularly in the upper portion. Above 1500 feet (458 m) the valley side slope flattens

N A G—K

off. On the west side the lower part fits best and then again the slope flattens more than the parabola from 500 feet (152 m) upwards. These two valleys show a similar asymmetry, which could possibly be related to post-glacial modification, although many more examples would be needed to confirm this.

Two valleys were measured in north Wales. One of these was the Nanttle valley, a diffluent glacial valley to the south-west of Snowdon, and the measured profile runs from Foel Rudd to Y Garn. The valley floor stands at 500 feet OD (152 m) and the hill crests rise to 2000 feet OD (610 m). The dimensions are very similar to those of the Lake District valleys. The south side of the valley is closely fitted by the parabola $y = 0.04x^2$. It is only near the top of the slope, that is to say above 1300 feet (400 m), that the parabola rises a little above the ground slope. The fit of the parabola to the profile derived from the map contours is good over the first 800 feet (244 m). The north side of the valley, however, rises more steeply from the centre line and the fit is not so good. The other valley was Nant Ffrancon. This valley has a flat floor, so an attempt was made to fit the curves to the sides and to extrapolate them down to the centre. The valley walls, however, are too straight to provide a good fit on either side over any considerable length of the valley. This valley, therefore, does not appear to have a well developed parabolic profile despite the fact that much ice must have passed through it.

The other valley that was tested was Bishopdale in the central Pennines. This valley received diffluent ice from Wharfedale at its head and has one of the best glacial cross-profiles in the Dales. The ice has lowered it considerably below the neighbouring valleys. The valley floor is at 600 feet OD (183 m) where it was measured with the walls rising 1600 feet OD (490 m) on either side. The valley runs from south-west to north-east. The north-west-facing slope is closely fitted by a parabola of the form $y = 0.02x^2$. The sides of the valley are considerably less steep than those of the other mountain areas, but the form is similar. The south-east-facing side of the valley, however, is not so smoothly parabolic. Beneath a scar the slope is steeper and then it flattens, finally steepening again towards the valley bottom. The recorded changes in slope could be accounted for by major slumping on the valley side, where shales underlie the limestone of the scar. This slumping is a common phenomenon in the dales and is a process which at times gives rise to reverse slopes on the hillsides. The south-facing slopes are more liable to this type of deformation, although it is not entirely restricted to them.

Parabolic curves have been fitted by least-squares to three of the glacial valleys already considered. Those that appeared to fit best by the approximate graphical method have been chosen for logarithmic fitting, using the side of the valley that showed the best developed form with least subsequent modification. The Y values (log y) for the height above the valley floors were taken at each 100 feet (30 m) contour interval, and the distance of this point to the centre of the valley was recorded.

The log values were used for a product-moment correlation analysis. This provided both the correlation coefficient and the regression coefficient, the former indicating the closeness of fit of the parabolic curve to the observed points, and the latter giving the exponent, which shows how closely the curve fits to a parabola. The value of log *a*, the regression constant, can also be calculated and converted to numbers from logarithms; the equation can then be written as a power relationship.

The graphs in Figure 13.2 show the results of plotting the values logarithmically with the regression lines inserted. In working out the regression coefficient, the distance from the valley centre, x, is taken as the independent variable, and the height, y, as the dependent variable. There is, however, no causal relationship involved in this case. The values

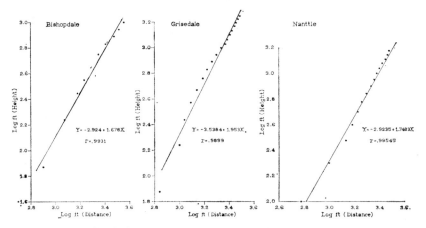

Fig. 13.2 The fit of glacial valley cross-profiles to logarithmic equations in north Wales, the Lake District and Bishopdale, Yorkshire.

for Bishopdale have been worked out for the south-east side of the valley. The value of *r* is 0·9931, which shows a very close correlation, and therefore a close logarithmic relationship that can be converted into a power function. The regression line reduces the square of the distance to the points in a vertical direction to a minimum. The equation for the regression line is:

$$\log y = -2 \cdot 924 + 1 \cdot 6785 \log x$$

which can be written:

$$y = 839 \cdot 54 x^{1 \cdot 6785}$$

The exponent is fairly close to 2, but it is sufficiently different to show that the best fitting curve is not a true parabola. The height range covered in this analysis was from 600 (183 m) to 1600 feet (488 m), so that the curve was fitted to ten points.

This valley may be compared with the south side of the Nanttle valley in north Wales. The slope here rises from 500 feet OD (152 m) to the

summit of Y Garn at 2000 feet OD (610 m), and 15 points were thus used for curve-fitting. In this case $r = 0.995$, which shows a good fit. The curve was not, however, a true parabola, as the equation is:

$$Y = -2.9235 + 1.748X$$

or:

$$y = 838.0x^{1.748}$$

The exponent of this curve (1.748) is rather closer to 2, and the slope therefore approximates rather more closely to a parabola than in the case of the Bishopdale example. The value of a is very similar in both equations and very much lower than the value derived from fitting the true parabola graphically. This difference may be due to the fact that the trend of the points, when plotted on a logarithmic scale, tends to flatten near the base of the curve. The points also get closer near the upper part on the logarithmic scale so that the lowest points do not influence the fit so much as the upper ones. In the graphical method the fit is generally better at the base of the slope where the curve is gentler in gradient.

The third profile that has been fitted by the method of least-squares is for the south-east side of Grisedale near Helvellyn. The slope at this point rises from 725 feet OD (220 m) to 2500 feet OD (760 m), which has been divided into 18 100 feet divisions. The horizontal distance covered in this vertical rise was 3100 feet (945 m). The value of r in this analysis is 0.9899, which is slightly lower than the other values of r obtained in the two examples considered above. Nevertheless it does indicate a good fit of the logarithmic curve to the measured points. The equation for the regression line is:

$$Y = -3.5384 + 1.953X$$

which can be stated also as:

$$y = -3450.05x^{1.953}$$

The exponent in this equation approaches closely to 2, and hence the best-fit curve is nearly a parabola. The value of a (in the general form $y = a x^b$) is lower in this curve, because the exponent, b, is rather larger. The points on this slope are similar to those in Bishopdale in that the upper and lower ones fall below the regression line and the central ones lie above it. In the Nanttle valley the upper points are slightly higher from the line and the lower ones lie below it, with the exception of the last point. The analysis makes it clear that in the Lake District and the Dales the middle to lower part of the slope has been oversteepened, while the upper and lowest part is rather flatter than the best-fit curve. The results of the analysis do show, however, that there is a good fit of a power equation to the slope of a glaciated valley. The value of the exponent and constant provide a quantitative value for comparison with other areas.

These examples of fitting a simple parabola, $y = a x^b$, to valley side slopes in glaciated valleys relate to data obtained from the 1 : 25,000 Ordnance Survey maps. Such data are not sufficiently accurate to warrant the application of a more elaborate curve-fitting procedure, although they

do illustrate well the general parabolic form of strongly glaciated valleys. A rather more elaborate curve-fitting technique can be applied to profiles surveyed in the field. A case study is provided by data obtained in the Front Range of the Rocky Mountains of Colorado. Three slopes were measured (Fig. 13.3A). Two of them together form a high-level col through which ice possibly passed—across the continental divide eastwards over the high interfluves—in an earlier and more extensive glaciation than that which formed the most conspicuous glacial features of the area (see Fig. 15.8 and discussion on p. 344). The third profile (inset to Fig. 13.3A)

Table 13.1 Correlation coefficients for glaciated slope curve fitting

| Number | Transformation of | | Correlation coefficient and rank order for the slope of | | | | | |
| | | | Kiowa | | Albion | | Rainbow Lakes | |
	Height	Distance	r	rank	r	rank	r	rank
1	Linear	Linear	0·9771	4	0·9923	5	0·9917	2
2	Linear	Log	0·7146	9	0·9125	10	0·8329	9
3	Log	Linear	0·9981	1	0·9972	2	0·9549	4
4	Log	Log	0·6442	10	0·9497	9	0·9386	6
5	Linear	Inverse	0·3125	11	0·7371	14	0·4763	15
6	Inverse	Linear	0·9778	3	0·9855	6	0·8190	10
7	Inverse	Inverse	0·1961	13	0·8449	13	0·7598	11
8	Linear	Square $^{-1}$	0·3120	12	0·5695	16	0·3484	16
9	Square $^{-1}$	Linear	0·9280	6	0·9604	8	0·7243	12
10	Square $^{-1}$	Square $^{-1}$	0·1541	16	0·7369	15	0·7180	13
11	Linear	Square	0·8890	7	0·9944	3	0·9747	3
12	Square	Linear	0·9285	5	0·9724	7	0·9503	5
13	Square	Square	0·8029	8	0·9994	1	0·9954	1
14	Log $^{-1}$	Linear	0·9949	2	0·9943	4	0·8771	7
15	Linear	Log $^{-1}$	0·1756	15	0·8459	12	0·6226	14
16	Log $^{-1}$	Log $^{-1}$	0·1947	14	0·9110	11	0·8412	8
			$n = 34$		$n = 14$		$n = 17$	

Note: the symbol $^{-1}$ refers to the inverse value

is of the side of a small valley that probably contained a cirque glacier during the last major ice advance, but now only contains a small semi-permanent snow patch.

In order to assess the transformation of the data that gives the best-fit curve, the observed heights and distances, ordered cumulatively, were used as the independent x and dependent y values in a curve-fitting computer program written by Dawson (1969). 16 different transformations were fitted to the data and the results are set out in Table 13.1 where the correlation coefficients between x and y are given for each of the three slopes. The Spearman rank correlation test has been used to assess the

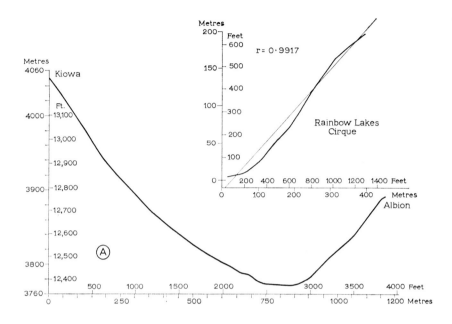

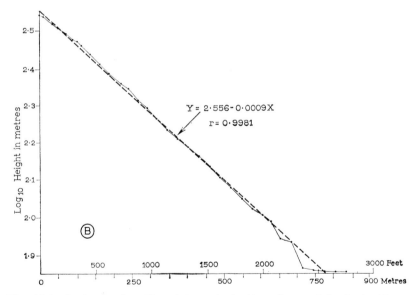

Fig. 13.3 **A**: Surveyed profiles of slopes in the Front Range of the Rocky Mountains of Colorado; **B**: Transformed log linear version of the Kiowa Slope.

similarity of slope forms. The order of best fit for each of the transformation combinations has been ranked from 1 to 16, and the Spearman test applied to these rank orders. The results are set out below:

Albion and Kiowa slopes   —Spearman rho 0·802, significant at 99%
                           level for n = 16
Albion and Rainbow Lakes—Spearman rho 0·809, significant at 99%
                           level for n = 16
Kiowa and Rainbow Lakes—Spearman rho 0·580, significant at 95%
                           level for n = 16

The results show that there is a general similarity in the form of the three surveyed glaciated valley sides, a similarity that is strong both between the two sides of the probable diffluent col and between the Albion and Rainbow Lakes slopes; the latter have similar dimensions, being about half the height of the Kiowa slope.

The best-fit curve for the Kiowa slope is that which necessitates a log-transformation for the height and an untransformed distance value (Fig. 13.3B), giving an r value of 0·9981. This form is related to a parabola, in that it indicates a steepening of the slope steadily away from the valley centre. The parabola, $y = a x^b$, can be written in logarithmic form as $\log y = \log a + b \log x$, whereas the best-fit curve found to apply to this data has the form $\log y = a + b x$. The values for a and b in the example are 5·886 and −0·0021 respectively. A double log-transformation does not fit any of the three slopes well. The best-fit transformation for the Albion slope is in the form $y^2 = a + b x^2$, giving an r-value of 0·9994. This is closely followed by the same $\log y = a + b x$, curve that gave the best fit to the Kiowa slope which gives an r-value of 0·9972. Both these transformations have the effect of contracting the scale upwards along the slope, thus straightening the curve in which the gradient increases progressively up-slope. The equations for the two best-fit transformations for the Albion slope are as follows:

$$\log y = 1·778 + 0·0014x \quad (r = 0·9972)$$

$$y^2 = 4023·986 + 0·2394x^2$$

The same double square transformation also fits the Rainbow Lakes cirque valley slope best, giving the equation:

$$y^2 = −1565·1066 + 0·27645x^2 \quad (r = 0·9954)$$

The linear form of the correlation fits the Rainbow Lakes slope slightly less well than the double square transformation, but the high correlation both for this slope and for the Albion slope shows that they do not depart far from a linear form, the respective correlation coefficients being 0·9917 and 0·9923. The reasons for the approximate linearity of the slopes may be due to different causes. In the Rainbow Lakes valley the glacier was a small cirque glacier and the slope was measured near the head of the valley. The ice, therefore, was probably relatively ineffective in modifying

the valley side as it would not be very fast moving or thick. The Albion slope, on the other hand, was glaciated much earlier and there has been considerably more time for the slope to have been modified subsequently by processes of mass movement. The slope is covered in rock debris and shows signs of periglacial activity and nivation processes that could modify the parabolic glacial form.

The results of these analyses suggest that the main characteristic of glaciated slopes is a steady up-slope steepening, but that this need not be exactly parabolic in form. All the transformations that fit well have the effect of contracting the vertical scale relative to the horizontal one.

Data of the type analysed in this section is similar to time-series, in that the values are not independent of those adjacent to them. This means that *autocorrelation* may occur. This happens when one value is not independent of the ones that precede and follow it (section 10.4). The presence of autocorrelation can be detected by a simple test, in which a value k is calculated from $k = m^2/s^2$, where s^2 is the variance and m^2 is the sum of the differences between each value and the preceding one squared and divided by $(n-1)$ (see worked example below). If k is small then there is positive autocorrelation, but if k is larger than a value given in tables by B. I. Hart (1942) then there is negative autocorrelation. If the k value lies within the upper and lower limits given by Hart (1942) then the values can be considered as independent of each other. It seems likely that on a slope there would be autocorrelation between adjacent portions, which means that both x and y values must change continuously unless there are overhangs. The test can also be used to test the residuals from the regression for autocorrelation. This is of value in the present study in that the residuals are a measure of the irregularity of the slope in this instance, or of its departure from a smooth curve according to the transformation used.

As an example of testing residuals from regression for autocorrelation, the Albion slope for a logarithmic height transformation and a linear distance scale is given. The r-value for this fit is 0·9972. The data are set out in Table 13.2. The residuals multiplied by 100 are listed in the first column and squared in the second. The consecutive differences are listed in column 3 and squared in column 4. The sums of columns 2 and 4 are respectively $u^2 = 17·20$, and $(u_{i+1} - u_i)^2 = 19·32$. The value of m^2, which is given by:

$$m^2 = \frac{\sum_{1}^{n-1} (u_{i+1} - u_i)^2}{n-1} = \frac{19·32}{13} = 1·49$$

The value of s^2, which is given by:

$$s^2 = \sum_{1}^{n} (u_i - \bar{u})^2 = \frac{17·20}{14} = 1·23$$

The value of k is now found:

$$k = \frac{m^2}{s^2} = \frac{1 \cdot 49}{1 \cdot 23} = 1 \cdot 21$$

This value lies between the upper and lower significance levels for the 95% and the 99% confidence level. (NB for n = 10 and at the 95%

Table 13.2 Autocorrelation of residuals: Albion slope

u × 100	(u × 100)²	$(u_{i+1} - u_i)$	$(u_{i+1} - u_i)^2$
2·2	4·84	3·7	13·69
−1·5	2·25	0·7	0·49
−2·2	4·84	1·1	1·21
−1·1	1·21	1·6	2·56
0·5	0·25	0·2	0·04
0·7	0·49	0·3	0·09
1·0	1·00	0·4	0·16
0·6	0·36	0·0	0·00
0·6	0·36	0·4	0·16
0·2	0·04	0·2	0·04
0·4	0·16	0·2	0·04
0·2	0·04	0·8	0·64
−0·6	0·36	0·4	0·16
−1·0	1·00		
−6·4 +6·4	Σ = 17·20		Σ = 19·32
Σ = 0			

$\Sigma u = 0$ $\bar{u} = 0$ n = 14

level the upper limit is 3·26 and the lower limit 1·18; for the 99% level the upper limit is 3·61 and the lower limit 0·84.) It may, therefore, be concluded that there is no autocorrelation in the residuals. They represent random fluctuations of the surveyed data points around the fitted curve. In this example the semi-log-transformation provides a very close fit to the surveyed profile.

13.3 Corrie form

Studies of the detailed distribution of corries have been made which pay particular attention to variation in their altitude and aspect, as these variables appear to be interrelated. The data can be examined by means of an analysis of variance test (section 8.2) to consider the significance of these and other variables in influencing the distribution of corries in detail in any one area. Seddon (1957) recorded the elevation and aspect of the

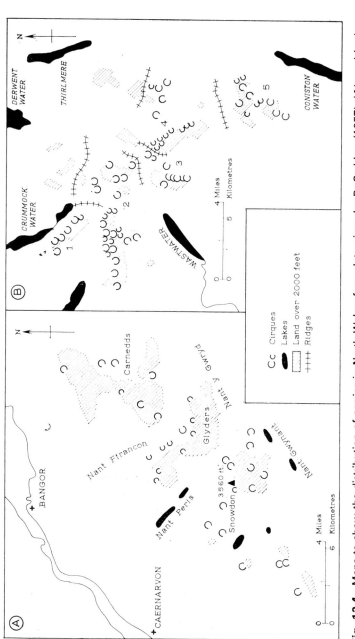

Fig. 13.4 Maps to show the distribution of corries in North Wales, from data given by B. Seddon (1957) (**A**), and in the western Lake District from data given by P. H. Temple (1965) (**B**).

corries in three mountain groups in north Wales, namely the Snowdon groups, the Glyders and the Carnedds (see Fig. 13.4A). These corries, in part at least, are the result of the last ice to affect north Wales. The groups lie approximately in a belt running from south-west to north-east. The Snowdon group, which is the most south-westerly, has the maximum precipitation, and the Carnedds group, which is the most north-easterly, has the lowest precipitation. In each group there is a variation in the aspects of the corries and in their altitude. The observations collected

Table 13.3 Corrie altitudes (feet) in north Wales

Aspect	North-east	North-west	South-east	Means
Group				
Snowdon	1200	1700	1300	1400
Glyders	1500	2000	2000	1833
Carnedds	2000	2100	2200	2150
Means	1566	1933	1833	1777

by Seddon can be studied by analysis of variance techniques to test the significance of the observed variations in altitude and aspect in the three different groups. As three variables are involved, a two-way analysis of variance design is required. The data will first be analysed by a simple two-way analysis of variance and then replications will be added to show how the effect of interaction can be worked out. The data need to be

Table 13.4 Adjusted corrie altitudes in north Wales

Aspect	North-east	North-west	South-east	Totals
Group				
Snowdon	−5	0	−4	−9
Glyders	−2	3	3	4
Carnedds	3	4	5	12
Totals	−4	7	4	7

slightly adjusted to provide equal numbers of replications in each cell of the matrix.

In the simple test the data are averaged to give a mean altitude for each group for each aspect (see Table 13.3). In order to simplify the calculations the data are adjusted by dividing each entry by 100 and deducting 17 from the result. The table can now be re-written as in Table 13.4. From this it will be seen that the correction factor is $7 \times 7/9 = 5.44$ and N is 9; the between 'aspect' sum of squares is $27 - 5.44 = 21.56$;

the between 'group' sum of squares is $80.33 - 5.44 = 74.89$; the total sum of squares is $113.00 - 5.44 = 107.56$. The analysis of variance table can now be set up with these values (see Table 13.5). The residual value is used as the denominator in working out the F-ratio. The F-ratio for the aspect is $10.8/2.78 = 3.92$, and F 2, 4, 0.05 is 6.94 (see section 8.2). The F-ratio for the groups is $37.45/2.78 = 13.46$ and F 2, 4, 0.025 is 10.6. In this analysis the residual value is used for the denominator because it represents that part of the total height variation which is explained neither by the aspect nor by the groups. The null hypotheses are set up to state that there is no significant difference between the altitude of the corries either with respect of their aspect or to their group. The result of the test shows that with regard to aspect there is no significant difference in the altitude of the corries. The corries facing north-east are lower than those facing in other directions, but the difference is not sufficiently large to be statistically significant, even at the 95% level. With regard to the different mountain groups the F-value is high enough to allow the null hypothesis to be rejected with 97.5% confidence. From

Table 13.5 Analysis of variance table for the data in Table 13.2

Source of variation	Sum of squares	df	V	F
Between aspect	21.56	2	10.8	3.92
Between group	74.89	2	37.45	13.46
Total	107.56	8		
Residual	11.1	4	2.78	

this analysis it may be concluded that precipitation, which in this area decreases north-eastwards, is of greater significance than aspect in determining corrie altitude.

It is worth considering the analysis that can be carried out if replications are used, as these allow an interaction effect to be calculated. The interaction is affected by the relationships between the variables. An example of interaction will make the concept clearer. In most mountain groups in the British Isles the north-east-facing corries are at a lower elevation than those facing in other directions. However, in one group the relief may be such that wind drives snow off the north-east-facing slopes, so that in this group the north-east-facing corries are higher. In this instance there is an interaction between one of the groups and the other variable (the corrie elevation) causing its value to be anomalous in the general pattern. One of the assumptions in the analysis of variance test is that there is no interaction between the variables. It is therefore useful to be able to test for this effect, for if there is interaction it may reveal interesting relationships that previously escaped notice. The test using replications will be applied to a comparison of corrie elevations

in the Lake District and Snowdonia with reference to their aspects, using the different groups within these areas to provide the replications.

First, however, the Lake District data must be analysed separately. Data suitable for this analysis have been collected by P. H. Temple (1965) for the west and central part of the Lake District. The measured altitude of the corries was used and the two criteria analysed were the aspect of the corrie and the group of mountains where it was situated. The groups of mountains (Fig. 13.4B) were

1   the hills south of Buttermere
2   the hills of Pillar and Great Gable
3   the Scafell group of hills
4   the Langdale Fells
5   the Coniston Fells.

The rather uneven spread of data necessitated linking some of the nearest groups together. The data were analysed in three groups by combining (1) and (2), and (3) and (4). The resultant groups formed compact areas arranged in a line from north-west to south-east. There is, therefore, not nearly so strong a variation in precipitation associated with this grouping as in north Wales, where to a certain extent the groups farther to the

Table 13.6   Corrie altitude Lake District

	(Absolute values of altitude, feet)			
	NE	SE	NW	SW
Groups (1) and (2)	1600	1700	1600	1800
Groups (3) and (4)	2100	1900	2000	1400
Group (5)	1800	1700	1800	1900

Table 13.7   Adjusted values of corrie elevation

(Adjusted values found from: absolute values divided by 100 minus 17)

	NE	SE	NW	SW	Totals
Groups (1) and (2)	−1	0	−1	1	−1
Groups (3) and (4)	4	2	3	−3	6
Group (5)	1	0	1	2	4
Totals	4	2	3	0	9

Correction factor is $81/12 = 6.75$
Between 'aspect' sum of squares is $9.67 - 6.75 = 2.92$
Between 'group' sum of squares is $13.25 - 6.75 = 6.50$
Total sum of squares is $50.00 - 6.75 = 43.25$

north-east lie in the rain-shadow of those farther south-west. The pre-valent rain-bearing winds come from the south-west at present, and this pattern probably applied to the snow-bearing winds when the corrie glaciers were active. The data for the three Lake District groups are set out in Table 13.6. Both F-values in Table 13.8 are too low to allow the

Table 13.8 Analysis of variance table for corrie altitudes in the Lake District

Source of variation	Sum of squares	df	V	F
Between aspect	2·92	3	0·97	0·17
Between groups	6·50	2	3·25	0·58
Total	43·25	11		
Residual	33·83	6	5·64	

null hypothesis to be rejected, and it must be concluded that neither the aspect nor the groups makes a significant contribution to the variation in corrie elevation.

The other test is now carried out for the frequency with which corries

Table 13.9 Corrie frequency in Lake District

Frequency	NE	SE	NW	SW	Totals
Group (1)	6	0	2	0	8
Group (2)	11	2	13	2	28
Group (3)	4	5	2	0	11
Group (4)	10	3	1	1	15
Group (5)	7	3	0	1	11
Totals	38	13	18	4	73

fall into different aspect groups, regardless of elevation. All the groups can be included separately in this analysis, the data for which are set out in Table 13.9. The F-value (Table 13.10) for the variation in frequency between different aspects is significant at the 97·5% level as F 3, 12,

Table 13.10 The analysis of variance table, using adjusted figures for corrie frequency

Source of variation	Sum of squares	df	V	F
Between aspect	124·15	3	41·38	4·94
Between areas	62·30	4	10·57	1·39
Total	286·55	19		
Residual	100·10	12	8·35	

0·025 is 4·49, but there is no significant difference in the frequency of corries between the different areas. The result shows that the pattern of distribution in aspect is significantly different but that the numbers do not vary significantly in the different groups. The preponderance of north-east-facing corries is a noticeable feature of most mountain corrie groups in western Britain. It reaches extreme proportions in western Kerry, where in the Dingle Peninsula all but one of 36 corries face north-east. This characteristic appears to increase in intensity towards the more marginal areas of corrie development.

The two-way analysis of variance test with replications can now be used

Table 13.11   Corrie altitude N. Wales and Lake District

	(Elevation in hundreds of feet)			
	NE	NW	SE	District totals
Lake District	16 21 18	16 18 20	17 19 17	162
Subtotals	55	54	53	
North Wales	12 15 20	17 20 21	13 20 22	160
Subtotals	47	58	55	
Totals	102	112	108	
Grand total				322

The correction factor is $\dfrac{322^2}{18} = 5760·2$, N is 18, K is 3, L is 2

to examine the null hypothesis that there is no difference in elevation of corries in north Wales and the Lake District, nor in the elevation with respect to aspect. The data used for the replications are the mean values for elevation of corries in the three groups already used separately in each of the mountain areas. The data can be set out as in Table 13.11. The error sum of squares is found by deducting the between-subclass sum of squares from the total sum of squares: 131·8 − 22·5 = 109·3. The inter-action sum of squares is found by the difference between the sum of the row and column sum of squares and the between-subclass sum of squares: 22·5 − (8·5 + 0·2) = 22·5 − 8·7 = 13·8. The test for interaction is made by calculating the F-value for interaction, which is the interaction variance divided by the error variance, 6·9/9·1 = 0·76. The F-value for

interaction is not significant, so that this term can be added to the error term to test the significance of the between-column and between-row variation in the next analysis of variance table. The residual $df = (N - 1) - (K - 1) - (L - 1)$. Both F-values are so small that the null hypothesis must be accepted, with the conclusion that there is no significant difference in the corrie heights in the two mountain groups of north Wales and the Lake District. The elevations are in fact extremely similar. Even the differences in aspect are not significant. This result might have been

Table 13.12   Analysis of variance table: test of interaction

Source of variation	Sum of squares	df	V	F
Total	131·8	17		
Between subclass	22·5	5	4·5	
Between column (aspect)	8·5	2	4·25	
Between row (area)	0·2	1	0·2	
Interaction	13·8	2	6·9	0·76
Error	109·3	12	9·1	

expected from the lack of significance in this value in both the Lake District and north Wales when they are analysed separately. The only significant effect on the elevation of corries which the analysis has confirmed is that which exists between the groups in north Wales, where the rain-shadow effect of the successive hill masses on each other is marked.

Table 13.13   Analysis of variance table, test for between-column and between-row variation

Source of variation	Sum of squares	df	V	F
Between column (aspect)	8·5	$(K - 1) = 2$	4·25	0·49
Between row (area)	0·2	$(L - 1) = 1$	0·2	0·02
Residual	123·1	$(N - 1) - (K - 1)$ $- (L - 1) = 14$	8·7	

It is noteworthy that in north Wales the hills are highest to the west, and so would have a greater effect than in the part of the Lake District analysed.

In the Lake District the hill groups are almost equally open to precipitation-bearing winds. Even though aspect does not influence the elevation of corries, it has been shown that it does have a marked effect on their frequency.

Correlation of the relationship between corrie elevation and the height of the mountain groups where they occur for seven Lake District hill groups gave an insignificant value of $r = +0.29$. The heights of the mountains do not, therefore, appear to account for the variations in corrie elevations in the different mountain groups of the Lake District that were used in the analysis.

In a study of a wider area Linton (1959) has shown that the corries of Scotland rise in elevation eastwards. They are lowest in the wetter western highlands and highest in the drier eastern highlands, such as the Cairngorms. The distribution repeats, on a much larger scale, the relationship that has been analysed in the three mountain groups in north Wales. It appears that precipitation plays the largest part in determining corrie altitude. Aspect is more important in controlling the frequency of corries, with the north-east aspect strongly favoured in the British Isles, especially and to an increasing extent as conditions become more marginal for corrie formation.

References

ANDREWS, J. T. 1965: The corries of the northern Nain-Okak section of Labrador. *Geogr. Bull.* **7,** 129–36.

DAWSON, J. A. 1969: Curve fitting by least-squares methods. *Univ. of Nottingham, Computer Applications in the Natural and Social Sciences* **3.**

HART, B. I. 1942: Significance levels for the ratio of the mean square successive differences to the variance. *Ann. of Math. Stats.* **13,** 445.

LEWIS, W. V. 1947: Valley steps and glacial valley erosion. *Trans. Inst. Brit. Geogr.* **14,** 19–44.

1947: The cross sections of glaciated valleys. *J. Glaciol.* **1,** 37–8.

1954: Pressure release and glacial erosion. *J. Glaciol.* **2,** 417–22.

LINTON, D. L. 1959: Morphological contrasts of east and west Scotland. In Miller, R. and Watson, J. R., editors, *Geographical essays in memory of A. G. Ogilvie.*

1963: The forms of glacial erosion. *Trans. Inst. Brit. Geogr.* **33,** 1–28.

NYE, J. F. 1965: The flow of a glacier in a channel of rectangular, elliptical or parabolic cross-section. *J. Glaciol.* **5,** 661–90.

SEDDON, B. 1957: Late-glacial cwm glaciers in Wales. *J. Glaciol.* **3,** 94–9.

SVENSSON, H. 1959: Is the cross-section of a glacial valley a parabola? *J. Glaciol.* **3,** 362–3.

TEMPLE, P. H. 1965: Some aspects of cirque distribution in the west central Lake District, northern England. *Geogr. Ann.* **47,** 185–93.

14 Forms of glacial and fluvioglacial deposition

14.1 Distribution of drumlins
  [*test for random distribution,* t-*test, chi-square test, area under the normal curve*]
14.2 Drumlin morphology
  [*ellipse, ellipsoid, lemniscate loop, rose curve, analysis of variance,* t-*test*]
14.3 Moraine spacing
  Terminal moraines [*regression analysis, product-moment correlation coefficient*]; Ice-cored moraines [*chi-squared test, null hypothesis, degrees of freedom, regression analysis*]
14.4 Moraine morphology
  [t-*test, analysis of variance,* F-*value*]
14.5 Esker morphology
  [*topology, nodes, test for randomness*]

14.1 Distribution of Drumlins

One of the variable characteristics of a drumlin field is the number of drumlins per unit area, i.e. drumlin density. In part this is a function of the size of the drumlins. Numerical analysis can be applied to a study of the distribution of drumlins in order to test for differences in drumlin density and to see if individual drumlins are distributed at random. Three areas have been selected for analysis:

1   The Appleby area and the Eden valley immediately north of Appleby, where the drumlins are fairly rounded and the ice flow was not vigorous.

2   The Solway lowlands around Wigton to the west of Carlisle, where the ice flow was both constant and vigorous.

3   An area in central western New York State, around the town of Newark, which is used for comparison as the ice conditions here were different from those in the Eden valley area. It lay far from the ice centre and its relief is generally lower. This drumlin field is one of the most extensive in the world, containing an estimated 10,000 drumlins and covering an area 32 × 130 miles between Lake Ontario and the Finger Lakes.

The pattern of drumlin distribution in these three areas may be compared by assessing the frequency of drumlin occurrence in units of uniform area. In the Eden valley area the map prepared by S. E. Hollingworth

(1931), on which all the drumlins are shown, was used to record the drumlin frequency distribution, and in the New York State area the 1 : 62,500 map of the United States Geological Survey was used. This map, with a 20 feet contour interval, shows the drumlins clearly. The size of unit which gave a reasonable number of drumlins per square unit was a square with a side of 1·5 miles. The values used here are the number of drumlins per 2·25 square miles and 60 squares have been counted for each of the three areas covered. The pattern of distribution in the Eden valley and Solway is shown in Figure 14.1. The mean and standard

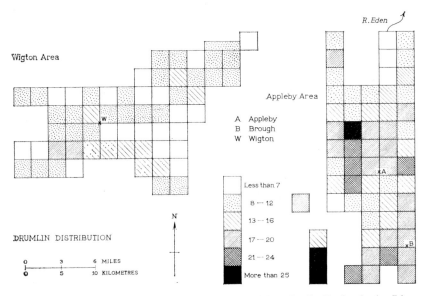

Fig. 14.1 Diagram to show variations in density of drumlin distribution in the Eden valley around Appleby and the Solway around Wigton.

deviation of the drumlin frequency in calculated below for each area, these values enabling a *t*-test to be carried out to compare the density of drumlins in the different areas. The chi-square test is also used to ascertain whether the distribution of drumlins within the area studied is significantly different from a random distribution. The results of the chi-square test will be considered first.

The expected frequency of drumlins per unit square, assuming a regular distribution, is found by dividing the total number of drumlins recorded by the number of squares (60 for each area), which gives the mean number of drumlins per square. For the Appleby area there are 946 drumlins and the mean value of drumlins per 2·25 square mile is 15·77; the total number of drumlins in the Wigton area is 560, giving a mean frequency of 9·33; and in the New York State area a total of 422 drumlins gives a mean frequency of 7·03.

As the expected frequency in any one area is the same for each square, the value of chi-square can be worked out by using the equation:

$$\chi^2 = \sum \frac{(o - e)^2}{e}$$

(See section 10.2.) The value for this equation, in the Appleby area (1) is:

$$\chi^2 = 1954 \cdot 42 / 15 \cdot 77 = 123 \cdot 933$$

There are 59 degrees of freedom: this is more than the maximum number entered in significance tables, so the significance can be worked out by using the value of Z (the area under the normal curve—see section 10.1), where:

$$Z = \sqrt{2\chi^2} - \sqrt{2df - 1}$$

For the values in this area $Z = \sqrt{248} - \sqrt{117}$ which is 4·931. This gives an area under the normal curve equal to 0·49999, a result significant to 0·00001 (i.e. 0·5 − 0·49999 = 0·00001). It may be concluded from this result that the distribution within the Appleby area is far from regular. In parts of the area there are as many as 29 and 34 drumlins in one unit area, while elsewhere the number falls to 4, 5 and 6. The larger numbers occur in the more marginal parts of the belt, and in isolated areas around Orton and Ravenstonedale. In addition, high density values occur around Brough, where the ice was beginning to move upwards to pass over the Stainmore Pass and also just around Appleby, in what was the vicinity of the basal ice-shed. Density decreases to less than the mean northwards along the Eden valley, where ice flow became more concentrated. It is also noticeable that where drumlin density is high drumlins are small. Within the Appleby area, therefore, there is a significant variation in the distribution of drumlin density, which is associated with the nature of the ice flow—which was also variable within the area. The drumlins are more numerous in those areas where conditions were marginal for their formation and where the ice moved less consistently and slowly, as at the basal ice-shed, or where it was being forced to flow uphill to escape over the col of Stainmore, which acted as a diffluent gap at this stage.

The pattern in the Wigton area west of Carlisle is different. The expected number of drumlins is 9·33 per unit area. The chi-square value is 667·40/9·33 = 71·533, the significance of which may be worked out as in the previous example to give:

$$Z = \sqrt{143} - \sqrt{117} = 11 \cdot 958 - 10 \cdot 817 = 1 \cdot 141$$

The area under the normal curve is 0·3729. This gives a probability of 0·5000 − 0·3729 = 0·1271, which is not a significant result. There is no reason, therefore, to reject the null hypothesis that the distribution of the drumlins is anything but regular. In this area it can be assumed that the ice forming the drumlins was behaving in a uniform way and drumlins were formed regularly. (The density pattern may be seen in Figure 14.1.) The setting and relief of the second area supports this conclusion, as there

are no obstacles to prevent a uniform flow of ice and the relief is much more subdued than in the Eden valley. The form of the drumlins, with their greater elongation, also suggests rapid and consistent ice flow.

In the area around Newark in New York State the drumlins have been counted on the map, and the 60 2·25 square mile units, used for the two earlier case studies, have been maintained for comparability. The expected frequency per square being 7·03 drumlins:

$$\chi^2 = 154/7{\cdot}03 = 21{\cdot}906$$

This value is much lower than that obtained for each of the other two areas, and it is definitely not significant: H_0, the null hypothesis, cannot be rejected. The drumlins in this area, therefore, show a great degree of uniformity and regularity in their distribution. The ice appears to have behaved in a regular manner, and, as in the Wigton area, the relief in the immediate vicinity is not great.

The t-test (see section 8.7) may be applied to the results for the three areas to ascertain whether the density of drumlins is significantly different in the three areas. The data required for the analysis are set out in Table 14.1. The t-test between X and Y gives a t-value of $6{\cdot}44/0{\cdot}86 = 7{\cdot}4884$. There are 118 df and t 120, 0·001 is 3·291, so that the calculated t-value is significant at the 99·9% level. It may be concluded confidently that there

Table 14.1 Drumlin density in three areas

Appleby area, X		Wigton area, Y		Newark area, Z	
n.	60	n.	60	n.	60
ΣX	946	ΣY	560	ΣZ	422
ΣX^2	16,872	ΣY^2	5894	ΣZ^2	3122
$\bar{X}$	15·77	$\bar{Y}$	9·33	$\bar{Z}$	7·03
$V = \dfrac{16{,}872}{60} - (15{\cdot}77)^2$		$V = \dfrac{5894}{60} - (9{\cdot}33)^2$		$V = \dfrac{3122}{60} - (7{\cdot}03)^2$	
	= 32·61		= 11·12		= 3·56
St. dev.	5·7	St. dev.	3·33	St. dev.	1·89

is a significant difference in the drumlin density in the Appleby area and the Wigton area, both of which occur in the same drumlin field. The t-value between Y and the Newark area, Z, was calculated to be $2{\cdot}30/0{\cdot}5 = 4{\cdot}60$. This value is also significant at the 99·9% level because, although the difference between the means is considerably less, so also is the difference between the standard deviations, particularly in Z, where the drumlin distribution has already been shown to be very regular. There is also a significant difference in the density of the drumlins in these two fields.

The density of the drumlins seems to be related to the nature of the ice flow. An attempt is now made to see if the density is also associated with

the size of the drumlins. It is possible that densely spaced drumlins are smaller and sparsely spaced ones larger, but this need not necessarily be so. The three areas have been sampled to study this relationship. 20 drumlins were chosen randomly from each area, using random numbers for the purpose, and the long-axis of the sample was measured on the map. It was not, however, possible to measure this value very accurately in this way but at least some indication of size has been obtained. The resulting values may be compared with each other by use of the t-test (Table 14.2).

Table 14.2 Drumlin size in three areas

Appleby area, X		Wigton area, Y		Newark area, Z	
n.	20	n.	20	n.	20
Σ X	1492	Σ Y	2302	Σ Z	3159
Σ X^2	117,300	Σ Y^2	272,720	Σ Z^2	514,275
$\bar{X}$	74·6	$\bar{Y}$	115·1	$\bar{Z}$	157·9
Variance	300	Variance	411	Variance	749·7
St. dev.	17·3	St. dev.	20·27	St. dev.	27·37

The t-test between Y and Z gives $t = 42 \cdot 8/7 \cdot 8 = 5 \cdot 487$, significant at 99·9% level, and between Y and X $t = 40 \cdot 5/6 \cdot 12 = 6 \cdot 6176$, which is significant at the 99·9% level. The three different areas have thus significantly different sized drumlins and these values are inversely related to the density of the drumlins. The relationships are shown by the values given below:

	Density	Size
Appleby area (X):	15·77	746 m
Wigton area (Y):	9·33	1151 m
Newark area (Z):	7·03	1579 m

There is therefore a tendency for areas with large drumlins also to have a lower drumlin density. This is not in fact as obvious as it at first appears, for although large drumlins could not, because of their size, provide high density values, small drumlins could, by being spread sparsely, result in low density values. However, it seems that the forces allowing very regular, but fewer, drumlins to form also allow them to grow larger. In marginal or changeable conditions the drumlins only grow to small dimensions and are more irregularly spaced, being smaller where they are crowded together.

14.2 Drumlin morphology

The morphology of drumlins indicates past ice movement and can also give useful information concerning the mechanism of drumlin formation, which is a process by no means yet fully elucidated. The shape of a

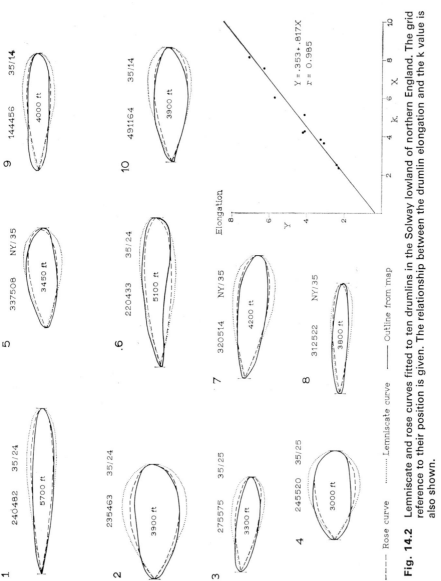

Fig. 14.2 Lemniscate and rose curves fitted to ten drumlins in the Solway lowland of northern England. The grid reference to their position is given. The relationship between the drumlin elongation and the k value is also shown.

drumlin can be assessed in several ways. A simple measure of shape is the elongation ratio, which is the maximum length divided by the maximum width. An alternative method is to fit specific shapes to drumlin outlines. Ellipses, ellipsoids and lemniscate and rose curves have been used for this purpose (Chorley, 1959; Reed *et al.*, 1962). The ellipse and lemniscate loop are two-dimensional shapes, while the ellipsoid also takes the third dimension into account.

The ellipse can be fitted by using the equation for an ellipse which is given by:

$$x^2/\mathbf{a}^2 + y^2/\mathbf{b}^2 = 1$$

x and y are chosen for different coordinate points and $\mathbf{a}$ is the long axis of the drumlin and $\mathbf{b}$ is its intermediate axis or width. The solid form of the ellipsoid is given by:

$$x^2/\mathbf{a}^2 + y^2/\mathbf{b}^2 + z^2/\mathbf{c}^2 = 1$$

where z is the third coordinate and $\mathbf{c}$ is the short axis or height of the drumlin. The equation for the lemniscate loop is:

$$r = \sqrt{\mathbf{a}^2 \cos k\,\theta}$$

where r defines the form of the loop, $\mathbf{a}$ is the long axis length, and k is a dimensionless constant defining the elongation of the loop. The value of k is calculated from the length of the drumlin and its area A, by the equation

$$k = \mathbf{a}^2\pi/4\mathrm{A}$$

The value of k can be used as an alternative to the elongation ratio of $\mathbf{a}/\mathbf{b}$. However, if a simple measure of elongation is required, the $\mathbf{a}/\mathbf{b}$ value is quicker and easier to measure. Figure 14.2 shows that the $\mathbf{a}/\mathbf{b}$ elongation correlates closely with k. The difference between the lemniscate curve and the rose curve is also worth mentioning. The rose curve has the equation

$$r = \mathbf{a} \cos k\theta$$

On Figure 14.2 both the lemniscate and the rose curves are shown for ten drumlins from the Solway lowland area. The rose curves, which are thinner than the lemniscate curves, provide a much closer fit to the observed drumlin form. In Table 14.3 the values required to plot the rose curve and the lemniscate curve are set out. The top row lists sufficient angles to cover one half of the symmetrical curve, as the last value lies below and beyond the origin, which represents the pointed end of the drumlin. The first point represents the limit of the rounded stoss end. The values of $k\theta$ are entered in the second row, and in the third these values are multiplied by the cosine value for the angles listed in the top row. In the fourth row the values in the third row are multiplied by $\mathbf{a}$, the length of the drumlin. These values give the lengths of the vectors at the angles given in the top row. The third-row values are multiplied by the square of the drumlin length and the square root of the result is

Table 14.3 Computation of rose curves and lemniscate curves

$\theta = 0$	2	5	10	15	20	22	25
$k\theta = 0$	7·64	19·1	38·2	57·3	76·4	84·0	95·5
$\cos k\theta = 1\cdot0$	0·991	0·945	0·785	0·540	0·218	0·105	−0·096
$\mathbf{a}\cos k\theta = 3\cdot9$	3·86	3·68	3·06	2·103	0·850	0·410	−0·374 = r for rose curve
$\sqrt{\mathbf{a}^2}\cos k\theta = 3\cdot9$	3·88	3·78	3·45	2·76	1·82	1·26	−1·19 = r for lemniscate curve

A = Area; **a** = length; **b** = width; **a** = 3900 feet; **b** = 1200 feet; A = 3,130,000 square feet. $k = \mathbf{a}^2\pi/4A$. $k = \dfrac{3900 \times 3900 \times 3\cdot14159}{4 \times 3,130,000} = 3\cdot82$

found giving the values for the lemniscate curve, entered in the bottom row. The calculations listed in Table 14.3 are for drumlin 10 in Table 14.4. Chorley fits rose curves to drumlins, according to the equation he gives, although they are referred to as lemniscate curves.

The ten drumlins measured in the Solway area of the Eden valley drumlin field have dimensions as set out in Table 14.4. The value of k and elongation correlate very closely, so that it is reasonable to use the elongation to assess regional differences in drumlin shape, as this measure

Table 14.4 Drumlin dimensions in the Solway area

Drumlin	Length feet	Width feet	Area × 10,000 square feet	k	Elongation
1	5700	800	313	8·17	7·1
2	3900	1600	475	2·52	2·4
3	3300	800	200	4·27	4·1
4	3000	1300	300	2·36	2·3
5	3400	1100	250	3·64	3·1
6	5100	900	338	6·05	5·7
7	4200	1000	325	4·26	4·2
8	3800	600	150	7·57	6·3
9	4000	700	244	5·16	6·1
10	3900	1200	313	3·82	3·25

is more readily obtained. The measurements may be made from closely contoured maps, such as the 1 : 25,000 Ordnance Survey or the 1 : 62,500 United States Geological Survey, or on air photographs. The accuracy of the maps, however, is such that the simpler method of estimating drumlin shape is more suitable when many drumlins are being measured.

In the fitting of shapes to drumlins the value of the lemniscate and rose curves in particular, lies in their form and their relationship to flow phenomena (Chorley, 1959). The best examples of the curves in nature

are found in pure streamlined forms, such as sand and snow drifts, which form in the shelter of an obstacle: there is a small accumulation in front of the object and a long tapering accumulation in its lee, where material can collect as long as the wind, which is responsible for these forms, is constant in direction. A similar shape is used for forms such as aeroplane wings that must offer as small a resistance to the wind as possible. These curves offer a minimum of resistance to the flowing medium in which the object (drumlin) is situated. In the same way the close approximation of many drumlins to rose curves suggests that they owe their shape to flow in the ice by which they are formed. The variation of the k-values is a measure of the stress of the ice on the drumlin. The rounded end of the drumlin faces the direction from which the greatest pressure came. The degree of tapering is indicated by the value of k. For a highly tapering drumlin the value of k is small and the two ends of the drumlin differ considerably in shape. For a more elliptical form k is larger and there is less difference between the two ends of the drumlin. Chorley states that a greater k-value indicates a small resistance to the flow, so that on this argument large k-values should be associated with maximum ice pressure. The large k-values are also associated with greater elongation, so the more elongated drumlins are likely to occur in areas where the ice pressure was greatest on the developing drumlin. It is also necessary to take constancy of flow direction into account. A more constant direction of flow is likely to produce a longer drumlin than a variable direction of flow—by the analogy with developing sand shadows, which are much longer when the wind direction is constant. If the flow is constant then the stoss end of the drumlin will protect a long stretch in its lee in which material can accumulate or remain uneroded, according to whether the drumlin is forming by deposition or erosion. If on the other hand the flow is variable then only an almost circular area of deposition can result. Thus elongation of the drumlin is likely to provide a measure of the constancy of flow direction. An application of this principle to the Eden valley drumlin field will now be discussed.

20 drumlin elongations have been measured in each of five separate areas of the Eden valley.

1   The first area, near Appleby, has an elongation value which is relatively small, the mean of the 20 elongation values being 2·55. This area is situated near the basal ice-shed where the ice divided to flow north down the Eden valley, and east up and over the Stainmore Pass.

2   The second area is a little farther north where the ice flowed consistently north down the Eden valley between the Penrith sandstone scarp and the Pennine fault scarp. Here the drumlin elongation has a mean value of 3·03.

3   The third area is situated around Carlisle, where the ice again diverged at another basal ice-shed to flow east over the Tyne gap

and west around the northern side of the Lake District into the Irish Sea. The Lake District ice also met the ice coming south from Scotland at this point so that flow in this area would have been confused. The mean drumlin elongation value in the third area is 1·92.

4   The fourth area is situated around the Tyne gap, where flow was once again concentrated and the Lake District ice was augmented by Scottish ice, both ice masses moving east through the gap. The mean elongation value in this area is 2·85.

5   The fifth area occurs on the Solway lowlands. It is in this area around Wigton, to the north of the Lake District, that the Scottish and Lake District ice masses were merged to flow rapidly and constantly westwards towards the Irish Sea. Here, therefore, both the constancy of flow and the volume of the ice, and hence the pressure, would have been greatest. This area should have the greatest elongation from theoretical considerations, and in fact does have the largest elongation value of the five areas—4·66.

The significance of the differences between these five areas may be tested by using the analysis of variance test (section 8.2) and the t-test (section 8.7). The tabled value of F 4, 95, 0·001 is about 5·1 so that the

Table 14.5 Analysis of variance table for drumlin elongation

Source of variation	Sum of square	df	V	F
Between-sum of squares	82·9	4	20·7	21·03
Within-sum of squares	93·5	95	0·984	

null hypothesis of no significant difference between the areas can be confidently rejected, and the differences between elongation values may be taken to indicate significant differences in drumlin morphology in these five parts of the Eden valley drumlin field.

The differences between the pairs of values are also of interest and their significance can be calculated by working out the least significant difference (LSD). This is given by:

$$\text{LSD} = t\,0\cdot01 \sqrt{\frac{2s^2}{n}} = 2\cdot63 \times \sqrt{\frac{2 \times 0\cdot984}{20}} = 0\cdot826$$

Where $s^2 = V$. The value for the 95% confidence level is 0·52. The differences between the mean value for each pair of observations may be compared with these values to assess the significance of the differences between the five areas. The results are set out in Table 14.6. The areas that are not significantly different in elongation are (1) and (2) (which are adjacent to each other), (1) and (4), and (2) and (4). The three areas involved are (1), (2) and (4). These areas have a number of flow characteristics in common, for in each case the ice was constrained within a

valley. (1) and (4) are situated where ice was flowing uphill towards a gap which would tend to reduce pressure, and these have the greatest similarity, there being only one difference significant at the 95% level which applies to the two basal ice-shed areas. Area (5) is significantly

Table 14.6   Differences in drumlin elongation

Areas	Differences between means	Significance	Order of significance
1 and 2	0·48	NS	(8)
1 and 3	0·63	0·05	7
1 and 4	0·30	NS	(9)
1 and 5	2·11	0·01	2
2 and 3	1·11	0·01	5
2 and 4	0·18	NS	(10)
2 and 5	1·63	0·01	4
3 and 4	0·93	0·01	6
3 and 5	2·74	0·01	1
4 and 5	1·81	0·01	3

different from all the others at the 99% level. This is the most open of the areas on the lowest ground, where ice flow was most consistent and powerful. Area (3) is also significantly different from all the others, mainly at the 99% level. Both pressure and constancy of flow appear to be involved in explaining these differences, the latter being probably the more significant factor.

14.3 Moraine spacing

There is a wide variety of moraine types, but the number of moraines of any one type is rarely large enough to necessitate sampling in any one area. Also the controls in different areas are not often sufficiently similar for the analysis of one area to be of value in a neighbouring area, except when the analysis concerns certain types of moraines. The kind of information that can be gained from an analysis of moraine distribution depends on the type of moraine. Two types will be mentioned.

1   *Terminal moraines* can be correlated with periods of ice advance and these in turn with climatic fluctuations. Space can be correlated with time in an analysis of this type of moraine. However, the behaviour of individual glaciers can vary, even in the same general area, so that the correlation between one valley and the next is not necessarily close, despite the fact that the major climatic episodes are reflected by all the glaciers. The differences between valleys in the same area may be due to the development of surges in the ice (J. F. Nye, 1965a, b).

2   The distribution of *ice-cored moraines* can also be related to climate
and glacier activity, and can thus provide information of value in
a study of climate over a wide area, as opposed to the detailed
climatic changes that may be investigated in one small area.

Terminal moraines

A well-recorded series of terminal moraines which can be correlated with
known dates of formation occurs in Austerdal, in Norway (Fig. 14.3). The
moraines nearer the present glacier snout are younger than those further
away. The moraines, which have equally steep proximal and distal slopes,
were formed in periods of minor advance during the general deglaciation
following the historic maximum in about 1750. The slope of the line
(Fig. 14.3) linking the plotted points is proportional to the mean rate of
retreat between the moraine-forming phases. The calculated rates of

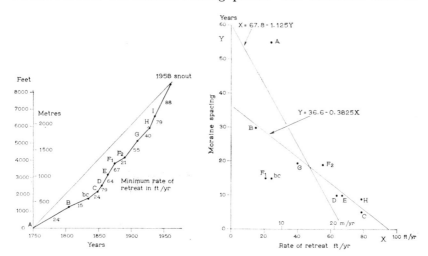

Fig. 14.3   The spacing of moraines in Austerdal, Norway, between the 1750
moraine and the glacier snout in 1958, and the relationship between
moraine spacing and rate of ice retreat for the same data.

retreat entered on the graph are minimum rates, as the periods of moraine
formation during which the snout was stationary or advancing are not
taken into account. The irregularity in the retreat rate is a measure of the
variations in climate over the 200 year period during which the 11
moraines were formed. The mean rate of retreat over the period is about
8500/200 = 42·5 feet per year. The actual minimum rates vary considerably around this value. The rate of retreat was at first slow, only
slightly exceeding 20 feet per year during the first 100 years, or about
half the mean rate. This was followed by a phase of rapid retreat between
1850 and 1875 when at times the rate was almost double the mean. After
this there were 15 years of slow retreat at 21 feet per year. The final period

between 1937 and 1958 was one of very rapid retreat at more than double the mean rate. This phase followed the last period of moraine building in the 1920s when records show that for a few years the glacier snout was in fact advancing. These records confirm that the moraines are formed during periods of advance.

There is a negative relationship, which can be assessed by correlation and regression analysis, between the time interval recorded for the building of successive moraines and the rate of retreat between moraine-building periods. The correlation coefficient, r, is -0.67, which is significant at the 95% level for the ten periods available. The regression equations are:

$$X = 67.8 - 1.1257Y$$
$$Y = 36.6 - 0.3825X$$

where X is the rate of retreat (feet per year) and Y is the time between moraine formations (years). It is not easy to determine which is the dependent variable, as both depend on the climate rather than on each other. The moraines tend to be more closely spaced when the retreat rate is greatest. The spacing depends on the frequency of colder and/or wetter spells of weather, which cause advances. The rapid retreat phases, on the other hand, depend upon warmer and/or drier spells of weather. The relationship between the two suggests that phases of variable climate, with warm/dry spells alternating rapidly with cold/wet periods, are replaced by phases when the cold/wet phases are spaced further apart and the warm/dry periods between them are less intense. This would lead to slower retreat between longer periods of colder, wetter weather. These phases of variable and less variable weather are not arranged systematically in chronological order. Periods of slow retreat and wide spacing occur near the beginning of the period, but after this the phases become rather irregular and often change markedly between consecutive moraine-building phases. The climatic changes that are reflected both in the moraine timing and in the rate of retreat appear to be complex, with periods of greater variability and extremes following periods of lesser extremes and slower changes. There is no correlation between the rate of retreat and the spacing of the moraines in distance, which depends on both rate of retreat and occurrence of cold spells during which moraines form. The long time interval is counteracted by the slow rate of retreat and so does not differ systematically from the shorter time gap with rapid retreat: the two distances will be comparable. The relationship between the time which elapses between the building of successive moraines and the rate of retreat is, therefore, more useful. The regression line:

$$Y = 36.6 - 0.3825X$$

shows that the maximum spacing in distance occurs within the central part of the data where the time interval is about 20 years and the retreat rate about 43 feet per year. In this analysis the regression line is not used for prediction because the whole population of moraines is available for

study and has been included in the calculations. The relationship established by the correlation does, however, throw light on the control of the
moraine spacing in time and on the rate of retreat, both of which depend
on the climate.

Ice-cored moraines

The distribution of ice-cored moraines in Norway has been studied
intensively by G. Østrem (1959, 1964, 1965). The pattern he has established
reflects the variations in size and activity of the glaciers. The ice of the
ice core consists to a large extent of snow that has been banked in drifts
against the terminal moraine and which has been slowly consolidated to
form ice and become buried beneath morainic debris. The character of
the ice can be seen to differ from glacier ice, when it is examined under a
microscope, and can be recognized as originating as snow. The ice-cored
moraines will form most readily where winter snow drifts do not melt,
especially where the snow drifts can accumulate in the same place for many
years in succession. The snow gradually becomes covered by debris
brought to it by the glacier, thus forming the ice-cored moraines. This
type of moraine can be recognized on the ground or air **photographs** by its
large size, relative to the glacier forming it, and by its steep slopes. They
are particularly conspicuous around small glaciers because these are
usually not so mobile as larger ones and respond more slowly to climatic
change, whereas they do not have a chance to form around rapidly
fluctuating glaciers, such as Austerdalsbreen, an outlet glacier from the
Jostedalsbreen ice-cap. They will, however, form around small, slow-
moving cirque glaciers, and once they have formed will create a barrier
that the ice cannot readily override. The ice will instead increase in thickness and the increase in elevation of the ice surface will allow debris to slide
down over the frontal snow banks. The ice-cored moraines are as a result
more common in those areas where the climate is continental. The long,
cold and relatively dry winters allow snow drifts to accumulate in sheltered
areas, such as on the leeward side of the terminal moraine. The glacier
accumulation is not, however, very great, so that the glaciers are not very
active. This results in a more or less static ice front, which gives time for the
ice-cored moraine to form. The great age of many of these moraines testifies
to the long periods of glacier stagnation in this type of situation. Oceanic
conditions provide larger gross glacier budgets and more rapid flow, and
hence ice-cored moraines cannot form so effectively. The distribution of
ice-cored moraines thus reflects the occurrence of the more continental
type of climate, and is dependent upon the character of the climate
rather than on the fluctuation of climate through time.

   G. Østrem's data may be used to analyse the distribution of ice-cored
and non ice-cored moraines in different parts of Scandinavia in order to
give quantitative evidence of the distribution of ice-cored moraines relative to climatic type. The data are given in Table 14.7. The first area is
close to the sea and, although it is in the highest latitude, the maritime

climate of this coastal zone ensures a predominance of non ice-cored moraines. The Kebnekaise and Sarek areas are similar with regard to continentality and ice-cored moraines predominate. The Jostedal and the Jotunheim areas lie farther south, in the same latitude, but the influence

Table 14.7   Number of ice-cored moraines in parts of Scandinavia

Area	Degrees north	Degrees east	Ice-cored		No ice core		Total no. of moraines
			obs.	exp.	obs.	exp.	
1 Narvik	70	20	28	43·7	90	74·4	118
2 Kebnekaise	68	18	68	49·7	66	84·4	134
3 Sarek	67	17	65	44·5	55	75·5	120
4 Jostedal	61·5	06	4	34·7	87	57·3	91
5 Jotunheim	61·5	09	43	36·3	55	61·6	98
Totals			291		270		561 = N

of continentality is clearly shown in the higher proportion of ice-cored moraines in the Jotunheim, which is higher and farther east and hence more continental. There are few ice-cored moraines in Jostedal.

The significance of the differences of these values can be tested by the chi-square test (section 10.2). The expected value is found by multiplying the row and column totals together and dividing by the grand total, N, which is 561. The value in the first cell is

$$\frac{291 \times 118}{561} = 43\text{·}7$$

The chi-square value may be found from:

$$\sum \frac{(o - e)^2}{e}$$

There are $(r - 1)(c - 1)$ degrees of freedom (where r is the number of rows and c the number of columns). Thus $df = 4$ in this case, and $\chi^2 = 78$. This value may be compared with the tabulated value for 99·9% level of significance, which is 18·465. The value is highly significant and the null hypothesis of no significant difference between the number of ice-cored and non ice-cored moraines in the five areas is rejected with a high degree of confidence.

The effect of continentality is related to the temperature and precipitation. This can be demonstrated in the Jotunheim, where Østrem has shown that there is a steady rise eastwards of the glacial limit. The glacial limit is itself related to the elevation of the seasonal snow-line, which rises north-eastwards in a similar way. The percentage of ice-cored moraines to total moraines has been counted between each 100 m isopleth of the glacial limit elevation. The results are given for an east–west traverse across the

Table 14.8 Ice-cored moraine in relation to glacial limit

Glacial limit m.	Number of moraines with ice core	Total number of moraines	Percentage
2200	5	5	100
2100	15	29	52
2000	6	13	46
1900	9	28	32
1800	2	13	15
1700	0	4	0

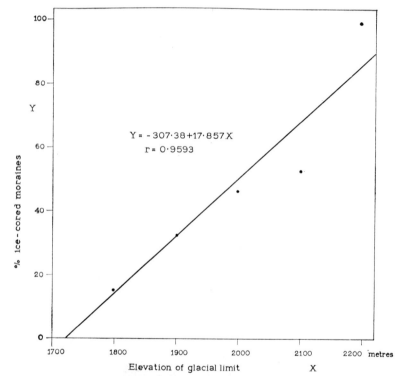

Fig. 14.4 The relationship between the percentage of ice-cored moraines and the elevation of the glacial limit in Norway from data given by G. Østrem (1964).

Jotunheim (Table 14.8). These values have been correlated by product-moment correlation and regression techniques. If X is taken as the independent variable, being the elevation of the glacial limit (Fig. 14.4) and Y as the percentage of ice-cored moraines, then:

$$Y = -307 \cdot 38 + 17 \cdot 857X$$
$$r = +0 \cdot 9593$$

which is significant at the 99% level. The relationship, therefore, supports the view that the distribution of ice-cored moraines is closely related to the general climate of the area. The numerical analysis therefore supports the ideas presented by Østrem concerning the pattern of distribution of the Scandinavian moraine types.

14.4 Moraine morphology

The term 'moraine' covers a wide variety of forms. In this section a series of moraines surveyed in east Baffin Island will be analysed: these provide a case study to show how *t*-tests (section 8.7) and analysis of variance (section 8.2) may be used to compare one set of moraines with another. The surveyed moraines show an interesting variation from very young moraines around the snouts of the present glaciers, to moraines along the outer coast which are about 40,000 years old. The moraines have been formed by different types of glaciers. The older ones, on the outer coast, have been formed by large ice sheets that extended along the fjord and pressed inland as fairly uniform ice masses. An intermediate series was formed by more restricted lobes pushing inland from the major glaciers in the fjords, where the fjord walls had been broken by valley embayments. The youngest of this set of moraine loops dates from about 7000 to 8000 years ago. The most recent moraines, associated with the present valley glaciers, occur in the inland valleys into which glacier tongues still protrude.

The youngest moraines consist of two types. There are those associated with the present glacier snouts and margins. These moraines are ice-cored and hence differ considerably in character from the moraines in the same area that have lost their ice cores. The form of the active moraines of three glaciers, namely 'Curlew', 'Linnet' and 'Magpie' glaciers, Baffin Island (Fig. 14.5), have been measured by surveyed profiles, and the surveys revealed some interesting differences. The moraines of 'Curlew' glacier are much smaller than those of the other two glaciers. The lateral moraine of 'Curlew' glacier is steep but small: the moraine crest is only 27·88 feet (8·5 m) above the ice surface and 31·16 feet (9·5 m) above the valley side; it has a slope of 30·5° on the proximal side adjacent to the ice and 24·5° on the distal side. 'Curlew' glacier has only a discontinuous terminal moraine, as its snout ends in a moraine-dammed lake along part of its length. The glacier is a large piedmont lobe, whose meltwater drains down two different valleys from different parts of its snout. The small size of its moraines, which are exceptionally unstable, could be accounted for by a recent advance of the glacier: this would mean that a shorter time would have been available to accumulate moraine. It has a large accumulation area which supports this suggestion.

The moraines of 'Linnet' glacier, which is a valley glacier protruding from a tributary valley into the main valley, are more typical of the

moraines of this area. They are steep, ice-cored, and show a strong asymmetry. The lateral moraine crest rises 19·68 feet (6 m) above the ice on the proximal side, but is 177 feet (54 m) high on the distal side. The slopes are 16·5° and 24·5° respectively. The slope at the terminal moraine is 27·75° on the proximal side, which is 52·5 feet (16 m) high, and 29·75° on the distal side, which is 210 feet (63 m) high. The moraine has some vegetation, but there is no lichen on the boulders. The very marked difference in height on either side of the moraine is due to its ice core. This property also accounts for its steepness, particularly on the long distal side. The length of the slope renders it more unstable as slides

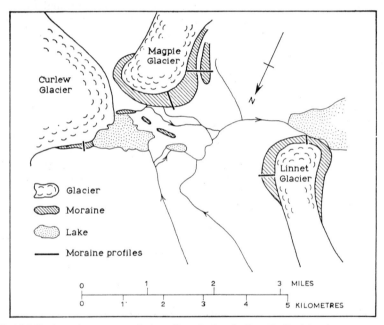

Fig. 14.5 Location map to show valley glaciers in East Baffin Island.

will gather momentum and remove more material. The temperature will also be slightly higher on the distal side, which is sheltered from the cold down-glacier winds, and this will facilitate melting through insolation warming the bare stones. This process will also aid the maintenance of the steep slope. The slow forward movement of the ice will also steepen the slope on the distal side of the moraine.

The steepness of the proximal and distal slopes may be compared, by using a *t*-test, to see if there is a significant difference between their means. On the proximal slope the mean gradient is 16·75°, while on the distal slope it is 25°. As there are only four sets of observations, the *t*-value for the significance of this difference is only 1·76, the variances being 1·07 and 0·28 respectively for proximal and distal slopes. The *t*-value is thus

significant only at the 80% level, but the results indicate a tendency that would probably be significant if more values were available for testing. However, this relationship may not hold good for all moraines: it may only apply to those developed by cold arctic glaciers. In many temperate areas ice-cored ridges are usually steeper on the proximal side and often show pronounced fluting. Down-melting of the ice surface is much more active in temperate areas where ice flow and insolation play a more important part.

The moraines of 'Magpie' glacier are particularly interesting in that their morphology shows clearly the static nature of these arctic glaciers compared with more mobile temperate glaciers. The terminal moraine of 'Magpie' glacier shows two steps with a very steep slope, 141 feet (43 m) high, up to the first step. The gradient is 31°. At the top of the first step a dense lichen growth covers the rocks. The size and density of the lichens indicates that this part of the moraine must have been static for several hundred years at least. The lower front of the moraine, is, however, formed of fresh rocks and its gradient indicates that the ice core within the moraine is still present. The second step of the moraine rises to a total elevation of 270 feet (82 m) above the valley level. It has smaller lichens on the rocks, indicating a younger age. The proximal side of the moraine is only 69 feet (21 m) high and slopes at 16·5° to the glacier surface. The morainic debris continues under the clean ice edge, where the margin is visible in a small ice cave. The double step suggests a renewed phase of moraine formation. The ice cannot, however, have had sufficient force to enable it to override its own earlier deposits of terminal moraine. It therefore added an inner ridge behind the main ridge.

The survey of the lateral moraine of 'Magpie' glacier confirms the double nature of the moraine. The glacier must have shrunk in width, if not in length, between the advancing phases. Two distinct lateral moraines occur. The inner one has a crest height of 56 feet (17 m) above the ice and a distal slope 256 feet (78 m) high. The gradients are 12·5° and 17·5° respectively, again showing the steeper distal slope. There are lichens on this moraine, indicating its relatively great age. A stream section, however, revealed its ice core beneath about 10 feet (3 m) of stones and debris. The outer moraine appears to have lost its ice core as it is a nearly symmetrical ridge. The proximal side rises to a height of 118 feet (35·5 m) and slopes at 22°. The distal slope is 138 feet (42·5 m) high and has a gradient of 13°. The outer moraine is about half the height of the inner, ice-cored one.

The ice core in the active moraines causes a marked difference between the height of the proximal and distal sides of the moraines. The differences in height of the active, ice-cored moraines may be compared with the height differences for the older moraines, most of which have probably lost their ice cores. The values are compared by use of the t-test. Calculated t is 2·26 and t 11, 0·05 is 2·201, so that the difference is significant at the 95% level. If the anomalous moraines of 'Curlew' glacier are

omitted, the calculated t reaches 3·67, which is significantly different at the 99% level for ten degrees of freedom. Asymmetry is a feature of the youngest moraines and is due to their ice core and the long stillstand at the present snout position, which has allowed the large bulk of material to accumulate at the snout and cover the ice in this part of the glacier. These arctic glaciers differ from many temperate ones in the asymmetry

Table 14.9 Height differences between proximal and distal sides of moraines, Baffin Island

	Mean height difference	n	Variance
Ice-cored moraines	45·6 m	5	576
Old moraines	16·4 m	8	348

of their moraines, owing to the much more rapid ablation in the latter where these are decaying by down-wasting. In temperate glaciers, such as the Tasman, Hooker, Mueller and Cameron glaciers in the Southern Alps of New Zealand, the slope down to the ice on the proximal side of the moraine is usually much steeper and longer than the distal slope.

The recent ice-cored moraines of east Baffin Island differ in two respects

Table 14.10 A comparison of moraine heights

Group 1 Active ice-cored moraines		Group 2 Moraines in inland valleys		Group 3 Fjord lobe moraines		Group 4 Outer coast moraines	
metres		metres		metres		metres	
9·5	n 5	42·5	n 3	14	n 4	5	n 4
63	m 56·7	15	m 24·5	20	m 28·75	13·5	m 9·55
54	v 741·8	16	v 162·2	34	v 114·2	8·7	v 10·0
82				47		11	
75							

where n is the number, m is the mean, and v is the variance

from the older moraines of this area. In order to assess these differences the heights and gradients of the moraines have been compared. A simple analysis of variance test can show whether there is a significant difference in the moraine heights. The heights of the distal sides have been used as these are normally the greater, especially on the most recent moraines. The values have been divided into four groups. The data are presented in Table 14.10. The analysis of variance is given in Table 14.11. The tabled value for F 3, 12, 0·05 is 3·49 so that the differences are significant

at the 95% level, but not at the 99% level because of the small number of observations available. The individual pairs have been tested using the *t*-test. Some of the results are significant. Groups (3) and (4) differ significantly at the 95% level, and (1) and (4) at the 98% level. Owing to

Table 14.11   Analysis of variance table for the data in Table 14.10

Source of variation	Sum of squares	df	V	F
Between groups	5427	3	1809	4·98
Within groups	4347	12	362	
Total	9774	15		

the large variances and small samples, despite large mean differences, groups (1) and (2) and groups (1) and (3) differ significantly at only the 80% level and (2) and (3) are not significantly different at all.

The gradients for both sides of the moraines in the four groups have

Table 14.12   Moraine gradients

Group 1	Group 2	Group 3	Group 4
m 2·168	m 3·83	m 6·39	m 15·69
n 10	n 6	n 6	n 6
v 1·05	v 0·69	v 5·63	v 123·17

also been analysed. The data are summarized in Table 14.12 and the analysis of variance table is given in Table 14.13. The mean values in Table 14.12 are given as cotangents of the angle of slope, which is equivalent to the gradient. The value of F is significant at the 99% level and very nearly significant at the 99·9% level, as F 3, 24, 0·001 is 7·55. A

Table 14.13   Analysis of variance table for the data in Table 14.12

Source of variation	Sum of squares	df	V	F
Between groups	736·46	3	245·49	7·468
Within groups	789·22	24	32·88	
Total	1525·68	27		

t-test applied to each pair of groups gives the results shown in Table 14.14. The groups are all significantly different at least at the 95% level, with the exception of groups (3) and (4), which show a significant difference at only the 90% level, owing to the large variance of group (4). The variance values are themselves of interest, showing that the moraines

appear to become less homogeneous with increasing age. The variability, as well as the other differences between the groups, reflects the processes by which these different moraines have been formed.

All the moraines in the inner valleys—groups (1) and (2) have been formed around the margins of land-based, mainly valley glaciers. Their gradients reflect this homogeneity of formation by the small variance value and their consistently steep slopes. The height differences are accounted for by the presence or absence of an ice core. The moraines of the third group were formed around the margins of ice lobes protruding into embayments along the fjords during a re-advance. There are four moraines in each of three embayments. These become progressively higher in absolute altitude inland. Some of them are very large, one being 420 feet (128 m) high on its proximal side. They do, however, vary considerably in height and slope. This variation is due to variation in the time of ice stillstands, the second moraine of the sequence from the fjord being the most massive. Another factor which has affected the slope of the older

Table 14.14 t-test results for the groups listed in Table 14.12

Groups	Mean differences	t	df	Significance level
1 and 2	1·66	3·13	14	0·01
1 and 3	4·22	4·60	14	0·001
1 and 4	13·52	3·57	14	0·01
2 and 3	2·56	2·28	10	0·05
2 and 4	11·86	2·38	10	0·05
3 and 4	9·33	1·84	10	0·1

moraines of this sequence is the considerable quantity of fine material they contain. This has led to marked solifluction activity, which has reduced the gradients of these moraines relative to the younger ones, which are stonier.

The moraines of the outer coast are the lowest, gentlest in mean gradient and most variable in slope, as well as being the oldest. The variation in slope is due in part to differences of material and mode of formation. Some of the moraines consist mainly of stratified marine material, containing shells, with only a thin cover of drift and stones. The shells in the moraine at Cape Henry Kater have been dated at 40,000 ± 2000 B.P. The moraines were probably deposited in the sea, and this would account for their fairly low height. Their relatively steep slopes could be accounted for by later river erosion. On the other hand, some of the moraines of the outer coast are formed of relatively coarse material, and these have suffered less subsequent modification of gradient.

The variation in moraine morphology in this area can be accounted for by the presence or absence of an ice core, by type of material, by

type of glacier or ice sheet, by age and subsequent modification dependent on age, by type of deposition, whether it be on land or in water, and by time of ice stillstand. The actively forming moraines of these arctic cold glaciers differ in several respects from similar features of many temperate glaciers.

14.5 Esker morphology

Eskers can have a variety of forms, and these are frequently the result of different methods of formation. Their form can be expressed either by a statement of their overall plan or by a measurement of their cross-profiles and long profiles. The spatial relationships of esker ridges is also important. The eskers of west Baffin Island display a variety of types and this area may once again be used to provide a case study to illustrate another use of numerical techniques. The eskers in west Baffin Island are small features unlike the very long esker ridges of other parts of Canada, and of Scandinavia. All the eskers of the area, which lies around Eqe Bay and Lake Gillian, are associated with dead-ice in large spreads of fluvioglacial deltaic outwash.

There are two main types of esker ridges in this area, though in fact the first type are not true eskers. They are found within the large kettle holes. They are short ridges, some of which bifurcate and rejoin in complex patterns. True eskers form the second type. These are longer and may be sinuous or irregular in form and, although they are associated with kettle-hole lakes, their plan is rather simpler and they do not bifurcate. They are mostly associated with delta surfaces with which they merge at their distal ends. At times they are connected with deltas by kame terraces. 'Eskers' of the first type are probably not representative of true eskers in that they are not formed by flowing water. The second type, on the other hand, do reveal evidence of flow.

An example of the first type is located among the kettle holes in the dead-ice area near Tikerarsuk Point. The pattern of ridges is complex and a *topological* representation is shown in Figure 14.6. The pattern of ridges can be described in topological terms related to simple graph theory. The pattern is shown to contain 14 *arcs* or lines between nodes, **a**, representing individual ridges. There are 14 *nodes*, which can be subdivided into seven terminal nodes **n**, one two-node at F, five three-nodes where three ridges converge from different directions, and one four-node where two ridges join the main ridge at the same point. The seven terminal nodes show that the whole system splits up into seven separate ridges at its margins, where the ridges diverge in all directions. There are two *regions*, **r**, in the system. One is small and enclosed by the arcs joining nodes K, L and M. This region contains a small kettle-hole lake. The other is the external region around the whole system.

The whole system conforms to the general topological relationship, in

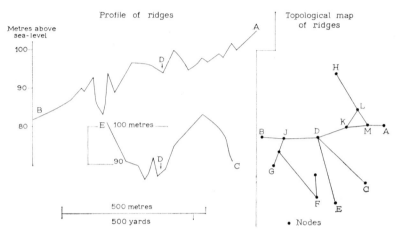

Fig. 14.6   A topological map of an esker in west Baffin Island and the profile of the main ridges.

that $\mathbf{a} + 2 = \mathbf{n} + \mathbf{r}$, because $14 + 2 = 14 + 2$, where $\mathbf{a}$ is 14, $\mathbf{n}$ is 14 and $\mathbf{r}$ is 2. The connectedness of the graph of the ridges may be described by the mu (μ) index, which is given by $\mathbf{a} - \mathbf{n} + \mathbf{p}$. For this ridge system the value is $14 - 14 + 1 = 1$, the $\mathbf{p}$-value being the number of sub-graphs. As the whole system is linked into one pattern of connecting ridges there is only one sub-graph, so $\mathbf{p}$ is 1. Between the ridges there are a number of more or less circular kettle holes, so that the ridges round them are arcuate in pattern, although the major ridge AB is aligned almost parallel to the general slope of the delta surface in which the dead-ice area lies.

The description of the plan-form of the ridges in topological terms gives an indication of the complexity of the pattern. The form of the ridge height along the ridge crests also indicates a complex pattern. The profiles shown in Figure 14.7 indicate that the ridge crest consists of peaks and troughs of an amplitude of 5 to 10 m. Along the profile of the main ridge AB, on which there are 23 surveyed points, there are 16 turning points. The *test for randomness* in the number of turning points can be applied to the series, although the test should ideally be carried out when there is a larger number of points. The expected number of turning points, e, is given by:

$$2/3(\mathbf{n} - 2)$$

In this case $\mathbf{n}$ is 23 and $e = 14$. The variance is given by:

$$16\mathbf{n} - 29/90 = 3{\cdot}76$$

This value is multiplied by $1{\cdot}96$ (i.e. $3{\cdot}76 \times 1{\cdot}96 = 7{\cdot}3696$) to give the 95% confidence range within which randomness cannot be rejected. In this case the confidence range is $14 \pm 7{\cdot}3696$, and the actual value of 16 turning points lies within it. This means that the pattern of turning

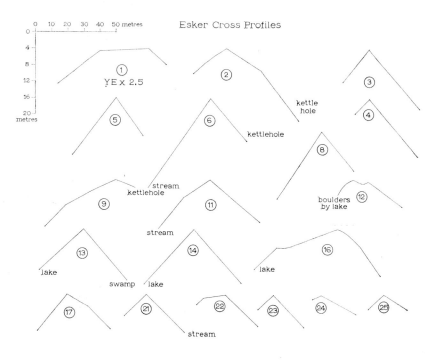

Esker Cross Profiles

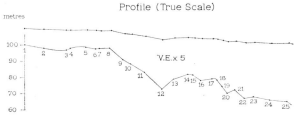

Profile (True Scale)

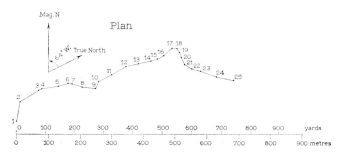

Plan

Fig. 14.7 The cross-profiles, longitudinal profile, and plan of an esker in west Baffin Island.

points could have occurred randomly, although the small value of **n** in this series means that the result is not very reliable.

The conclusion reached from the results of these analyses is that the pattern of the eskers, both in plan and in elevation, is complex. The system as a whole shows little trend, apart from the general lowering of the main ridge as it is followed from the delta surface at 344 feet (105 m) to its distal end where it disappears into the kettle-hole lake at 269 feet (82 m). This is the direction in which the whole delta slopes. The ridges have steep slopes at the angle of rest of the coarse material of which they are composed. Their irregularity, both in plan and profile, indicates that they are not formed by flowing water, but that they represent the filling of crevasses and other cracks in the dead-ice by material being accumulated from the delta surface above in elongated troughs in the ice. The method of accumulation, by slow down-washing into the hollows from the delta surface to the north, would account for the diminution of ridge height towards the distal end of the ridge. The steep slope of the ridges would be the result of slumping as the enclosing ice slowly melted away to form the kettle holes.

One of the greatest contrasts between the complex ridge pattern of the kettle-hole eskers just described and the longer esker ridges associated with kettle-hole lakes is the trend of the ridge crest. The first ('untrue esker') type slopes down generally in the direction of delta slope, indicating slow accumulation. The second ('true esker') type trends generally upwards in the direction of former flow to reach a maximum at the distal end, where the ridge merges into a delta surface. The second type may be illustrated by examining an esker situated at the edge of Reflection Lake near Lake Gillian. In plan the esker ridge is simple, although it is not straight. It is just over 840 m long and runs in a generally southerly direction to merge at its distal end with a kame terrace, which slopes down to the south to merge into a large delta surface. The supply of material to build the ridge must have been derived from the north. There are many kettle holes in this area, pointing again to the presence of dead-ice while the esker ridge was accumulating.

The morphology of the ridge is shown in Figure 14·7. The slopes of the ridge are at about 25° and represent the slumping of material as the kettle holes developed with the melting of the dead-ice. The esker crest rises from 210 feet (64 m) at its proximal end, where it is only 13 feet (4 m) above the ground around, to 328 feet (100 m) where it merges with the kame terrace at its distal end. Its maximum height above the surrounding flatter ground is 72 feet (22 m). The esker consists of a mixture of material from fine sand and gravel to large boulders. The latter are irregularly distributed in size along the esker, although there is some tendency for size to decrease with the height of the esker crest. This tendency is not statistically significant as the correlation coefficient, r, is -0.44, and is based on only 11 observations. The mean size of the boulders along the esker crest varies between 12 cm at the distal end and 24 cm at the

proximal end, but, in between, some large boulders reach a mean of 59 cm. The largest boulders were probably not carried by the subglacial stream, but were accumulated by down-washing from the surrounding ice as the esker melted out. The finer material forming the core of the esker was probably carried, at least in part, by the subglacial stream, which deposited more and more of its load as the open end of the channel was approached. This would account for the upward slope of the esker crest towards its distal end. Other eskers in the area show a similar rise of crest elevation as the kame terraces or deltas into which they feed are approached. Others have a smoothly sinuous form suggesting flow in a confined and swinging channel by a very vigorous stream.

The morphological criteria which can be used to differentiate between the two types of dead-ice ridges, defined at the beginning of this case study, include first of all the direction of slope of the ridge crest. One type slopes down towards its distal end and the other slopes up. The pattern of the ridges provides a second criterion. The complex pattern of the first type has a higher μ index (p. 317) which was 1 in the given example. The value of μ for the second ridge was zero as:

$$\mathbf{a} - \mathbf{n} + \mathbf{p} = 1 - 2 + 1 = 0$$

References

CHORLEY, R. J. 1959: The shape of drumlins. *J. Glaciol.* **3,** 339–44.

FAEGRI, K. 1948: On the variation of western Norwegian glaciers during the last 200 years. *Gen. Ass. Int. Un. Geol.* (Oslo) **2,** 293–303.

FLINT, R. F. 1928: Eskers and crevasse fillings. *Am. J. Sci.* **235,** 410–16.

GRAVENOR, C. P. 1953: The origin of drumlins. *Am. J. Sci.* **251,** 674–81.

HOINKES, H. C. 1968: Glacier variations and weather. *J. Glaciol.* **7,** 3–19.

HOLLINGWORTH, S. E. 1961: The glaciation of western Edenside and adjoining areas, and the drumlins of Edenside and the Solway Basin. *Quart. J. Geol. Soc.* **87,** 281–359.

HOPPE, G. 1967: Case studies of deglaciation patterns. *Geogr. Ann.* **49**A, 204–12.

KING, C. A. M. 1959: Geomorphology in Austerdalen, Norway. *Geogr. J.* **125,** 357–69.

KUPSCH, W. O. 1955: Drumlins with jointed boulders near Dollard, Saskatchewan. *Bull. Geol. Soc. Am.* **66,** 327–38.

LEWIS, W. V. 1949: An esker in process of formation, Boverbreen, Jotunheimen, 1947. *J. Glaciol.* **1,** 314–19.

NYE, J. F. 1965a: A numerical method of inferring the budget history of a glacier from its advance and retreat. *J. Glaciol.* **5,** 589–608.

1965b: The frequency response of glaciers. *J. Glaciol.* **5,** 567–88.

ØSTREM, G. 1959: Ice melting under a thin layer of moraine and the existence of ice cores in moraine ridges. *Geogr. Ann.* **41,** 228–30.

1961: A new approach to end moraine chronology. *Geogr. Ann.* **43,** 418–19.

1964: Ice-cored moraines in Scandinavia. *Geogr. Ann.* **46,** 282–337.

1965: Problems of dating ice-cored moraines. *Geogr. Ann.* **47**A, 1–38.

PRICE, R. J. 1966: Eskers near the Casement Glacier, Alaska. *Geogr. Ann.* **48**A, 111–25.

REED, B., GALVIN, C. J. and MILLER, J. P. 1962: Some aspects of drumlin geometry. *Am. J. Sci.* **260,** 200–210.

SMALLEY, I. J. and UNWIN, D. J. 1968: The formation and shape of drumlins and their distribution and orientation in drumlin fields. *J. Glaciol.* **7,** 377–90.

STOKES, J. C. 1958: An esker-like ridge in process of formation, Flatisen, Norway. *J. Glaciol.* **3,** 286–90.

VERNON, P. 1966: Drumlins and Pleistocene ice flow over the Ards Peninsula, Strangford Lough area, Co. Down, Ireland. *J. Glaciol.* **6,** 401–9.

15 Analysis of glacial and related deposits

15.1 Glacial sediment size analysis
Stone sizes [*regression analysis, means, standard deviation, coefficient of variation, t-test*]; Fine sediments [*cumulative frequency curves, probability paper, histogram, skewness, kurtosis*]
15.2 Stone shape
[*parametric and non-parametric tests, factor analysis, product-moment correlation matrix, factor loadings, weightings on factors, Kolmogorov–Smirnov tests, analysis of variance, F-value, t-test, discriminant analysis, Kruskal–Wallis test, trend-surface analysis*]
15.3 Orientation of surface stones
[*orientational data, Tukey chi-square test, Rayleigh test of significance*]

15.1 Glacial sediment size analysis

The deposits associated with glaciation can be broadly divided into the stratified fluvioglacial deposits and the unstratified tills. The size range of particles in both types of sediment may be very large, ranging from clay particles to large boulders. The aim of this section is to illustrate some of the numerical methods of sediment size analysis by reference to examples from Baffin Island and a drumlin in Yorkshire.

Stone sizes
An analysis of stone sizes will be illustrated first by reference to measurements made in west Baffin Island. The stones come from a variety of glacial and fluvioglacial environments. These include outwash deltas, eskers, ice-contact deposits, as well as moraine and lag (see below) deposits. For purposes of comparison stone sizes have also been measured on some raised beaches within the same area. The difference of size of stones in these different environments can be seen by plotting the mean size of each sample against its standard deviation (section 2.3), as shown on Figure 15.1. Each sample consists of the measured lengths of the long axis of 50 stones, which were sampled randomly. The relationship between the mean values and their standard deviation is linear, the value of r, the coefficient of correlation, being 0.929. The equation for the regression line is:

$$Y = 2.26 + 0.57X$$

where Y is the standard deviation and X is the mean size. The standard

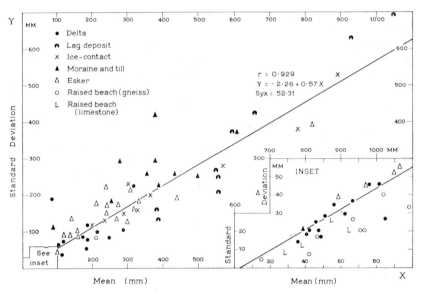

Fig. 15.1 Graph to relate mean size of samples of 50 stone lengths from various features in west Baffin Island to the standard deviation of the samples.

deviation, therefore, increases with the mean. There are, however, some differences in the ratio of the mean to the standard deviation among the various groups of stones. Table 15.1 gives the mean, the standard deviation, the *coefficient of variation*, and the number of samples for which the values are calculated for the six environments. The coefficient of variation

Table 15.1 Stone sizes in West Baffin Island

Environment	Mean cm (m)	St. dev., cm (s)	Coeff. of var. (s/m × 100)	No. of samples (n)
Moraines	38·6	25·1	65·0	10
Ice-contact	44·0	24·7	56·2	9
Lag deposits	58·9	32·8	55·7	10
Eskers	23·7	13·8	60·0	22
Deltas	12·1	5·9	48·8	24
Raised beaches	7·9	2·8	35·4	9

Total no. 84

is found by dividing the standard deviation by the mean and multiplying the result by 100. These figures show that, relative to their size, beach deposits are better sorted than delta deposits. This is reasonable considering the sorting action of waves, which wash away the finer material and which could only reach the outermost, and therefore smaller, delta

stones in this particular area. The deltas are made up in general of smaller stones than the eskers. Relative to their size they show better sorting than the eskers, although the difference is not so great as that between the deltas and beaches. The better sorting of the delta stones is probably due to the greater distance they have been transported. The ice-contact slope stones are larger than those in the eskers, as the ice-contact stones would have been the first to be deposited at the ice margin. Their coefficient of variation is similar to the esker stones, which have also not been carried very far. Moraine stones are only slightly smaller than the ice-contact stones, their smaller mean size being due to a larger quantity of smaller material. They have a relatively large standard deviation and hence a high coefficient of variation (which adjusts the standard deviation for variations in the mean). The value of the latter is the largest for all the deposits, reflecting the completely unsorted nature of glacial till when compared with fluvioglacial deposits. The largest stones are concentrated at the ice-contact faces and the mean size of the stones in this environment is only exceeded by the lag deposits, which are found in abandoned stream channels, such as subglacial or marginal drainage channels, where the largest stones have become concentrated, the smaller material being washed away by the powerful but diminishing currents. The mean size of the lag deposits is very large, but their coefficient of variation is similar to those of the ice-contact and esker deposits, indicating the result of

Table 15.2   Analysis of stone size in Baffin Island

Pairs	Difference between means (cm)	df	t	Significance level
Moraines and ice-contact	5·4	17	0·55	NS
Beaches and deltas	4·2	31	1·19	NS
Deltas and eskers	11·4	44	3·16	0·01
Deltas and ice-contact	31·9	31	5·51	0·001
Eskers and ice-contact	20·3	29	2·74	0·02
Ice-contact and lag deposits	14·9	17	1·285	NS
Eskers and moraines	14·9	30	1·96	0·1

Pairs	Difference between standard deviations (cm)	df	t	Significance level
Beaches and deltas	3·1	31	1·00	NS
Deltas and eskers	7·9	44	4·16	0·001
Deltas and ice-contact	18·8	31	1·92	0·1
Eskers and ice-contact	10·9	29	2·80	0·01
Ice-contact and lag deposits	8·1	17	0·95	NS
Eskers and moraines	11·3	30	3·3	0·01

removal of finer material. Some of the lag deposits were found amongst dead-ice features, and probably represent moraines from which the finer material has been washed away. The largest sample of all, with a mean length of 106·5 cm, was found in a broad col in the hills where moraine had accumulated and boulders had been left after finer material had been washed away by temporary, but vigorous water flow.

The significance of the differences between the size of stones in these different environments has been tested by using the t-test. Some of the most important pairs are shown in Table 15.2. The tests show that the differences between the deposits are statistically significant in most cases in one or other of the properties tested. The differences between the deltas and beaches as regards size is not significant. Deltas and eskers, however, can be differentiated both by the mean and by the standard deviation of their stone sizes. Deltas and ice-contact deposits, as well as eskers and ice-contact deposits can be differentiated on the basis of both size and sorting. Ice-contact and lag deposits are not, however, significantly different. Eskers and moraines are better differentiated by their sorting than by their size.

A study of the size of the larger elements of these deposits, therefore, provides a means of differentiating them, if for some reason this cannot be done morphologically.

Fine sediments
The analysis of fine sediment from glacial deposits yields rather different information and must be treated by different methods. Five samples of fine material taken from a drumlin in Wensleydale will be used to illustrate the type of information that can be gained from a study of fine sediment. The results of the sieve and hydrometer analysis, which give the percentage of sediment falling within certain size ranges, mostly in $\frac{1}{2} \phi$ units in size, are plotted as a cumulative frequency curve on an arithmetic probability paper graph as shown in Figure 15.2. The ϕ units are related to mm by the relation $\phi = -\log_2 \text{mm}$. The ϕ units are therefore logarithmic, and if the cumulative frequency curve plots as a straight line then the distribution of sediment size is log-normal (section 2.4). This follows from the property of probability paper on which a normal distribution cumulative frequency curve plots as a straight line. The cumulative frequency curve can be used to read off the necessary percentile values from which four graphic moment measures (Folk and Ward, 1957) can be calculated. These measures are the mean value of the size of the material, its degree of sorting and the character of the frequency distribution curve (or histogram) of the material size as defined by the skewness (section 2.3) and kurtosis (or peakedness) of the curve. These graphic measures are defined by:

Mean, $M_z = (\phi_{16} + \phi_{50} + \phi_{84})/3$
Sorting, $\sigma = (\phi_{84} - \phi_{16})/4 + (\phi_{95} - \phi_{5})/6\cdot6$

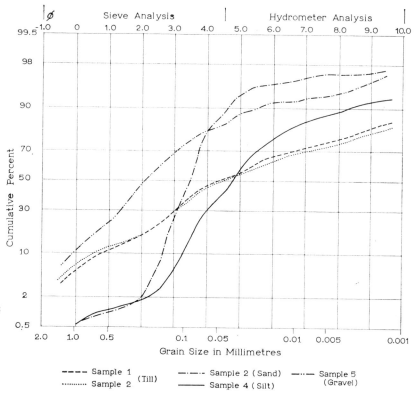

Fig. 15.2   Material size distribution of five samples of sediment from a drumlin in Yorkshire plotted on arithmetic probability paper.

Skewness, $\alpha = \dfrac{\phi_{16} + \phi_{84} - 2\phi_{50}}{2(\phi_{84} - \phi_{16})} + \dfrac{\phi_{5} + \phi_{95} - 2\phi_{50}}{2(\phi_{95} - \phi_{5})}$

Kurtosis, $\beta = (\phi_{95} - \phi_{5})/(2\cdot44(\phi_{75} - \phi_{25}))$

These measures may be used in the analysis of the drumlin sediments. The relevant values are given in Table 15.3.

The samples show that there is a considerable range of material within the drumlin. The first two samples are, however, very similar in character. Both were derived from the unstratified till in the central part of the drumlin, and are typical of clayey till: they contain 25% clay, 54% silt and clay combined, but 46·5% of the sediment is sand, of which 10% is coarse sand. There are also stones and boulders of various sizes. The sediment shows a very wide range of size distribution. This results in the very high sorting coefficient of 3·80. The inclusion of the larger stones would still further increase the sorting coefficient. The frequency distribution is approximately normal over the range analysed as the plotted points do not fall very far from a straight line. A much closer fit is achieved,

Table 15.3 Drumlin sediment size characteristics

Sample	ϕ_5	ϕ_{16}	ϕ_{25}	ϕ_{50}	ϕ_{75}	ϕ_{84}	ϕ_{95}	M_z	σ_0	σ_1	sk_i	% silt	% coarse sand	M_d (mm)
1	−0·3	1·9	2·7	4·5	7·9	9·3	—	5·3	3·80	—	+0·32	54	12	0·04
2	−0·1	2·0	2·7	4·4	7·5	9·3	—	5·2	3·65	—	+0·34	54	12	0·05
3	2·4	2·8	3·0	3·5	3·8	4·2	5·4	3·5	—	0·81	0	18	1	0·09
4	2·9	3·5	3·8	4·7	5·8	6·7	—	4·9	1·60	—	+0·25	69	1	0·04
5	−0·6	0·5	1·2	2·1	3·4	4·5	8·4	2·4	—	2·36	+0·20	19	22	0·12

NB. A modified sorting coefficient, σ_0, is used when the graph (Fig. 15.2) does not allow the 95% (cumulative per cent) value to be read off. This happens when the analysis stops at 9·6 ϕ, and 15% of the material finer than this still remains in the medium to fine clay grades.

however, if two straight lines are drawn. One line fits the silt and clay grade and another the sand grade, indicating a bimodal distribution. This could be related to differences in transport of the different sized material. These two samples also have the highest skewness of the set. The value is positive indicating a considerable tail to the histogram caused by the presence of much fine material. This is confirmed by the high percentage of clay relative to coarse sand. The very close similarity between these samples shows that the bulk of the drumlin is composed of uniform fine silty till.

The analysis so far considered deals only with the fine fraction of the sediment, i.e. that which is less than 2 mm in size. In fact the fine material constituted 76·4% of the total in one sample and 78% in the other. Thus about 25% of the whole sediment consists of material larger than this, including gravel, cobbles and larger stones. The total carbonate content of the till was 22·9%. The proportions of sand, silt and carbonate correspond fairly closely to the proportion of sandstone, shales and limestones in the area from which the till was derived. This suggests, therefore, that all the till is of local origin and was derived from the rocks to the west within about 12 miles of the drumlin.

The other three samples, from the same drumlin, show considerable differences both within themselves and with the true till or boulder clay. All three samples, however, have few really large particles and two of them have negligible amounts of clay. They were obtained from a small area of stratified deposits near the upstream stoss end of the drumlin, where sand overlies a finer silty layer, which in turn rests on a gravelly layer. The uppermost sand layer contained material 4·3% of which was coarser than 2·0 mm ($-6\ \phi$), and only 3·0% clay and 15·3% silt. The bulk of the sample was medium and fine sand. The material has a mean size of 3·5 ϕ, a low sorting value of 0·805, indicating very good sorting, and a zero skewness. 80% of the sample lies between 2 and 4 ϕ. Its size frequency distribution curve is, however, far from normal and it appears to have three peaks (trimodal). On this curve there are three separate straight line sections, indicating three distinct but normally distributed populations when examined separately. These are related to the silt and clay content, the amount of fine sand and the 2% residue of coarser sand respectively. The total carbonate content is lower in this sample, amounting to only 18·6%. The material is a clean, washed sand deposited by moving water.

The underlying silty layer contains nearly 60% silt and only 1·2% is coarser than 2 mm. The material is not quite so well sorted and has a positive skewness due to the tail of fines. It must have been deposited in quieter conditions. The relatively low carbonate content of 14% still indicates washing of the material. This layer is also not normally distributed.

The lowest stratified gravelly layer shows another change of depositional process. It has 23·5% coarser than 2 mm, but no really large stones

as in the first two samples. Of the finer fraction 81% is sand, with almost 50% of this being coarse and medium sand. The amount of clay is 6% and silt 13%. It is, therefore, the coarsest of the sediments. It has a positive skewness, which often indicates deposition in flowing water. The sorting is relatively poor.

These four very different sediments all come from within a few yards of each other in one drumlin. Their analysis can throw light on the different processes operating to build up the drumlin. The true till or boulder clay indicates that much of the drumlin was deposited directly from the ice. This is also indicated by its poor sorting and by other properties, such as the proportion of clay, sand and carbonates. The sand, silt and gravel, however, were stratified and they show that some of the sediment was deposited by flowing water. The intensity of flow must have varied with time. Starting with vigorous flow, the water then slowed down to allow silt to be deposited and after this conditions changed to allow local sand deposition. These sediments at the stoss end of the drumlin could be related to pressure melting where the impact of the ice on the growing mound of till was greatest.

15.2 Stone shape

The shape of stones can give valuable information concerning the nature of the feature in which they are found and also concerning the processes by which they have been transported and deposited. The first problem, however, is to obtain a quantitative value to describe the shape of the stone, which is usually an irregular solid. Shape can be assessed qualitatively on the basis of comparison with regular geometrical figures. The division of stones into rod, wedge, plate, disc, sphere and varihedroid, used by Holmes (1941) in his discussion of till fabrics, provides a nominal method of shape description. The degree of angularity can be assessed by comparison with a graded series of outlines, providing an ordinal method of shape description. Shape classes can be obtained, as by Zingg (1935), by using the ratios of the a, b and c axes of the pebbles. These are the long, intermediate and short axes, respectively. He gives four classes, called discs (I), spherical (II), blades (III) and rods (IV). These are found by calculating the ratios as shown:

Class I	$b/a > 2/3$, $c/b < 2/3$
Class II	$b/a > 2/3$, $c/b > 2/3$
Class III	$b/a < 2/3$, $c/b < 2/3$
Class IV	$b/a < 2/3$, $c/b > 2/3$

This method, although based on values measured on the ratio scale, only provides nominal classes. If parametric tests (see introduction) are to be applied a value given on the ratio scale is necessary. One such value is the sphericity value of Krumbein (1941). It is given by $\sqrt[3]{bc/a^2}$, where

a, b and **c** are as defined above. The value gives the cube root of the ratio of the stone volume to that of the circumscribed sphere. This measure only takes into account the shape of the pebble and is not concerned with its angularity.

The degree of roundness of a pebble (Cailleux, 1947) is a more useful measure, as it is directly related to the type of process by which the pebble has been moved. This measure can be obtained from:

$$2\mathbf{r}/\mathbf{a} \times 1000$$

where **r** is the minimum radius of curvature in the principle plane. The value of roundness, **R**, reaches a maximum of 1000 for a pebble which is completely circular in the principle plane. Cailleux has also developed a flatness value which is almost inversely related to Krumbein's sphericity formula. Cailleux's flatness value is given by $[(\mathbf{a} - \mathbf{b})/\mathbf{c}]100$. This value increases with increasing flatness and can exceed 1000 in very plate-like rocks, such as slate.

In order to consider the value of the different parameters in terms of their ability to characterize the deposits in which the stones occur, multivariate techniques such as factor analysis can usefully be applied. The application of factor analysis in this type of geomorphological study will be illustrated by reference to data obtained from 67 samples of 50 stone roundness and other shape measures studied in west Baffin Island. 14 variables have been derived from the data. The variables used are the first, second, third and fourth moment measures calculated for the 50 values in each sample of stones of the sphericity, flatness and roundness values. These figures provide twelve of the variables. The last two are the mean and standard deviation of the maximum length of the 50 stones in each sample. The moment measures in this analysis describe the distribution of the three shape measure values over the 50 stones sampled in each set of data. The stones have been selected, as far as size is concerned, to fall within a fairly narrow range in order to eliminate one important variable—size—in the analysis of shape. The first stage in the analysis of shape (see chapter 4) provides the product-moment correlation matrix between all the variables. The mean and standard deviation—the first two moment measures—of the size, roundness and flatness are all closely positively correlated. The sphericity values do not show this correlation and none of the other variables show any strong correlation, apart from a strong negative correlation between mean sphericity and mean flatness of -0.971. Since sphericity and flatness are the inverse of each other, as measures of shape, it is unnecessary to use both. A strong correlation exists between all the roundness measures, particularly the skewness and kurtosis, a relationship that is also strong in the flatness values. These strong relationships suggest that it is not necessary to use all the moment measures.

The strength of the relationships is also seen in the values of the factors and the factor loadings on the individual variables, which are derived

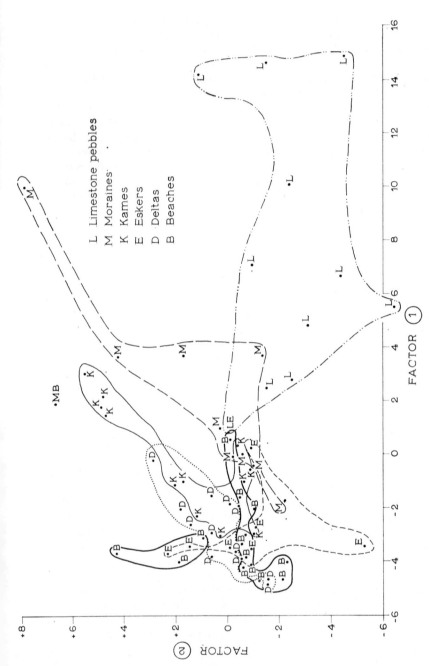

Fig. 15.3 The weightings on factor (1) of 67 samples of 50 stone shape measurements, made in west Baffin Island, plotted against the weightings on factor (2) to differentiate the different morphological features from which the samples were obtained. MB refers to the modern beach, while the beach samples were from raised beaches.

L Limestone pebbles
M Moraines
K Kames
E Eskers
D Deltas
B Beaches

from the correlation coefficients (chapter 4). The first factor accounts for 34% of the variability, the second and first together for 52·5% and the first three for 66%. The variables that load highly on the first factor are the measures associated with roundness and flatness. This is the result of their mutual intercorrelation. The mean sphericity also weights highly on the first factor on account of its strong negative correlation with flatness. On the second factor the strongest variables are the sphericity moments and the skewness and kurtosis of the flatness. It is only on the third factor that size has any weight on the factors. This is the result of its lack of correlation with the other variables, except its own moment measures. Since size does not load heavily on the first two factors, this shows that it does not affect the differentiation of the samples by shape—size of material having, in this case, been effectively eliminated as a discriminating variable. The factor loadings show that mean roundness provides a satisfactory measure of shape. Roundness is also easy to calculate and is diagnostic of the different depositional features.

The value of shape measurement for differentiating the different features can be shown by considering the weightings, on factors 1 and 2, of the individual samples (Fig. 15.3). The different features fall into distinct but overlapping groups among the pattern of points, the 67 samples representing 6 major classes of deposits from different features.

In order to assess the value of the mean roundness as a measure to differentiate between the environments, the mean roundness values have been correlated with the weightings on the first factor for the same groups. The correlation coefficient is $+0·978$, indicating a significant relationship between roundness and weightings on the first factor. The regression line is shown in Figure 15.4. The mean roundness value may, therefore, be used in the rest of the analysis to characterize the shape of the pebbles.

Another important variable that affects stone shape is the nature of the rock. In the western Baffin Island analysis nearly all the samples have been measured in granite-gneiss stones derived from the basement complex. The uniformity of rock type and in this case, stone size, means that any significant differences found in the roundness of the pebbles are likely to be due to differences in depositional processes. These processes will be revealed as differences in the land forms from which the samples have been derived. Thus a study of stone roundness should provide a diagnostic criterion in establishing the identity of deposits, and it should also give some information concerning the formational processes.

The analysis is, therefore, concentrated on studying the variations of roundness within the granite-gneiss pebble samples. The environments represented in these samples will be briefly mentioned. They include beach deposits and deltas formed of glacial outwash in close proximity to the retreating ice. The rear end is usually in the form of a steep ice-contact slope, and the front edge is modified in places by waves to form raised beaches. There are eskers and kames in the dead-ice areas associated with the delta spreads. The other deposits are morainic, some of the material having

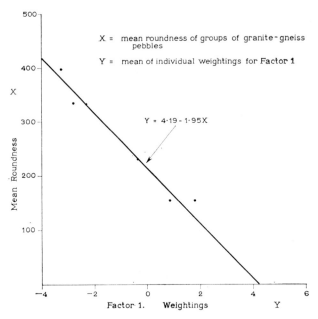

Fig. 15.4 The relationship between the mean roundness and factor (1) for the stone shape measurements from west Baffin Island.

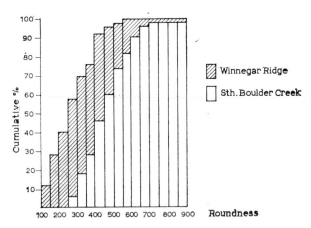

Fig. 15.5 Cumulative frequency graphs of two stone roundness observations to Ilustrate the application of the Kolomogorov-Smirnov test.

been moved subsequently to form solifluction deposits. The final category consists of rather doubtful deposits that, morphologically, do not fall clearly into any of the previous categories. The number of samples in each type are: Beaches, 24 (including one modern gneissic, 12 raised gneissic and 11 limestone beaches); Deltas, 12; Eskers, nine; Kames, six; Solifluction, three; Moraines, six; Doubtful, seven. The significance of the different roundness values for the 67 samples may be tested, first of all, by the non-parametric *Kolmogorov–Smirnov* test. This test is based on a comparison of two cumulative frequency percentage histograms. An example is illustrated in Figure 15.5, which refers to two samples of stone roundness of 50 stones, one obtained from an old high level deposit in the Rocky Mountains at Colorado, which is probably till, and the other from the present river bed in the same locality. The aim of the test is to establish whether the two sets of data are likely to have been drawn from the same population. The test is, therefore, a non-parametric version of the *t*-test. The values are divided into classes at 50 roundness value intervals. The maximum difference between any two classes is noted and compared with the critical differences given, for example, by Miller and Kahn (1962). In the example illustrated the maximum difference is 52%. This value is compared with the tabled value of 32% for the 99% significance level and 26·5% for the 95% significance level. These values depend on the number of stones in the samples so that in order to enter the tables it is necessary to calculate the value of N, where:

$$N = n_1 n_2 / (n_1 + n_2)$$

n_1 is the number in the first sample and n_2 in the second. In the example both values are 50, thus $N = 2500/100 = 25$. The two samples are shown to be highly significantly different. It can, therefore, be concluded that the stones in the older, higher deposit are significantly less rounded than those worn by fluvial erosion. This supports the suggestion that the older deposit is a till and not an old fluvial deposit.

The comparison of all the Baffin Island samples with each other is shown in matrix form in Figure 15.6. The samples have been divided into categories of decreasing roundness in the matrix. The pattern in the matrix reveals a number of unshaded triangles, which represent samples that are not significantly different. These mostly belong to one environment, for example, the eskers. Three main groups in which the roundness is significantly similar stand out. In the first group are most of the raised beaches. A second large group includes nearly all the deltas and eskers and a few of the raised beaches. The third group includes the kames and the doubtful deposits from the 'Heart Lake' valley. The till-moraine deposits and the solifluction deposits do not show any significant difference among themselves except for a few samples. The Kolmogorov–Smirnov matrix shows the significance of differences between individual pairs of samples and the broad groups into which these pairs fall. In order to assess the significance of the similarities and differences between the stone

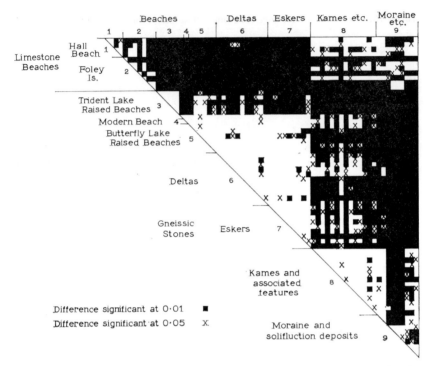

Fig. 15.6 The matrix of 67 samples tested against all the other samples by the Kolmogorov-Smirnov non-parametric test. The unshaded triangles are those that are not significantly different. The types of features from which the samples were obtained are indicated at the side and top. As the matrix is symmetric only half is reproduced.

roundness in the various environments the values for the morphological units have been collected together, and the groups differentiated by analysis of variance. There are six groups in the analysis with the following mean values:

Table 15.4 Means of stone roundness in each environmental group

	Beaches	Deltas	Eskers	Kames	Moraines	Solifluction
Mean	398	334	332	232	138	185
Number of samples	12	12	9	13	6	3

The analysis of variance is shown in Table 15.5. The null hypothesis of no significant difference between the groups can be rejected with a high degree of confidence as $F_{5, 49, 0.001}$ is 5·5. In the six groups analysed the doubtful deposits are included with the kames. The data were also analysed with these deposits in a separate category, and although the

Table 15.5 Analysis of variance for stone roundness in the six groups of Table 15.4

Source of Variation	Sum of square	df	V	F
Total	469,463	54		
Between groups	393,576	5	78,715	50·826
Within group	75,887	49	1,548·7	

value of F was rather smaller it was still significant at the 0·001 level. The limestone pebbles are excluded from the analysis as they are so different from the granite-gneiss pebbles that if these were to show a significant difference, the limestone ones would show an even greater difference.

It is also important to know which pairs of groups are significantly different, and the t-test may be used for this purpose. The results are summarized in Table 15.6. The pairs in the table are not all derived

Table 15.6 Stone roundness t-tests

t-test pairs	Mean roundness		Difference $(\bar{R}_1 - \bar{R}_2)$	t	df	Significance level
	first group $(\bar{R}_1)$	second group $(\bar{R}_2)$				
Deltas and eskers	334	332	2	0·14	19	NS
Kames and 'Heart Lake' deposits	238	227	11	0·65	11	NS
Butterfly Lake Beaches and Trident Lake beaches	436	360	76	3·01	10	0·02
Moraine and solifluction	138	185	47	1·68	7	NS
Beaches and deltas	398	334	64	3·32	22	0·01
Deltas and kames	334	238	96	5·65	16	0·001
Eskers and kames	332	238	94	6·86	13	0·001
Moraines and kames	138	238	100	5·44	10	0·001
Moraines and deltas	138	334	196	10·90	16	0·001
Moraines and eskers	138	332	194	12·60	13	0·001

directly from the analysis of variance groups as some of these have been subdivided: for example two sets of raised beaches have been differentiated. Only three of the pairs tested show no significant difference. These are the deltas and eskers, the kames and doubtful deposits, and the moraines and solifluction. The similarity of deltas and eskers is to be expected in that both are laid down by vigorous fluvioglacial streams, which are heavily loaded and hence have a considerable ability to abrade and round stones. It has already been shown (section 15.1), however, that the

two deposits can be differentiated by the mean size of their constituent stones.

The lack of significant difference between the kames and doubtful deposits in 'Heart Lake' valley is an indication that these features are similar to kames in at least some respects. The doubtful feature is a ridge of gravel, extending perpendicular to the valley wall, which was probably deposited very closely to the margin of the ice, when the ice was damming up a lake in this small tributary valley. The presence of spillways across cols in the valley sides supports this view.

The third similar pair are moraines and solifluction deposits. Although the solifluction stones are rounder they cannot, in this case study, be shown to be significantly more round than those within the moraines. This lack of significance is partly due to the small number of observations and partly due to the large variance. There is some indication that the slow down-slope movement in solifluction causes some rounding of the stones, but this effect is not marked.

The significantly greater roundness of the beach stones than those of the delta shores is due to the effectiveness of even fairly brief periods of wave action in the vigorous surf and breaker zone. The more exposed beaches at Butterfly Lake show a significantly greater degree of rounding than those at Trident Lake, but the difference is not so great as that between the beaches and deltas.

The considerable difference between eskers and deltas on the one hand and kames on the other, shows that even fairly short distance transport in an active and turbulent meltwater stream can effect very considerable rounding of the stones. The kames are also water deposited, but the stones have probably only been carried a very short distance before being deposited at the margin of the ice. The stones in the kames have usually been carried by slow-flowing meltwater, apart from those in kame terraces along which the water was flowing marginal to the ice. The kame stones are, however, significantly more rounded than the moraines. The difference indicates that even brief transport by moving water is enough to cause some rounding of the stones. The relatively low roundness of the moraine stones suggests that the cold-glacier ice is not a very efficient agent of glacial rounding in this area.

A further series of roundness measurements has been carried out in the very different environment of eastern Baffin Island, where moraines are better developed. The aim of the study required a different sampling plan, this aim being to establish, firstly, the difference of roundness values of stones in moraines of different ages and types, and, secondly, to see whether there is any significant difference in roundness in different parts of one moraine. For this purpose nine samples of 50 stones each have been measured, three on the distal side of the moraine, three on the crest and three on the proximal side. 11 different moraines have been sampled in this way, ranging from moraines around present glaciers to moraines about 40,000 years old on the outer coast. The mean values for

the roundness range between 166 and 233 for samples of 450 and 150 stones respectively.

The significance of the differences between these moraines has been tested by analysis of variance. There are ten moraines each with nine values, which are the means of 50 stone roundnesses. It is therefore possible to test whether the variation within the nine groups is greater or less than the variation between the groups. The Table 15.7 shows that the groups

Table 15.7   Moraine stone roundness analysis of variance

Source of variation	Sum of square	df	V	F
Total sum of squares	63,220	89		
Between groups	36,604	9	4067·11	12·22
Within groups	26,616	80	332·7	
F 9, 80, 0·001 = 3·9				

are significantly different at the 99·9% level. The least significant difference has been calculated to ascertain which groups differ significantly. The values of the least significant differences are as follows: 95% 17·2, 99% 22·8, 99·9% 30·0. The tabulated values for the differences between the means of the pairs of samples, Table 15.8, show that the moraines

Table 15.8   Moraine stone roundness, differences between means of 450 stones

	01	02	03	04	05	06	07	08	09	10
01	—						95% significant difference 17·2			
02	2	—					99% „ „ 22·8			
03	12	10	—				99·9% „ „ **30·0**			
04	13	11	1	—						
05	40	38	28	27	—					
06	46	44	34	33	6	—				
07	41	39	29	28	1	5	—			
08	57	55	45	44	17	11	16	—		
09	44	42	32	31	4	2	3	13	—	
10		50	40	39	12	6	11	5	8	—

Moraines   01 'Linnet' glacier ⎫
           02 'Magpie' and 'Curlew' glacier ridge ⎬ Young moraines
           03 Moraine knob recently formed ⎪
           04 'Curlew' glacier ⎭

           05 Isabella Bay glacier moraine ⎫ Fjord lobes
           06 Itirbilung fjord glacier moraine ⎬

           07 'Cache camp' moraine ⎫
           08 'Cache camp' moraine ⎬ Old, outer coast moraines
           09 'Haar' Lake, outer coast moraine ⎪
           10 Cape Henry Kater moraine ⎭

fall into two groups. There is no significant difference between the roundness of the moraine stones of the active glaciers. Nor is there a significant difference, on the basis of stone roundness, between any of the older moraines. The differences between all the active moraines and all the older ones are significant at the 99% level and most at the 99·9% level. This result could be explained in three ways, firstly, by the relatively short period during which the present moraines have been forming, or secondly, by the short distance that their stones have been moved, or thirdly, by the fact that the greater roundness of the older moraines may be the result of subsequent weathering. Probably all three effects have played a part, but in view of the considerable age range of the features, the first and second are probably the more important, especially as the present glaciers are very sluggish. They have not moved their snout positions for several hundred years.

When two or more variables can be used to characterize each observation, instead of one as in the analysis of variance, then discriminant analysis can be used to differentiate between two groups of data (section 11.3). Here discriminant analysis may be applied to two sets of moraine observations. The analysis is used first to differentiate the moraines of one of the present glaciers from those of the most recent ice advance, dating from about 8000 B.P. For both sets of nine observations on the two moraines, the mean stone roundness and the flatness of the samples of 50 stones give 36 values in all. The discriminant function (D^2) is calculated (sections 5.3 and 11.3), and from this the value of Hotelling's T^2 is found. For the first set of observations, D^2 is 8·499 and T^2 is 38·2455. From the T^2 value the F-value is calculated and this is 17·928. The tabled value of F 2, 15, 0·001 is 11·34, so these moraines are shown to be significantly different with 99·9% confidence.

The simple t-test does not show a significant difference between the oldest moraines on the outer coast and the fjord lobe moraines of intermediate age further up the fjords. Discriminant analysis (section 11.3) may be used to assess whether there is a difference if two variables are considered together. For this test the mean and standard deviation of the roundness are used for nine sets of data from two moraines. The old moraine is situated at Cape Henry Kater and the younger one is the Itirbilung moraine, which was used in the first discriminant analysis test (see Fig. 15.7). The mean values of the nine sets of data shown in the Table 15.9.

From these values the following calculations may be made (see Table 11.3 for another worked example):

$$d_1 = \bar{x}_1 - \bar{x}_2 = 216·3 - 212·9 = 3·4$$
$$d_2 = \bar{y}_1 - \bar{y}_2 = 110·56 - 90·67 = 19·89$$
$$d_1^2 = 11·56 \quad \text{and} \quad d_2^2 = 395·61$$

The difference between the means (d_1) is too small to be significant for discriminatory purposes, but when it is combined with the difference

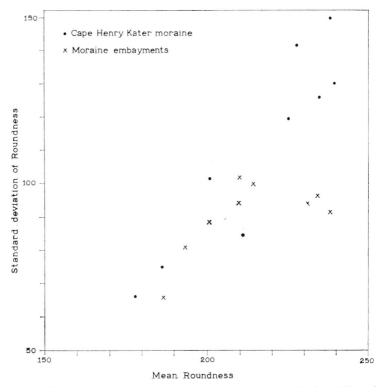

Fig. 15.7 The relationship between mean roundness and standard deviation of the roundness of moraines from the Cape and the embayment moraines of Henry Kater Peninsula, east Baffin Island.

between the standard deviations (d_2) the results are significant. The discriminant function (D^2) is (see section 11.3) found to be 1·616, from which T^2 is calculated as 8·06. The F-value is 3·73, which is significant at the 95% level. The use of the standard deviation as well as the mean of the stone roundness thus differentiates these moraines of different ages.

The geomorphological reason for this difference must be sought in both variables used. The older moraines have slightly rounder stones because most of the stones within them have probably been carried farther from their source. The higher standard deviation in the older moraines is brought about by the greater number of weathered stones, which have a low roundness, though better rounded stones must also be present to give the slightly higher mean roundness. Thus both glacial rounding and subsequent weathering have combined to give the stones their characteristic differences in shape in the two moraines.

Some information concerning the activity of the present glaciers can be derived from an analysis of the nine samples from the lateral moraine of

Table 15.9 Discriminant analysis of moraines

	Cape Henry Kater moraines		Itirbilung fjord moraines	
	Mean (x_1)	St. dev. (y_1)	Mean (x_2)	St. dev. (y_2)
	238	150	193	81
	226	119	201	88
	187	75	238	92
	228	142	210	104
	240	131	208	94
	236	126	232	94
	178	66	187	66
	212	84	214	100
	202	102	233	97
Total	1947	995	1916	816
Mean	216·3	110·56	212·9	90·67

$n_1 = n_2 = 9$ $n_1 + n_2 - 2 = 16$

Sum of squares

A	425,401·0	117,303·0	410,536·0	75,022·0
C	421,201·0	110,002·8	407,895·1	73,984·0
	4,200·0	7,300·2	2,640·9	1,038·0

Sum of products

A	220,331·0	174,760·0
C	215,251·7	173,717·3
	5,079·3	1,042·7

xx 6840·9   yy 8338·2   xy 6122·0
s_1^2 427·56 s_2^2 521·14 $s_{1 \cdot 2}$ 382·63
$D^2 = 1/76{,}412 \cdot 9(6024 \cdot 37 - 51{,}751 \cdot 47 + 169{,}147 \cdot 0) = 1 \cdot 616$
$T^2 = 81/18 \times 1 \cdot 616 = 8 \cdot 06$
$F = 15/32 \times T^2 = 0 \cdot 46875 \times 8 \cdot 06 = 3 \cdot 73$

F 2, 15, 0·05 is 3·68

'Linnet' glacier. These data allow a two-way analysis of variance with 50 replications to be carried out. Interaction can now be assessed and the difference between the proximal, distal and crest samples may be compared with sets of samples longitudinal to the moraine crest. The mean roundness of the nine samples is set out in Table 15.10. The table shows little difference in the means for the different direction, but a considerably greater variation between the positions relative to the ice. The results of the two-way analysis of variance are set out in Table 15.11, and the values confirm that the difference is significant but that interaction is not.

The variance ratio with the interaction and error terms combined for the 'between positions' and the combined 'error and interaction' term is F 11·717. This F-value is significant at the 99·9% level, as F 2, ∞, 0·001 is 6·91. The F-value for the 'direction and the error and interaction' term is 0·125, which is not significant. These results show that the stones on the proximal side of the moraine are significantly rounder than those on the crest and distal side.

The reason for the difference is probably related to the static nature of

Table 15.10 Moraine stone roundness

	Proximal	Crest	Distal	Means
West	205	150	153	169
Central	192	145	168	166
East	179	168	144	164
Means	192	154	155	166

this glacier. During the slow process of moraine formation the stones on the proximal side have probably been moved more than those on the distal side, which have been little disturbed since they were first deposited. There is no similar significant difference in roundness of the stones of 'Curlew' glacier, which has possibly advanced more recently. Therefore the time during which the 'Curlew' moraine stones on either side of the moraine have been subjected to glacial erosion is shorter and less disparate

Table 15.11 Moraine stone roundness, analysis of variance table

Source of variation	Sum of squares	df	V	F
Total sum of squares	2,923,484·920	449		
Between subclasses	188,241·360	8	23,530·17	
Between positions	147,110·414	2	73,555·207	
Between directions	1,568·814	2	784·407	
Interaction	39,562·132	4	9,890·533	1·595 NS
Error	2,735,244·729	441	6,202·369	

than the stones on either side of the moraine crest of 'Linnet' glacier. Some of the other moraines of the area show a similar difference in roundness on either side of their crests, but this variation does not occur in many of the older moraines that have been tested by this method.

The effectiveness of glacial erosion in producing significant rounding of stones in a short distance has been observed by measurements made on the moraines of Blea Water in the English Lake District. The mean roundness of one set of 50 stones from the moraine at the margin of the

tarn is 85·3, while about one mile down-valley another set of 50 stones have a mean roundness of 111·2. The variances were 1465 and 2586 respectively. The difference of 25·9 between the mean values is significant at the 99% level as indicated by the t-test. Other observations in the same vicinity also show significant differences in the roundness, with increasing values down the valley. These results point to the effectiveness of glacial transport in rounding stones—which in this area were of Borrowdale volcanic series rocks—in a distance of about one mile.

Another example of the use of stone roundness values is illustrated by reference to the roundness of moraine and till stones in the Front Range, Colorado. The landscape of part of the Indian Peaks in the Front Range shows two distinct phases of development. There is an older, high level

Table 15.12   Data on stone roundness for Kruskal–Wallis test

Old till		Young moraine		Periglacial		Fluvial deposits	
Roundness	Rank	Roundness	Rank	Roundness	Rank	Roundness	Rank
159	13·5	252	30	182	17	471	33
292	32	211	25	159	13·5		
289	31	206	23	126	9		
248	29	202	22	125	8		
241	28	194	21	121	7		
235	27	167	16	102	5·5		
222	26	156	12				
207	24	137	11				
192	19·5	135	10				
183	18	102	5·5				
192	19·5	90	4				
		87	2·5				
		87	2·5				
		74	1				
		161	15				

landscape of mature aspect, into which young glacial valleys are incised with associated deposits, and a younger low level landscape. There is some doubt in the former as to the extent of glaciation and glacial deposits. In order to attempt a solution of this problem a series of deposits have been sampled for stone roundness in both older and younger landscapes. The sets of 50 stone samples include material from till or possible till on the older surface, from the younger morainic deposits, and from periglacial deposits; one sample of modern river pebbles has been obtained for comparison. The difference between the four types of deposits can be examined by means of a non-parametric test known as the *Kruskal–Wallis test*, which can be applied if the data are not normally distributed. The data are set out in Table 15.12. The Kruskal–Wallis test is a non-parametric version of the analysis of variance test. It uses the ranked values of the variable

being tested. The values are all ranked as indicated on the table, and then summed separately for each set of data. The results are as follows: old till, 267·5, n = 11; young moraine 200·5, n = 15; periglacial 60, n = 6; fluvial 33, n = 1. A value, H, is now found from the equation:

$$H = \frac{12}{N(N+1)} \sum_{j=1}^{k} \frac{R_j^2}{n_j} - 3(N+1)$$

k is the number of samples, N is the total number of values, R_j is the sum of the ranks in the jth set. In the example under discussion:

H = 12/33(34) × 10,869 − (3 × 34)
                = (12/1122) × 10,869 − 102 = 11·6

If there are a large number of ties in the ranks the result can be adjusted by dividing H by

$$1 - \frac{\sum T}{N^3 - N}$$

T equals $t^3 - t$, where t is the number of tied observations in a tied group of scores. T is $(2^3 - 2) = 6$ where two observations are tied, $3^3 - 3 = 24$ for three observations, and so on. In the example there are eight ties of two observations each so that the correction is $1 - (48/35,937)$, which is $1 - 0·001335, = 0·998665$. The value is so near unity that it can be ignored. The value of H can be tested for small samples of less than 5 values for significance by means of tables given by Siegel (1956), using (k − 1) degrees of freedom. In the example k is 4, and there are 3 *df*. For large samples the chi-square tables can be used. For 3 *df* the chi-square value is 11·34, so the result is significant at the 99% level, indicating that the four types of deposit can be differentiated by means of the roundness value of the stones which they contain.

The range of values in the groups for the old and new glacial deposits given in Table 15.12 is considerable. *Trend-surface analysis* (section 9.1) may be applied to the data to see if this variability can be related to a systematic trend in distribution. The samples of glacial sediment round-ness have been collected over a considerable range of country, extending out from the corries near the continental divide to the foothills. The trend-surface analysis is performed to ascertain whether there is any systematic variation in the pattern of roundness values. The independent variables, U and V, are the Cartesian coordinates of the sample positions, and the dependent variable, Z, is the mean roundness of a sample of 50 stones. The sampling sites stretch in a north-west to south-east belt covering a total distance of 12·5 miles (20 km) from the continental divide, as shown in Figure 15.8, in which the residuals indicate the data points. The north-east and south-west corners of the area for which the trend maps have been drawn contain no data points: they are therefore unreliable and have been omitted from the figure.

The linear surface explains 72·2% of the total sum of squares. Its equation is Z = 52·93 + 1·36U + 0·498V. It shows an increase of round-

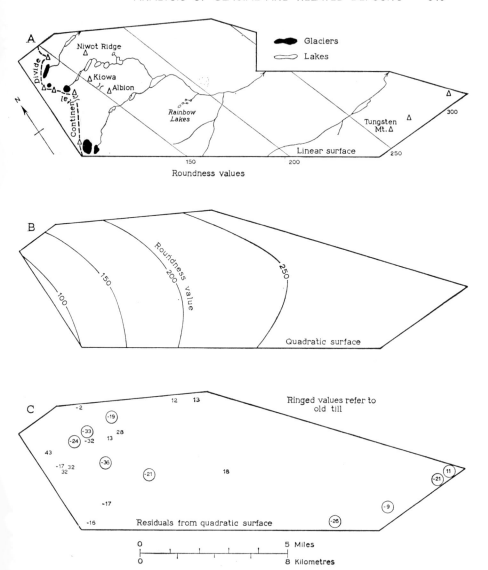

Fig. 15.8   A: Linear trend surface of stone roundness values of part of Indian Peaks area in the Front Range of Rocky Mountains, Colorado; **B:** Quadratic trend surface; **C:** Residuals from Quadratic surface.

ness away from the continental divide where most of the glaciers originated. The quadratic surface explains 84·2% of the sums of squares, an increase of 12% above the linear surface, and gives a good generalization of the increase in stone roundness away from the glacier source, as the contours show a rapid increase in roundness close to the source areas

and a slower one further away. It is not necessary to consider the cubic surface, which only raised the explanation a further 6·2% to 90·4%. Its form is very similar to the quadratic one. The trend equation for the quadratic surface is:

$$Z = 333·08 + 0·387U - 3·743V - 0·039U^2 + 0·146UV - 0·144V^2$$

The result of the analysis, therefore, illustrates well the effectiveness of glacial erosion in rounding stones, an effectiveness that decreases as the glaciers move further from their source. The samples included both older and newer deposits, thus supporting the view that the older, high level deposits are glacial in origin as they fit into the sequence.

The residuals also indicate the effectiveness of glacial erosion in rounding stones. Some of the positive residuals, indicating stones less round than the calculated values, occur where a small corrie joined the main valley at a point some distance down-valley from the main corries on the continental divide. On the other hand some of the older, high level deposits have higher roundness values than calculated, because the ice that laid them down probably came from across the continental divide, passing through the col analysed in section 13.2 and through another further west on the continental divide, shown in Figure 15.8.

15.3 Orientation of surface stones

A preferred orientation of the long axes of elongated stones on the surface has been observed on some glacial deposits (Andrews, 1965). The measurement of surface stone orientation of a sample of 50 stones is a very rapid method of obtaining some useful information about the nature of the deposits and past ice movements. If the stones show a preferred orientation this may well be related to the past direction of ice flow. A considerable number of observations have been made and they indicate that many deposits of till and moraines show a preferred pattern of orientation of the surface stones. In some instances this characteristic may help to identify morainic deposits, but the evidence should not be considered more than an indication because not all moraines show a preferred orientation. On the other hand, if a deposit does have a strong preferred orientation it is unlikely to have received this by random processes of mass movement, unless the slope is fairly steep. Some patterns of stone orientation emerge as a result of cryoturbation sorting processes, but these will not show a single preferred direction in most situations.

In this section some examples of orientation measurements made for the purpose of identifying the nature of a particular feature will be discussed. These observations will then be compared with the results of orientation studies on moraines in Baffin Island. Some of the observations have been made on the moraines of active glaciers, others on older moraines.

The first set of observations to be considered has been made on a lobate

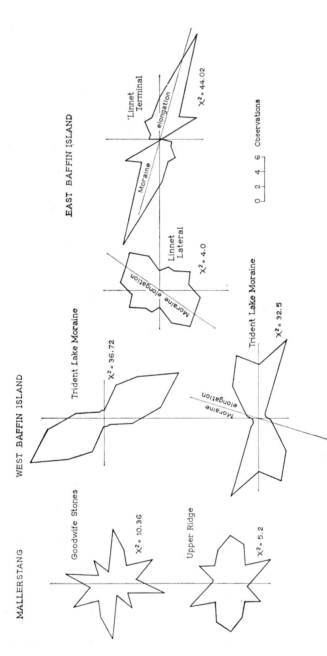

Fig. 15.9 Two-dimensional orientation diagrams for the long axis orientation of samples of 50 stones measured on solifluction deposits in Mallerstang, and moraines in west and east Baffin Island.

deposit of drift high on the hillside of Mallerstang in the upper Eden valley. This feature has been described by Rowell and Turner (1952) as a corrie moraine dating from the zone III cold phase, which is contemporaneous with the youngest moraines of Blea Water. The morphological character of the Mallerstang feature is, however, more consistent with the view that it is the product of nivation processes in association with massive periglacial solifluction slumping. The strata are conducive to this type of movement, as shales, which become soft and liable to flow when saturated, underlie massive grit strata. The grits have been broken by frost weathering and have moved down the slope to form a series of large lobate ridges. The ridges nearest the rock face are composed mainly of grit blocks, but the marginal part of the deposit consists of more isolated blocks of grit mixed with the shales.

A series of orientation measurements, made on different parts of this feature, is intended to give some evidence concerning the most likely of the two possible modes of origin (see Fig. 15.9). One set of observations has been made on elongated blocks of grits in the ridge nearest to the rock face. The data are divided into nine classes of $20°$ each to cover $180°$, as the orientation pattern is circular and non-directed, and may be analysed by a simple chi-square test (section 10.2) to assess the randomness of the distribution. The expected number of observations in each $20°$ class is $5 \cdot 6$, as there were 50 observations. The expected values may be compared with the observed frequency, and $\chi^2 = 6 \cdot 28$. Two more sets of values have been measured near the base of the lobate spread. The chi-square values for these two sets are $5 \cdot 14$ and $0 \cdot 777$. None of these three values allows the null hypothesis of random orientation to be rejected, and the conclusion to be drawn from these observations is that the orientation of the stones does not confirm the hypothesis that this feature is a moraine. The evidence, on the other hand, is not in conflict with the view that the feature is the result of nivation and extensive mass movement by solifluction. However, this evidence alone is not adequate to indicate which of the hypotheses is correct. It does, however, provide some useful information on the nature of the deposits and possible formative processes.

The ordinary chi-square test may be modified to provide an estimate of the preferred orientation as well as a test of significance by means of the *Tukey chi-square test*. The application of the test is exemplified here by using data on surface stone orientation measured in East Baffin Island. The data and calculations are set out in Table 15.13. The 50 observations have been divided into nine $20°$ classes. The observed frequency in each class is given in column two. The expected frequency is given, on the assumption that there is no preferred orientation, in the third column. The null hypothesis of the test is, therefore, that the orientation pattern shows no preferred orientation. The number of classes must be arranged so that none of the expected frequencies falls below 5. Thus if 100 stones had been measured, $10°$ classes could have been used. The values of observed minus expected frequencies are next entered in column 4, and

Table 15.13 Tukey chi-square test for orientational data

Degrees	Observed frequency	Expected frequency	O − E	x	Cos 2θ	x cos 2θ	Sin 2θ	x sin 2θ
0–20	11	5·6	+5·4	+2·28	+1·000	+2·280	0·000	0·000
20–40	15	5·6	+9·4	+3·98	+0·766	+3·220	+0·643	+2·560
40–60	6	5·6	+0·4	+0·16	+0·174	+0·028	+0·985	+0·158
60–80	5	5·6	−0·6	−0·25	−0·500	+0·125	+0·860	−0·214
80–100	0	5·6	−5·6	−2·37	−0·940	+2·220	+0·342	−0·810
100–120	1	5·6	−4·6	−1·95	−0·940	+1·830	−0·342	+0·667
120–140	3	5·6	−2·6	−1·10	−0·500	+0·550	−0·866	+0·954
140–160	1	5·6	−4·6	−1·95	+0·174	−0·339	−0·985	+1·920
160–180	8	5·6	+2·4	+1·04	+0·766	+0·796	−0·643	−0·669

$$\Sigma = 50 \qquad\qquad \Sigma = \begin{array}{r} +11{\cdot}049 \\ -0{\cdot}339 \\ \hline +10{\cdot}710 \end{array} \qquad \Sigma = \begin{array}{r} +6{\cdot}259 \\ -1{\cdot}693 \\ \hline +4{\cdot}566 \end{array}$$

$$x = O - E/\sqrt{E} \qquad \sqrt{E} = 2{\cdot}36$$

$$C = \frac{\Sigma\, x\cos 2\theta}{\sqrt{\Sigma\cos^2 2\theta}} \qquad S = \frac{\Sigma\, x\sin 2\theta}{\sqrt{\Sigma\sin^2 2\theta}} \qquad \tan 2\theta = S/C$$

$$\sqrt{\Sigma\cos^2 2\theta} = \sqrt{\Sigma\sin^2 2\theta} = 2{\cdot}117 \qquad C = \frac{+10{\cdot}710}{2{\cdot}117} = 5{\cdot}09$$

$$S = \frac{+4{\cdot}566}{2{\cdot}117} = 2{\cdot}16$$

$$\text{Tan } 2\theta = 2{\cdot}16/5{\cdot}09 = 0{\cdot}42 = 23° \qquad \theta = 11\tfrac{1}{2}° + 10° = 21\tfrac{1}{2}°$$

$\chi^2 = C^2 + S^2 = 25{\cdot}81 + 4{\cdot}66 = 30{\cdot}47$. This value for 8 degrees of freedom is significant at the 99·9% level.

Rayleigh test of significance:

	f cos 2θ	f sin 2θ
	+11·000	0·000
	+11·500	+9·650
$\Sigma f \sin 2\theta = A$	+1·042	+5·900
$\Sigma f \cos 2\theta = B$	−2·500	+4·300
	0·000	0·000
$\tan 2\theta = A/B = 0{\cdot}42$	−0·940	−0·342
$= 23°$	−1·500	−2·598
$\theta = 11\tfrac{1}{2}°$	+0·174	−0·985
preferred orientation $= 11\tfrac{1}{2} + 10°$	+6·930	−5·144
$= 21\tfrac{1}{2}°$	+30·706	+19·850
	−4·940	−9·069
	$\Sigma = +25{\cdot}706$	+10·781

$$R = \sqrt{A^2 + B^2} = \sqrt{660 + 111{\cdot}6} = 27{\cdot}8 \qquad L\% = R/n \times 100 \qquad n = 50$$
$$L\% = 2R = 55{\cdot}6$$

the x value, which is given by $O - E/\sqrt{E}$, is calculated in the fifth column. The values of $\cos 2\theta$ and $\sin 2\theta$ are fixed and they are entered in columns 6 and 8. Columns 7 and 9 contain the products of these values and the x values. These two columns are then summed, the signs being noted. The values of C and S are then calculated by dividing the sums by 2·117. This is the value of $\sqrt{\sum \cos^2 2\theta}$ and $\sqrt{\sum \sin^2 2\theta}$, which are equal. The preferred orientation is then found by $\tan 2\theta = S/C$. This value is halved and 10° is added to the result, to make allowance for the class size, giving $21\frac{1}{2}°$. The halving is necessary as the values have been calculated in terms of 2θ. The significance of the preferred orientation is checked in chi-square tables by calculating the value of $C^2 + S^2$, which in the example given is $25·81 + 4·66 = 30·47$. The chi-square tables are entered with eight degrees of freedom, which is the number of classes less one. The tabled value for 99·9% significance is given as 26·12, so as the calculated value exceeds this, the null hypothesis may be rejected at a very high significance level. It can thus be assumed that the pattern of stone orientation has been influenced by the consistent movement of glacier ice across the area where the stones were measured, the ice moving in a north-north-easterly direction.

There is also a rather simpler method of obtaining the same result by means of the *Rayleigh Test*. The calculations are shown in the lower part of Table 15.13. The frequency values are multiplied by the values for $\cos 2\theta$ and $\sin 2\theta$ as given in the upper part of the table, and summed, noting the signs. The tangent of the angle of preferred orientation is found by A/B, where $A = \sum f \sin 2\theta$ and B is $\sum f \cos 2\theta$, $\tan 2\theta = A/B$. The value must again be divided by 2° and 10° added to give the value corrected for the class boundaries. The same result is obtained by both methods. The significance of the result is tested by calculating R, where $R = \sqrt{A^2 + B^2}$, which in the example is $\sqrt{600 + 111·6} = \sqrt{771·6} = 27·8$. The L% value is then found from the equation $L\% = (R/n) \times 100$, where n is the number of observations, in this case $n = 50$, L% thus is equal to 2R, which is 55·6. The result can be checked against a graph given in Curray (1956, Fig. 4, p. 126). It is highly significant at a level well above 99·9%. For 50 observations the L value must exceed 32% to be significant at the 99% level, and 24% at the 95% significance level.

The aim of the analysis just described was to see if there is evidence for glacial action in the outer part of the Henry Kater Peninsula, where the observations were made on a thin stony till deposit situated on a rocky peninsula on the north side of Home Bay. The direction of ice movement could also be assessed, if there were a preferred orientation. The results of the analysis showed that there is in fact a very strong preferred orientation and as the site where the observations were made was relatively flat it can be assumed that moving ice was responsible for the pattern. The direction of preferred orientation suggests that ice from

the major glacier in the fjord was moving inland in this area, a conclusion that is supported by the presence of limestone erratics in the till near by. This rock could only have been derived from offshore in this area. Thus two lines of evidence confirm the assumed direction of ice movement.

A study of the orientation on a series of moraines in west Baffin Island has also been carried out (see Fig. 15.9). Ten observations have been made on moraines and the results may be tested by means of the chi-square test. Of the ten sets of data two are preferentially orientated at the 0·001 significance level, three at the 0·01 level, three at the 0·05 level and one at the 0·1 level, while three are not significantly orientated. The strongly orientated samples are arranged with the long axes of the stones at right-angles to the elongation of the moraines. In this area seven out of the ten moraines show a preferred orientation of the long axes of their stones.

The pattern of moraine stone orientation and that in other deposits is rather less systematic in east Baffin Island. Observations have been made on a variety of moraine types, some of them on moraines of active glaciers. The lateral moraines of Curlew and Linnet glaciers show no significant orientation, but the terminal moraine of Linnet glacier, which consists mainly of large boulders, shows a strong preferred orientation—with chi-square of 44·02, significant at 99·9% level—perpendicular to the moraine elongation (Fig. 15.9). The reason for this variation may be related to the movement of the ice, which could be upwards along thrust planes at the snout of the glacier. Flow would be in the direction of the stone orienta-tion. Along the margin of the glacier at the side of the valley the ice is moving parallel to the lateral moraine. The surface stones would tend to move more irregularly in this position and the pattern would change as the ice melted downwards and the stones moved randomly to their present position. The till at this site shows a strong preferred orientation a short distance below the surface, so only the surface stones have apparently undergone random final deposition.

The elongated rocks in a steep avalanche fan near the glaciers also show a strong preferred orientation—chi-square being 37·12—parallel to the slope. Farther from the present-day glaciers very bouldery till on a flattish hill summit shows strongly preferred orientation. Two samples have chi-square values of 40·19 and 36·44, but a third shows no significant pattern, probably due to later disturbance by solifluction as sorting has operated amongst the sediments in this position. Some of the moraines in the area of older moraine near the outer coast show strong preferred orien-tation, but others do not. The reason for this variation is not clear, but some of the moraines with no significant orientation appear to have been deposited into the sea, while others have been disturbed by cryoturbation and solifluction. These secondary processes have not influenced the west Baffin Island moraines so much, as their sediment is coarser.

The results of this type of observation must be treated with caution, but the analysis of the results by circular vectorial methods does give

some indication of the significance of the results from a statistical point of view. The geomorphological interpretation is rather more difficult. Later processes may destroy a preferred orientation but there is often no means of knowing if this has occurred or whether there ever was a preferred orientation.

References

ANDREWS, J. T. 1965: Surface boulder orientation studies around the north-west margin of the Barnes Ice Cap, Baffin Island, Canada. *J. Sed. Petrol.* **35,** 735–58.

ANDREWS, J. T. and KING, C. A. M. 1968: Comparative till fabrics and till fabric variability in a till sheet and a drumlin: a small-scale study. *Proc. Yorks. Geol. Soc.* **36,** 435–61.

ANDREWS, J. T. and SMITHSON, B. B. 1966: Till fabrics of the cross-valley moraines of north-central Baffin Island, N.W.T., Canada. *Bull. Geol. Soc. Am.* **77,** 217–90.

BATSCHELET, E. 1965: *Statistical methods for the analysis of problems in animal orientation and certain biological rhythms.* A.E.B.S. Mono.

CAILLEUX, A. 1947: L'indice d'emoussé: definition et premiere application. *C. R. Somm Geol. France* **13,** 250–52.

COOK, J. H. 1946: Kame complexes and perforation deposits. *Am. J. Sci.* **244,** 573–83.

CURRAY, J. R. 1956: The analysis of two-dimensional orientation data. *J. Geol.* **64,** 117–31.

DURAND, D. and GREENWOOD, J. A. 1958: Modification of the Rayleigh test for uniformity in analysis of two dimensional orientation data. *J. Geol.* **66,** 229–38.

FISHER, R. A. 1953: Dispersion on a sphere. *Proc. Roy. Soc. Lond.* **A217,** 295–305.

FOLK, R. C. and WARD, W. C. 1957: Brazos river bar: a study in the significance of grain-size parameters. *J. Sed. Petrol.* **27,** 3–27.

GALLOWAY, R. W. 1956: The structure of moraines in Lyngasdalen, N. Norway. *J. Glaciol.* **2,** 730–33.

GREEN, R. 1964: Available methods for the analysis of vectorial data. *J. Sed. Petrol.* **34,** 440–41.

HOLMES, C. D. 1941: Till fabric. *Bull. Geol. Soc. Am.* **52,** 1299–1354.

1947: Kames. *Am. J. Sci.* **245,** 240–49.

KAURANNE, L. K. 1960: A statistical study of stone orientation in glacial till. *Bull. Comm. Geol. Finlande* **188,** 87–97.

KING, C. A. M. 1969a: Glacial geomorphology and chronology of Henry Kater Peninsula, East Baffin Island, N.W.T. *Arctic and Alpine Research* **1** (3), 195–212.

1969b: Moraine types on Henry Kater Peninsula, East Baffin Island, N.W.T. Canada. *Arctic and Alpine Research* **1** (4), 289–94.

KING, C. A. M. and BUCKLEY, J. T. 1968: Analysis of stone size and shape in arctic environments. *J. Sed. Petrol.* **38,** 200–214.

KIRKBY, R. P. 1961: Movement of ice in central Labrador-Ungava. *Cah. de Geog. de Quebec.* **10,** 205.

KRUMBEIN, W. C. 1939: Preferred orientation of pebbles in sedimentary deposits. *J. Geol.* **47,** 673–706.

1941: Measurement and geological significance of shape and roundness of sedimentary particles. *J. Sed. Petrol.* **11,** 64–72.

LUNDQUIST, G. 1949: The orientation of block material in certain species of flow earth. *Geogr. Ann.* **31,** 335–47.

MANLEY, G. 1959: The late-glacial climate of north-west England. *Liv. and Manch. Geol. J.* **2,** 188–215.

MILLER, R. L. and KAHN, J. S. 1962: *Statistical analysis in the geological sciences.* New York: John Wiley. (483 pp.)

PINCUS, H. J. 1956: Some vector and arithmetic operations on two-dimensional orientation variates with application to geological data. *J. Geol.* **64,** 533–57.

ROWELL, A. J. and TURNER, J. S. 1952: Glaciation in the Upper Eden valley, Westmorland. *Liv. and Manch. Geol. J.* **1,** 200–207.

STEINMETZ, R. 1962a: Analysis of vectorial data. *J. Sed. Petrol.* **32,** 801–12.

1962b: Sampling and size distribution of quartzose pebbles from the New Jersey gravels. *J. Geol.* **70,** 56–73.

1964: Available methods for the analysis of vectorial data: a reply. *J. Sed. Petrol.* **34,** 441–2.

SWINZOW, G. K. 1962: Investigations of shear zones in the ice-sheet margin Thule area, Greenland, *J. Glaciol.* **4,** 215–29.

WATSON, G. S. and WILLIAMS, E. J. 1956: On the construction of significance tests on the circle and the sphere. *Biometrika* **43,** 344–52.

WEERTMAN, J. 1961: Mechanism for the formation of inner moraines found near the edge of cold ice caps and ice sheets. *J. Glaciol.* **3,** 965–78.

ZINGG, T. 1935: Beitrag zur Schotteranalyse. *Min. u. Petrog. Mitt.* **15,** 39–140.

Conclusion

The advance of the science of geomorphology must be based on the analysis of quantitative data, and not on mere description. Quantitative data fall into two main groups—one concerned with form and distribution, and the other with process. This book has concentrated on the former, although the latter has of necessity been frequently mentioned, as the two are intimately linked.

One reason for subdividing the book into major process studies has been to indicate the many diverse ways in which morphology may be studied: some of these have been explored. The range of methods must be especially wide when the variety of land forms examined is so great. At one end of the study has been the whole landscape in terms of drainage basins, while at the other have been individual features, such as moraines or eskers. The second and fourth parts of the book have been linked by the study of slopes in non-glaciated landscapes and in glaciated areas. These separate studies show how similar numerical techniques can differentiate between the major formative processes of slope development. In some sections it has been found valuable to analyse the form of the features through a study of their constituent materials. This has applied particularly in the study of beach forms and some glacial depositional features. Thus both erosional morphology and depositional morphology have been analysed by numerical methods at appropriate points. Nearly all the land forms considered have been analysed in numerical terms, which add precision to description and confidence to inference, whether the problem be simple or complex. Some of the problems analysed have been simpler than others, in that only one variable has been involved in the analysis. The pattern of stone roundness in glacial deposits exemplifies this type of analysis. Other problems have involved two variables, while yet others, more complex, have involved the analysis of many variables. The study of drainage basin morphometry illustrates the latter situation.

The book has not aimed to be a systematic study of geomorphology, but rather a collection of worked examples which it is hoped will elucidate some useful numerical methods of analysis, and also add to the sum of geomorphological knowledge.

Appendix:
Tables of statistics

One- and two-tailed tests

In general, to test whether, on the average, A is greater than B use a one-tailed test and to test whether A is different from B use a two-tailed test. In some instances, however, this is an over-simplified rule. For discussion see:

RATCLIFFE, J. F. 1967: *Elements of mathematical statistics* (2nd edition). London: Oxford University Press, 192–4

Table A Table of critical values of *t*

df	_	_	_	_	_	_
	Level of significance for one-tailed test					
	0·10	0·05	0·025	0·01	0·005	0·0005
	Level of significance for two-tailed test					
	0·20	0·10	0·05	0·02	0·01	0·001
1	3·078	6·314	12·706	31·821	63·657	636·619
2	1·886	2·920	4·303	6·965	9·925	31·598
3	1·638	2·353	3·182	4·541	5·841	12·941
4	1·533	2·132	2·776	3·747	4·604	8·610
5	1·476	2·015	2·571	3·365	4·032	6·859
6	1·440	1·943	2·447	3·143	3·707	5·959
7	1·415	1·895	2·365	2·998	3·499	5·405
8	1·397	1·860	2·306	2·896	3·355	5·041
9	1·383	1·833	2·262	2·821	3·250	4·781
10	1·372	1·812	2·228	2·764	3·169	4·587
11	1·363	1·796	2·201	2·718	3·106	4·437
12	1·356	1·782	2·179	2·681	3·055	4·318
13	1·350	1·771	2·160	2·650	3·012	4·221
14	1·345	1·761	2·145	2·624	2·977	4·140
15	1·341	1·753	2·131	2·602	2·947	4·073
16	1·337	1·746	2·120	2·583	2·921	4·015
17	1·333	1·740	2·110	2·567	2·898	3·965
18	1·330	1·734	2·101	2·552	2·878	3·922
19	1·328	1·729	2·093	2·539	2·861	3·883
20	1·325	1·725	2·086	2·528	2·845	3·850
21	1·323	1·721	2·080	2·518	2·831	3·819
22	1·321	1·717	2·074	2·508	2·819	3·792
23	1·319	1·714	2·069	2·500	2·807	3·767
24	1·318	1·711	2·064	2·492	2·797	3·745
25	1·316	1·708	2·060	2·485	2·787	3·725
26	1·315	1·706	2·056	2·479	2·779	3·707
27	1·314	1·703	2·052	2·473	2·771	3·690
28	1·313	1·701	2·048	2·467	2·763	3·674
29	1·311	1·699	2·045	2·462	2·756	3·659
30	1·310	1·697	2·042	2·457	2·750	3·646
40	1·303	1·684	2·021	2·423	2·704	3·551
60	1·296	1·671	2·000	2·390	2·660	3·460
120	1·289	1·658	1·980	2·358	2·617	3·373
∞	1·282	1·645	1·960	2·326	2·576	3·291

Source (tables A and B) : Siegel, S. 1965 : *Nonparametric statistics for the behavioral sciences.* New York : McGraw-Hill. *Adapted from original source* : Fisher, Sir R. A. and Yates, F. 1963 : *Statistical tables for biological, agricultural and medical research.* Edinburgh : Oliver and Boyd.

Table B Table of critical values of chi square

Probability under H_0 that $\chi^2 \geqslant$ chi square

df	0·99	0·98	0·95	0·90	0·80	0·70	0·50	0·30	0·20	0·10	0·05	0·02	0·01	0·001
1	0·00016	0·00063	0·0039	0·016	0·064	0·15	0·46	1·07	1·64	2·71	3·84	5·41	6·64	10·83
2	0·02	0·04	0·10	0·21	0·45	0·71	1·39	2·41	3·22	4·60	5·99	7·82	9·21	13·82
3	0·12	0·18	0·35	0·58	1·00	1·42	2·37	3·66	4·64	6·25	7·82	9·84	11·34	16·27
4	0·30	0·43	0·71	1·06	1·65	2·20	3·36	4·88	5·99	7·78	9·49	11·67	13·28	18·46
5	0·55	0·75	1·14	1·61	2·34	3·00	4·35	6·06	7·29	9·24	11·07	13·39	15·09	20·52
6	0·87	1·13	1·64	2·20	3·07	3·83	5·35	7·23	8·56	10·64	12·59	15·03	16·81	22·46
7	1·24	1·56	2·17	2·83	3·82	4·67	6·35	8·38	9·80	12·02	14·07	16·62	18·48	24·32
8	1·65	2·03	2·73	3·49	4·59	5·53	7·34	9·52	11·03	13·36	15·51	18·17	20·09	26·12
9	2·09	2·53	3·32	4·17	5·38	6·39	8·34	10·66	12·24	14·68	16·92	19·68	21·67	27·88
10	2·56	3·06	3·94	4·86	6·18	7·27	9·34	11·78	13·44	15·99	18·31	21·16	23·21	29·59
11	3·05	3·61	4·58	5·58	6·99	8·15	10·34	12·90	14·63	17·28	19·68	22·62	24·72	31·26
12	3·57	4·18	5·23	6·30	7·81	9·03	11·34	14·01	15·81	18·55	21·03	24·05	26·22	32·91
13	4·11	4·76	5·89	7·04	8·63	9·93	12·34	15·12	16·98	19·81	22·36	25·47	27·69	34·53
14	4·66	5·37	6·57	7·79	9·47	10·82	13·34	16·22	18·15	21·06	23·68	26·87	29·14	36·12
15	5·23	5·98	7·26	8·55	10·31	11·72	14·34	17·32	19·31	22·31	25·00	28·26	30·58	37·70
16	5·81	6·61	7·96	9·31	11·15	12·62	15·34	18·42	20·46	23·54	26·30	29·63	32·00	39·29
17	6·41	7·26	8·67	10·08	12·00	13·53	16·34	19·51	21·62	24·77	27·59	31·00	33·41	40·75
18	7·02	7·91	9·39	10·86	12·86	14·44	17·34	20·60	22·76	25·99	28·87	32·35	34·80	42·31
19	7·63	8·57	10·12	11·65	13·72	15·35	18·34	21·69	23·90	27·20	30·14	33·69	36·19	43·82
20	8·26	9·24	10·85	12·44	14·58	16·27	19·34	22·78	25·04	28·41	31·41	35·02	37·57	45·32
21	8·90	9·92	11·59	13·24	15·44	17·18	20·34	23·86	26·17	29·62	32·67	36·34	38·93	46·80
22	9·54	10·60	12·34	14·04	16·31	18·10	21·34	24·94	27·30	30·81	33·92	37·66	40·29	48·27
23	10·20	11·29	13·09	14·85	17·19	19·02	22·34	26·02	28·43	32·01	35·17	38·97	41·64	49·73
24	10·86	11·99	13·85	15·66	18·06	19·94	23·34	27·10	29·55	33·20	36·42	40·27	42·98	51·18
25	11·52	12·70	14·61	16·47	18·94	20·87	24·34	28·17	30·68	34·38	37·65	41·57	44·31	52·62
26	12·20	13·41	15·38	17·29	19·82	21·79	25·34	29·25	31·80	35·56	38·88	42·86	45·64	54·05
27	12·88	14·12	16·15	18·11	20·70	22·72	26·34	30·32	32·91	36·74	40·11	44·14	46·96	55·48
28	13·56	14·85	16·93	18·94	21·59	23·65	27·34	31·39	34·03	37·92	41·34	45·42	48·28	56·89
29	14·26	15·57	17·71	19·77	22·48	24·58	28·34	32·46	35·14	39·09	42·56	46·69	49·59	58·30
30	14·95	16·31	18·49	20·60	23·36	25·51	29·34	33·53	36·25	40·26	43·77	47·96	50·89	59·70

Table C Table of critical values of ρ, the Spearman rank correlation coefficient

N	Significance level (one-tailed test)	
	0·05	0·01
4	1·000	
5	0·900	1·000
6	0·829	0·943
7	0·714	0·893
8	0·643	0·833
9	0·600	0·783
10	0·564	0·746
12	0·506	0·712
14	0·456	0·645
16	0·425	0·601
18	0·399	0·564
20	0·377	0·534
22	0·359	0·508
24	0·343	0·485
26	0·329	0·465
28	0·317	0·448
30	0·306	0·432

Source: Siegel, S. 1965: *Nonparametric statistics for the behavioral sciences.* New York: McGraw-Hill.

Table D *F*-distribution ($F_{.95}$)

		Degrees of freedom for numerator									
		1	2	3	4	5	6	7	8	9	10
Degrees of freedom for denominator	1	161	200	216	225	230	234	237	239	241	242
	2	18·5	19·0	19·2	19·2	19·3	19·3	19·4	19·4	19·4	19·4
	3	10·1	9·55	9·28	9·12	9·01	8·94	8·89	8·85	8·81	8·79
	4	7·71	6·94	6·59	6·39	6·26	6·16	6·09	6·04	6·00	5·96
	5	6·61	5·79	5·41	5·19	5·05	4·95	4·88	4·82	4·77	4·74
	6	5·99	5·14	4·76	4·53	4·39	4·28	4·21	4·15	4·10	4·06
	7	5·59	4·74	4·35	4·12	3·97	3·87	3·79	3·73	3·68	3·64
	8	5·32	4·46	4·07	3·84	3·69	3·58	3·50	3·44	3·39	3·35
	9	5·12	4·26	3·86	3·63	3·48	3·37	3·29	3·23	3·18	3·14
	10	4·96	4·10	3·71	3·48	3·33	3·22	3·14	3·07	3·02	2·98
	11	4·84	3·98	3·59	3·36	3·20	3·09	3·01	2·95	2·90	2·85
	12	4·75	3·89	3·49	3·26	3·11	3·00	2·91	2·85	2·80	2·75
	13	4·67	3·81	3·41	3·18	3·03	2·92	2·83	2·77	2·71	2·67
	14	4·60	3·74	3·34	3·11	2·96	2·85	2·76	2·70	2·65	2·60
	15	4·54	3·68	3·29	3·06	2·90	2·79	2·71	2·64	2·59	2·54
	16	4·49	3·63	3·24	3·01	2·85	2·74	2·66	2·59	2·54	2·49
	17	4·45	3·59	3·20	2·96	2·81	2·70	2·61	2·55	2·49	2·45
	18	4·41	3·55	3·16	2·93	2·77	2·66	2·58	2·51	2·46	2·41
	19	4·38	3·52	3·13	2·90	2·74	2·63	2·54	2·48	2·42	2·38
	20	4·35	3·49	3·10	2·87	2·71	2·60	2·51	2·45	2·39	2·35
	21	4·32	3·47	3·07	2·84	2·68	2·57	2·49	2·42	2·37	2·32
	22	4·30	3·44	3·05	2·82	2·66	2·55	2·46	2·40	2·34	2·30
	23	4·28	3·42	3·03	2·80	2·64	2·53	2·44	2·37	2·32	2·27
	24	4·26	3·40	3·01	2·78	2·62	2·51	2·42	2·36	2·30	2·25
	25	4·24	3·39	2·99	2·76	2·60	2·49	2·40	2·34	2·28	2·24
	30	4·17	3·32	2·92	2·69	2·53	2·42	2·33	2·27	2·21	2·16
	40	4·08	3·23	2·84	2·61	2·45	2·34	2·25	2·18	2·12	2·08
	60	4·00	3·15	2·76	2·53	2·37	2·25	2·17	2·10	2·04	1·99
	120	3·92	3·07	2·68	2·45	2·29	2·18	2·09	2·02	1·96	1·91
	∞	3·84	3·00	2·60	2·37	2·21	2·10	2·01	1·94	1·88	1·83

Source (tables D and E): Merrington, M. and Thompson, C. M. 1943: Tables of percentage points of the inverted beta (F) distribution, *Biometrika* **33**, 73f.

Table E *F*-distribution ($F_{.99}$)

		Degrees of freedom for numerator									
		1	2	3	4	5	6	7	8	9	10
Degrees of freedom for denominator	1	4,052	5,000	5,403	5,625	5,764	5,859	5,928	5,982	6,023	6,056
	2	98·5	99·0	99·2	99·2	99·3	99·3	99·4	99·4	99·4	99·4
	3	34·1	30·8	29·5	28·7	28·2	27·9	27·7	27·5	27·3	27·2
	4	21·2	18·0	16·7	16·0	15·5	15·2	15·0	14·8	14·7	14·5
	5	16·3	13·3	12·1	11·4	11·0	10·7	10·5	10·3	10·2	10·1
	6	13·7	10·9	9·78	9·15	8·75	8·47	8·26	8·10	7·98	7·87
	7	12·2	9·55	8·45	7·85	7·46	7·19	6·99	6·84	6·72	6·62
	8	11·3	8·65	7·59	7·01	6·63	6·37	6·18	6·03	5·91	5·81
	9	10·6	8·02	6·99	6·42	6·06	5·80	5·61	5·47	5·35	5·26
	10	10·0	7·56	6·55	5·99	5·64	5·39	5·20	5·06	4·94	4·85
	11	9·65	7·21	6·22	5·67	5·32	5·07	4·89	4·74	4·63	4·54
	12	9·33	6·93	5·95	5·41	5·06	4·82	4·64	4·50	4·39	4·30
	13	9·07	6·70	5·74	5·21	4·86	4·62	4·44	4·30	4·19	4·10
	14	8·86	6·51	5·56	5·04	4·70	4·46	4·28	4·14	4·03	3·94
	15	8·68	6·36	5·42	4·89	4·56	4·32	4·14	4·00	3·89	3·80
	16	8·53	6·23	5·29	4·77	4·44	4·20	4·03	3·89	3·78	3·69
	17	8·40	6·11	5·19	4·67	4·34	4·10	3·93	3·79	3·68	3·59
	18	8·29	6·01	5·09	4·58	4·25	4·01	3·84	3·71	3·60	3·51
	19	8·19	5·93	5·01	4·50	4·17	3·94	3·77	3·63	3·52	3·43
	20	8·10	5·85	4·94	4·43	4·10	3·87	3·70	3·56	3·46	3·37
	21	8·02	5·78	4·87	4·37	4·04	3·81	3·64	3·51	3·40	3·31
	22	7·95	5·72	4·82	4·31	3·99	3·76	3·59	3·45	3·35	3·26
	23	7·88	5·66	4·76	4·26	3·94	3·71	3·54	3·41	3·30	3·21
	24	7·82	5·61	4·72	4·22	3·90	3·67	3·50	3·36	3·26	3·17
	25	7·77	5·57	4·68	4·18	3·86	3·63	3·46	3·32	3·22	3·13
	30	7·56	5·39	4·51	4·02	3·70	3·47	3·30	3·17	3·07	2·98
	40	7·31	5·18	4·31	3·83	3·51	3·29	3·12	2·99	2·89	2·80
	60	7·08	4·98	4·13	3·65	3·34	3·12	2·95	2·82	2·72	2·63
	120	6·85	4·79	3·95	3·48	3·17	2·96	2·79	2·66	2·56	2·47
	∞	6·63	4·61	3·78	3·32	3·02	2·80	2·64	2·51	2·41	2·32

Index of numerical techniques

Main references are printed in bold. Page numbers in italics refer to diagrams.

General index

Main references are printed in bold. Page numbers in italics refer to diagrams. Bibliographical references occur on the page numbers followed by the letter r (e.g. 109r). References to specific places and features are indexed under countries.